Abha Agrawal

Editor

# Patient Safety

## A Case-Based Comprehensive Guide

 Springer

*Editor*
Abha Agrawal, M.D., F.A.C.P.
Norwegian American Hospital and
    Northwestern University Feinberg School of Medicine
Chicago, IL, USA

ISBN 978-1-4614-7418-0          ISBN 978-1-4614-7419-7 (eBook)
DOI 10.1007/978-1-4614-7419-7
Springer New York Heidelberg Dordrecht London

Library of Congress Control Number: 2013941354

*To patients: they teach us everything.*
*To Mummy and Papa: who made it possible*
*to learn.*

# Preface

The fundamental premise of this book is the following: patient safety has always been at the core of medical professionals' ethic and value since Hippocrates and Florence Nightingale implored us to "do no harm." The newness in the patient safety movement of the last decade lies in a better understanding of the prevalence, causes, and potential solutions for medical errors.

Why is learning about patient safety critical to all healthcare professionals? We don't go to work to perform an operation or to administer medications; we go to work to treat, cure, and heal sick people. So of what value is the superb technical skill of a surgeon to a patient whose healthy leg gets amputated due to a trivial mistake in patient identification by a team of surgeons and nurses in a hurry? What good is the advanced skill and training of a specialized physician if a patient dies after receiving 100 times the dose of an anticoagulant caused by a trivial error in labeling the bag of intravenous medication? How do you console the mother of a newborn baby who dies due to an unwarranted and inexplicable delay in performing a Cesarean section caused by a breakdown in teamwork and communication between the obstetrician and the nurse? Death is binary; your patient is either alive or dead. And once someone is dead there is no coming back. Therefore, if delivering good outcomes for patients is at the heart of our profession, we have as much professional obligation to learn about the adverse events—the diseases of healthcare delivery system, as we have to learn about biological diseases—diseases of human body system.

The purpose of this book is to engage front-line clinicians and move patient safety from the boardroom to the bedside because only by practicing patient safety, will we be able to make a difference in the lives of our patients and their families.

## Error in Medicine

The evidence is now incontrovertible that many patients suffer serious harm due to *avoidable* adverse events in health care such as medication errors, hospital-acquired infections, surgical complications, and delays in necessary treatments.

These adverse events happen in every setting—clinic, hospital, emergency room, rural, urban, community center, academic hospital—across the globe to patients of all age groups, ethnicities, and socioeconomic backgrounds. They could happen to your patients and mine.

And this is not new. In 1964, Schimmel reported that 20 % of patients admitted to a university hospital suffered iatrogenic injury and that 20 % of those injuries were serious or fatal [1]. A 1981 report found that 36 % of patients admitted to the medical service in a teaching hospital suffered an iatrogenic event, of which 25 % were serious or life threatening [2]. In 1991, Leape et al. reported the results of a population-based study conducted in New York and found that 3.7 % of patients had "disabling" injuries as a result of medical treatment and that "negligent care" was responsible for 28 % of them [3]. Another 1991 study found that 64 % of cardiac arrests at a teaching hospital were preventable [4].

In spite of a multitude of reports, much of the discussion of error in medicine remained confined to the academic journals until the landmark 1999 report, "To Err is Human" catapulted the issue of preventable patient harm from academia into public discourse. The report estimated 48,000–98,000 deaths per year in US hospitals from medical errors and shocked the world by equating these deaths with the graphic analogy of one jumbo jet crashing per day [5].

More recently, a 2010 analysis of Medicare beneficiaries found that at least 13.5 % of hospitalized patients suffer an adverse event and almost half of these are preventable. The report concluded that about 15,000 patients (from the Medicare population alone) die in US hospitals *every month* as a result of potentially preventable adverse events [6].

These findings led healthcare experts to conclude that health care in the USA has an appalling problem of "waste, danger, and death"—words used to describe the grave condition of America's highway systems by President Eisenhower in a 1954 speech.[1]

Although the aforementioned reports are from the USA, a similar concern about adverse events has been found in hospitals around the world. Two widely quoted studies based on retrospective review from British hospitals found that approximately 10 % of patients experience adverse events; a third to half of these are preventable and often lead to disability and death [7, 8]. Similar findings have been reported from hospitals in Canada [9], Sweden [10], Brazil [11], Australia [12], and the Netherlands [13]. In a report from Israel, clinicians in a medical–surgical intensive care unit of a university hospital made 554 errors over 4 months or 1.7 errors per patient per day [14]. A recent 2012 report evaluating the extent of adverse events in developing countries (Egypt, Jordan, Kenya, Morocco, Tunisia, Sudan, South Africa, and Yemen) found that 8.2 % of the medical records showed at least one adverse event. Of these events, 83 % were judged to be preventable, and about 30 % were associated with death of the patient [15]. The report concluded that "unsafe

---

[1]President Dwight D. Eisenhower 1954 speech available at http://www.fhwa.dot.gov/interstate/audiogallery.htm. Last accessed Dec 30 2012.

patient care represents a serious and considerable danger to patients in the hospitals that were studied, and hence should be a high priority public health problem."

This irrefutable evidence of error and harm has spurred the healthcare community to action and there is now a global conversation about patient safety. Over the last decade, patient safety has become a focus of attention of healthcare leaders, quality experts, journalists, and concerned citizens. The Federal Government of the USA passed the Patient Safety and Quality Improvement Act of 2005 to create a network of patient safety organizations and to promote a culture of safety in health care. The World Health Organization created the World Alliance for Patient Safety to foster global awareness. The 2009 American Recovery and Reinvestment Bill (ARRA) provides for approximately $36 billion in incentive payments to hospitals and office practices who demonstrate "meaningful use" of electronic health records; improvement of quality and safety is a core component of the "meaningful use" criteria defined by the federal law. Patient safety is moving to the forefront of the strategic priorities agenda of most hospitals, regulatory agencies, improvement organizations, as well as legislative bodies.

The Institute of Medicine **defines patient safety** simply as "freedom from accidental injury [5]." Moreover, patient safety is also now an emerging scientific discipline—a field of both inquiry and action. Experts have defined it as *"a discipline in the health care sector that applies safety science methods toward the goal of achieving a trustworthy system of health care delivery. Patient safety is also an attribute of health care systems that minimizes the incidence and impact of, and maximizes recovery from, adverse events [16]."* Implicit in this definition is the understanding that with concerted systematic efforts, much of the harm from medical errors can be prevented.

## Why This Book?

Despite a flurry of activities in patient safety, many of my fellow practicing clinicians on the front line—physicians, nurses, ancillary professionals—remain disengaged if not disenfranchised from this important conversation. While administrators and leaders convene and deliver lectures at patient safety conferences, many clinicians believe they are too busy taking care of patients to learn this "new thing called patient safety" which is often viewed as one more activity imposed by their administrators. Although the evidence is clear, many of us believe that adverse events and medical errors happen at *other* institutions or in *other* departments or to *other* people's patients—*not ours*. We also feel that acknowledging medical errors is an affront to our skills, our education, our craft, and our fundamental commitment to our patients to "do no harm." This book aims to engage and educate practicing clinicians to challenge these long-held but no longer tenable values because changing them is a matter of urgency for our patients as well as our profession.

A unique feature, and I believe, a significant strength of the book is the use of the case-based learning format: clinical cases are described and analyzed to illustrate various types of medical errors and to propose systems-based solutions for the prevention of adverse events. Patient safety concepts such as "systems improvement," "cognitive biases," "heuristics," "human factors engineering," and "just culture" are by no means a routine part of most clinicians' vocabulary. Didactic lectures on patient safety do not engage many clinicians. They find the content, medical errors, threatening and the solutions, systems improvement, and baffling. Furthermore, compared to the long history of the scientific foundation of biological diseases based on anatomy, physiology, and molecular biology, the scientific foundation of patient safety is evolving only recently and much of the understanding of patient safety is based on narrative only; hence the value of case-based learning.

Case-based learning has been a vital tool in medical education but it is even more important for a new discipline like Patient Safety. I believe that harnessing the unique power of real-world clinical scenarios rich with the complexity of clinical experience and narrative will spark greater clinician enthusiasm in learning patient safety.

## Principles of Patient Safety

Traditionally, an unexpected adverse event was equated with an error. An error, in turn, was equated with incompetence or even negligence. Consequently, punishing individuals was considered to be the only method to improve safety of patients. However, this "name, blame, and shame" approach has a toxic effect. Not only does it not improve safety, it also continues to push the issue of medical errors into secrecy.

The discipline of Patient Safety acknowledges that risk is inherent in medicine and error is inherent in the human condition. Prominent theologian Saint Augustine declared over 1200 years ago "fallor ergo sum" or "I err, therefore I am." Savielly Tartakower, the famous Russian chess player wisely proclaimed, "The mistakes are all there, waiting to be made."

Based on this principle, the foundational contribution of the patient safety movement has been to propagate the insight that medical error is the result of "bad systems," not "bad apples" and CAN BE REDUCED by redesigning systems and improving processes so that caregivers can produce better results [17].

One thing is clear—while the discipline of Patient Safety is rooted in other high hazard industries, such as aviation, nuclear power, and manufacturing, the uniqueness of health care must not be lost. Health care is more unpredictable, complex, and nonlinear than the most complex of the airplanes and nuclear power plants. Machines respond in a predictable way to a set of commands and processes; patients don't—their response to medications and clinical interventions is far more variable and unpredictable. Machines don't have families, emotions, culture, language barriers, or psychosocial issues; patients do. While it is vitally important for us to learn techniques and lessons from other industries, health care must produce leaders and champions from within the clinical community to face up to this challenge and devise solutions unique to the clinical environment.

This patient safety text is founded on three propositions.

## *"The Soil, Not the Seed"*

The most fundamental intellectual contribution of the decade-long progress in patient safety is the seemingly simple yet profound insight that most errors are caused by bad systems, not bad people. Wrong-site and wrong-patient surgeries happen not because of incompetent surgeons but because of unreliable processes of patient identification and surgical site marking. Medication errors happen not because of inattentive nurses but because of a needlessly complicated multistep system of medication management from prescribing to dispensing to administration. As first proclaimed by the nineteenth century French chemist, Louis Pasteur, "it is the soil, not the seed" [18]. The patient safety discipline proposes that the fertile ground for medical errors is the "soil" of the healthcare delivery system and not the "seed" of the clinician.

Using the analysis of various clinical cases of adverse events, the book provides "real-world" examples of shifting the focus away from blaming and punishing individual clinicians to improving systems and processes.

## *From "I" to "We"*

The second quintessential underpinning of the Patient Safety discipline is that safer care is a function of good teams, not good individuals acting alone. This is because the technological sophistication of the last century has introduced unprecedented *complexity* and *fragmentation* in health care. The number and complexity of medical and computer equipment in an operating room or an intensive care unit has reached beyond the human capacity to safely monitor and operate them without great attention and team coordination. This complexity introduces an inherent risk of error lurking in what has been called "the bloody crossroads where complex technical systems meet human psychology."[2] In Medicine, poor management and coordination on this bloody crossroads cost patients their lives.

Nothing in clinical care is linear or predictable. There are frequent interruptions, shift changes, and discontinuity in care. Care has also become fragmented—a typical patient in an intensive care unit is the recipient of some 178 "activities" per day performed by tens of different types of professionals [19]. In The Emperor of All Maladies, the course of a cancer patient's illness exemplifies the complexity and fragmentation of modern patient care [20].

> *Eric's illness had lasted 628 days. He had spent one quarter of these days in a hospital bed or visiting the doctors. He had received more than 800 blood tests, numerous spinal and bone marrow taps, 30-X-rays, 120 biochemical tests, and more than 200 transfusions. No fewer than 20 doctors—hematologists, pulmonologists, neurologists, surgeons, specialists, and so on—were involved in his treatment, not including the psychologist and a dozen nurses.*

---

[2]David Brooks. Op-ed: Drilling for Uncertainty. The New York Times. May 27, 2010.

The relationships among these innumerable professionals are multidimensional and evolving. They are ever more subspecialized in their specific domains but not trained to work together as a team. The notion of teamwork may appear almost intuitive to lay people, but for medical professionals, this can initially appear as almost unnecessary to somewhat intrusive. Physicians have traditionally been thought of as the "captain of the ship." The professional training and practice model in medicine has been based on the competence and the accountability of the individual. By contrast, the discipline of Patient Safety rejects the notion of "I" in favor of "we." It proposes that the only possible way to deliver safe and efficient care in such a complex, fragmented system is for various professionals to work together as a coordinated team. No matter how obvious it is theoretically, bringing this notion to practice will require a recalibration of the role of various members in the interdisciplinary patient care team. Cases and the analyses in the following chapters illustrate the value of team work and provide evidence about the urgent necessity in making the cultural adjustments in how we view ourselves and our colleagues in the ecosystem of health care.

## "Just Culture"

The concept of "just culture" is based on the following three premises.

First, advances in patient safety are dependent on our ability to learn from adverse events and therefore, on the willingness of the clinical staff to report near-misses as well as patient harm events. The staff must believe and feel that the reporting is primarily for the purposes of learning and not for punishment. Given the current status of reporting, this is not a trivial issue. According to a recent report by the Office of the Inspector General, hospital incident reporting systems captured only an estimated 14 % of the patient harm events experienced by Medicare beneficiaries [21]. In his testimony to the U.S. Congress in 1997, Dr. Lucian Leape, a renowned patient safety expert, stated, "The single greatest impediment to error prevention is that we punish people for making mistakes."[3] David Marx, a noted author and expert in human error, explained in a 2001 report, "Few people are willing to come forward and admit to an error when they face the full force of their corporate disciplinary policy, a regulatory enforcement scheme, or our onerous tort liability system [22]." So our only hope for improving systems and processes of care lies in providing a fair and nonpunitive environment for reporting errors.

Second, the shifting of focus away from blaming and punishing individual clinicians has allowed us to recognize and acknowledge that even the most competent, skilled, and caring clinician is not exempt from human error and that human fallibility is inevitable. James Reason, the author of the Human Error, famous for the Swiss

---

[3] Leape LL. Testimony, United States Congress, House Committee on Veterans' Affairs; 1997 Oct 12.

Cheese model of medical errors observed, "Human fallibility is like gravity, weather, and terrain, just another foreseeable hazard [23]." In his book, Whack-a-Mole, David Marx writes, "Just as tornados and lightning strikes are unavoidable, predictable components of the weather, I know that human fallibility, my own included, is an unavoidable, predictable component of being human [24]." The Just Culture model proposes that since human condition cannot be changed, the only hope for safer care lies in a relentless focus on improving systems of care.

Third, the above two principles must be balanced with the need for accountability because no organization, no society can afford to offer a "blame-free" system where acts of gross misconduct or of reckless disregard for patient safety are not subject to appropriate disciplinary action. Just culture addresses this need to reconcile the "no-blame" approach to facilitate learning and reporting with "accountability" that is judicious, appropriate, and takes into account the type and magnitude of human error. Just culture provides a framework of shared accountability: healthcare institutions are responsible for providing systems and environment that are optimally designed for safe care and staff are responsible for their choices of behavior and for reporting system vulnerabilities.

The just culture model distinguishes between different type of errors and behaviors and provides guidance for potential disciplinary courses of action. The first type is the "human error," inadvertently doing other than what should have been done and includes errors such as a slip or a lapse. This is considered an inevitable part of human fallibility and should be managed through designing systems that are more error-proof and error-tolerant. The second is "at-risk behavior," behavioral choices that increase risk where risk is not recognized such as staff using workarounds to established processes. Such behavior should lead to coaching of the staff concerned regarding the consequences of their actions in addition to systems improvement. The final is "reckless behavior," behavior to consciously disregard a substantial and justifiable risk. For example, a surgeon refusing to sign the operative site or to participate in time-out process will be considered reckless behavior and will be worthy of punitive action [25]. Fortunately, such instances are rare and most errors fall into the category of human error or at-risk behavior.

The book illustrates the concepts of just culture through numerous case studies and includes a separate chapter on the culture of safety that discusses in details other elements that constitute a safe culture in a healthcare organization.

## What's in the Book?

Patient Safety is an evolving field. This text provides case-based discussions on various patient safety topics organized in four sections. The first section, Concepts, covers topics that are of universal application such as patient identification, teamwork and communication, and hand-off and care transition. The second section, Examples, provides analysis of root causes and best practices for preventing common complications of health care, e.g., medication errors, falls, and pressure ulcers.

The third section, Special Considerations, covers special patient safety issues relevant to specific fields such as Pediatrics, Radiology, and Behavioral Health. The fourth and final section, Organizational Issues, discusses topics around building a patient safety program from an organizational perspective, e.g., the culture of safety and error disclosure.

Each chapter provides an analysis of clinical cases based on the root cause analysis (RCA) methodology—a structured method relatively new to health care but with a long and successful track record in analyzing accidents in other high hazard industries [26]. The central tenet of RCA is to identify underlying systems problems that increase the likelihood of errors (called "latent errors") while avoiding the trap of focusing on mistakes by individuals (called "sharp-end errors"). The RCA process is designed to answer three basic questions: what happened, why did it happen, and what can be done to prevent it from happening again? While systems and processes often need to be tailored to local institutions, the basic principles of systems improvement are generalizable and therefore lessons learned from our cases are widely applicable.

# Book Chapters

**Chapter 1 on Patient Identification** begins with a discussion of the prevalence and causes of misidentification errors. The first case study takes place in an outpatient setting where various clinicians did not use proper identification procedures leading to the wrong patient being examined. The second case study describes an inpatient scenario when the blood is drawn from the wrong patient due to suboptimal processes of patient identification and specimen labeling at the bedside leading to a near miss event of mismatched transfusion. The chapter describes the RCA of the two cases using the "five rules of causation" and discusses corrective actions including the relative strength of the various actions in fixing the systems issues. Various patient, culture, and environment-related factors leading to misidentification are described. Key lessons emphasize the importance of double identifiers, active identification processes, "write-down" and "read-back" and the role of technology in facilitating patient identification.

**Chapter 2 on Teamwork and Communication** describes two illustrative case studies to emphasize the vital role of teamwork and communication in safe delivery of health care in an increasingly complex environment. The chapter begins with a discussion of the definition of team and teamwork, benefits of a team-based approach, and special interprofessional issues around nurses and physicians. It provides a comprehensive literature review on the contribution of poor teamwork and communication and disruptive behavior as root causes of adverse events. In the first case, the patient suffers a respiratory arrest when a paralytic agent is administered inadvertently before intubation due to poor teamwork and communication within the surgical team. In the second case, the patient suffers a pulmonary embolism due to a delay in the ordering and administration of heparin. The chapter discusses the adoption of the aviation industry's Crew Resource Management (CRM) methodology into health care and the TeamSTEPPS© program. Various practical strategies to improve communication, such as "SBAR" and critical language, e.g., "CUS," are described using clinical examples.

**Chapter 3 on Handoff and Care Transitions** examines two cases of adverse events to illustrate that transition of care and attendant handoffs are points of special vulnerability for patient safety. In the first case, the patient suffers death from poor management of postpartum hemorrhage due to ineffective handoff and communication between the operating room and the recovery room. In the second case, a patient with head injury suffers respiratory depression when excess dosage of opioids are prescribed due to poor handoff and communication between teams of neurosurgery and anesthesia (involving attendings and residents) during multiple shift changes. The chapter categorizes transitions of care into five points (1) interhospital, (2) interdepartmental, (3) inter-shift, (4) interprofessional, and (5) intra-team for a clearer understanding of the handoff issues. Barriers to effective handoffs include diversity of teams, time and resource constraints, as well as issues pertaining to the presence of residents in teaching hospitals. Various improvement strategies include standardization of handoff communication using written (e.g., SBAR, sign-out templates) and verbal (e.g., SBAR, read-back) methods, information technology-based solutions, and greater attending physician supervision when delegating care to a less experienced practitioner. The chapter emphasizes that effective sign-outs should generate a shared mental model, i.e., a common understanding of the patient's clinical condition.

**Chapter 4 on Graduate Medical Education and Patient Safety** discusses the evolution of regulatory and policy changes related to residents duty hour restrictions and their impact on patient safety (largely beneficial but concerns remain regarding increased discontinuity of care and handoffs). It describes various patient safety issues pertinent to resident supervision through the lens of two case studies. In the first case, the patient's condition deteriorates necessitating intubation and transfer to ICU due to poor supervision and failure to call for expert help. In the second case, a patient with do-not-resuscitate (DNR)/do-not-intubate (DNI) orders is inadvertently intubated due to poor communication during shift change facilitated, arguably, by duty hour restrictions. The chapter dissects issues around balancing the need for greater supervision for patient safety with the need for resident autonomy for adequate training. Tools to measure clinical supervision as well as best practices to improve supervision and communication including the innovative SUPERB/SAFETY model are described in details. The key lessons presented will be helpful to healthcare organizations in designing strategies for safe supervision in other types of teaching programs such as supervision of mid-level providers and nursing and pharmacy student trainees.

**Chapter 5 on Electronic Health Record and Patient Safety** discusses that while there are demonstrated benefits of health information technology tools such as electronic health record (EHR) and computerized physician order entry (CPOE), these systems can also introduce new safety hazards. In the first case study, a patient with an indwelling epidural catheter for postoperative analgesia is pre-scribed an anticoagulant using CPOE system. Since the system is not configured to detect a drug (anticoagulant)—route (epidural) interaction, the error goes unde-tected potentially exposing the patient to the risk of a spinal hematoma. In the second case, the CPOE system allows the patient's weight to the entered in pounds or kilograms. Consequently, the staff makes the error of entering the weight as 88 lb instead of the intended 88 kg. This leads to a substantial undercalculation of the weight-based dosing of unfractionated heparin. The chapter discusses emerg-ing evidence regarding safety concerns and unintended consequences of EHRs. The "sociotechnical model" is discussed as a framework for analyzing and solving EHR-related safety issues.

**Chapter 6 on Clinical Ethics and Patient Safety** describes that patient safety and ethics are interrelated concepts. Clinical ethics is similar to other clinical practices and can be evaluated and improved using basic quality improvement principles. In addition, promoting patient safety rests on core ethical principles ubiquitous in medicine—the professional duties to provide benefit and prevent harm. The first case study describes a 93-year-old patient with end-of-life decision making issues where multiple family members are in conflict regarding the plan to withdraw life-sustaining treatment. The analysis includes the intersection of ethics and law, ethics and patient safety, evaluation of decision-making capacity, and the role of DNR and the emerging POLST (Physician Orders for Life-Sustaining Treatment) protocol. The second case study describes disruptive physician behavior where an eminent cardiologist declines to comply with the hand-washing practice potentially contrib-uting to the *Clostridium difficile* outbreak in the hospital. The chapter analyzes the issue of professionalism and describes that discussing patient safety issues in terms of ethical responsibilities has the potential to motivate clinicians to improve quality and safety within their individual practices.

**Chapter 7 on Medication Error** describes that the medication errors occur in all clinical settings and are a source of substantial *preventable* harm to patients. The chapter elucidates various classification schemes for medications errors based on the level of patient harm, on the five stages of the medication management process, and on the root cause of errors. In the first case study, a patient with metastatic cancer suffered respiratory failure due to the inadvertent prescribing of opioids, a common source of adverse drug events. The analysis elucidates various practical measures for safe usage of opioids including assessment and reassessment of pain and accu-rate equivalence calculations for different types/routes of opioid administration. In the second case, a patient with breast cancer suffered from severe complications after receiving the wrong chemotherapeutic agent—Taxotere instead of Taxol. This is a case of look-alike, sound-alike, and spell-alike drugs. Measures to mitigate the risk of error from such medications include the use of "Tall Man lettering," color-coded storage bins, and the use of electronic systems such as the bar-coded medication administration. The five essential strategies in improving medication safety include: (1) the role of information technology, (2) addressing health literacy and engaging patients and families, (3) preventing risk from "high-alert" medications, (4) medication reconciliation, and (5) the vital role of pharmacists' collaboration on inpatient teams. Finally, an action plan is charted for various health team members including the prescriber, the pharmacist, the nurse, the patient, and the caregiver.

**Chapter 8 on Medication Reconciliation Error** defines medication reconciliation and its role as a key safety practice to prevent medication errors across the continuum of care using two case studies. In the first case, the patient is readmitted with digoxin toxicity when the digoxin is inadvertently continued as a home medication despite high digoxin levels during hospitalization. This adverse event illustrates the failure of appropriate medication reconciliation upon discharge. In the second case, the patient suffers from pulmonary embolism after hip fracture surgery when her anticoagulant therapy is inadvertently omitted upon transfer to a rehabilitation facility due to poor reconciliation of medications. The chapter provides practical strategies to reduce reconciliation errors at all points of transition (e.g., the role of pharmacists and nurses in obtaining a good medication history and the importance of the review of the electronic claims data) and discusses the role of electronic medication reconciliation systems in improving medication safety.

**Chapter 9 on Retained Surgical Items** discusses the problem of retained surgical items (RSI) from a perioperative safety perspective. According to the Joint Commission sentinel events database analysis, this has become the commonest surgical safety "never event" surpassing wrong-site surgery. The author emphasizes that although much of the current literature continues to focus on the traditional patient and surgical procedure-related risk factors for RSI, RSIs occur primarily due to suboptimal communication practices among multiple OR stakeholders. Three case studies are described to illustrate the three types of RSI events: No Count Retention Case (NCRC), Correct Count Retention Case (CCRC), and an Incorrect Count Retention Case (ICRC). This classification is valuable because of distinct prevention strategies for each type: implementation of a rigorous count policy for NCRC, improved and standardized sponge counting methodologies for CCRC, and improved communication with multiple stakeholders including radiologists, if needed, for ICRC. The chapter also discusses prevention strategies for retention of "small miscellaneous items" such as broken needles, instrument parts, or guidewires. The emerging technological adjuncts such as bar-coded sponges and radiofrequency identification (RFID) tagged sponges are also described.

**Chapter 10 on Wrong-Site Surgery** provides a detailed analysis of the incidence, etiology, and impact of wrong-site/wrong-patient surgery procedure—one of the most commonly reported sentinel events to the Joint Commission. Using two case studies, the chapter elucidates a chain of systems vulnerabilities that lead to this highly undesirable outcome. In the first case study, the patient underwent the amputation of the wrong side lower limb with devastating consequences both for the patient as well as the operating team due to a lack of Universal Protocol implementation in the operating room. In the second case study, a resident inserted a central venous catheter in the wrong patient who unfortunately died from a resultant pneumothorax, a fatal complication of a procedure she wasn't supposed to have. The chapter highlights the fact that such errors occur both inside and outside the operating room as well as across multiple specialties; hence the Universal Protocol must be utilized throughout an institution. Avoidance of these errors requires aggressive education of all staff, clinical and nonclinical, in the risk factors and root causes for these events.

**Chapter 11 on Transfusion-Related Hazards** describes in details the various process failures that led to two events involving blood and stem cell transfusions, along with possible solutions. In the first case study, the incorrect labeling of a unit of stem cells during the preparation and freezing process, and the lack of verification upon thawing and preparation for infusion led to the release of a pooled unit which was appropriately labeled, but which may have contained a unit from another patient. Discussion includes human fallibility and the tendencies in health care to blame, shame, and/or train in response to an error, as well as issues of safety culture, the value of verification and second-person checks, high reliability, and normal accident theories, form design, and other human factors. The second case study involves the repercussions of a misperceived verbal handoff leading to the selection of an incorrect patient in a hospital's computerized physician order entry (CPOE) system that nearly resulted in the mis-transfusion of red blood cells. Discussion in this case comprises communication and hand-off issues, resident duty hours, interruptions, and computer interfaces and alerts. The causal tree building method of RCA is described and illustrated in both of the case studies, including the classification of causes that leads to solution discovery.

**Chapter 12 on Hospital-Acquired Infections (HAIs)** summarizes a historical background of the infection control movement beginning with the nineteenth century physician Ignaz Semmelweis and continuing onto the current focus on the prevention of HAIs. The first case study describes an incident of *C. difficile* outbreak on a hospital floor due to multiple breakdowns in the infection control practices on the unit. The second case study describes a central line blood stream infection with MRSA due to a line placed in an emergency that the subsequent care team failed to notice and remove in a timely fashion. The discussion illustrates the role of infection control as a team-based enterprise including the success of the Comprehensive Unit-based Safety Program (CUSP), the role of video surveillance in promoting hand-washing, and the concept of device utilization ratio to measure the incidence of HAIs. A separate discussion section is provided on the prevention strategies for each of the HAIs including *C. difficile* antibiotic-associated diarrhea (CDAAD), central line-associated bloodstream infection (CLABSI), catheter-associated urinary tract infection (CAUTI), ventilator-associated pneumonia (VAP), and surgical site infections (SSI). The key lessons include the facts that HAIs are unacceptable at any level, many HAIs can be avoided by the consistent use of bundled checklists, and that the most effective way to prevent device-associated HAIs is to remove them as soon as possible.

**Chapter 13 on Hospital Falls** begins with a definition of the falls as a serious patient safety concern. Falls are common during hospitalization and are often associated with adverse outcomes such as fractures, head injury, and even death. In the first case study, an elderly man with multiple medical problems and on multiple cardiovascular medications, sustains a fall with an intertrochanteric fracture while trying to get up to grasp the water pitcher on the bedside table. In the second case study, the patient is an elderly woman with early Alzheimer's disease and delirium who sustains a fall and subdural hematoma while trying to avoid calling the nurse for assistance in getting out of the bed. The authors argue that while RCA is a common tool used to understand the underlying causes of adverse events, an expansion of this tool, aggregate RCA, can be more useful in analyzing high volume frequent events, such as falls, to identify trends and systemic issues across similar occurrences. A high-level process map of a hospital's experience related to falls prevention including various risk assessment tools is presented. The benefits and limitations of falls prevention interventions such as bed alarms, low beds, frequent patient rounding, and increased ambulation are discussed. Multifactorial interventions and addressing systems issues such as improving handoff and communication and improving skill and knowledge related to fall risk and prevention are most effective ways to prevent falls in hospitalized patients.

**Chapter 14 on Pressure Ulcers** begins with a discussion of the classification, staging, and epidemiology of pressure ulcers and highlights that as a "never event" their prevention as a serious patient safety issue. In the first case study, an elderly patient with complex medical conditions develops "suspected deep tissue injury." The detailed RCA of the case revealed the importance of various prevention measures including assessment and reassessment of skin integrity, nursing care including regular turning and positioning in the bed, adequate nutrition, and communication and information management between physicians and nurses. In the second case study, a young patient with a gunshot wound causing an unstable cervical/thoracic spine fracture undergoes prolonged life-saving surgery for ten hours and develops pressure ulcers to the occiput and the sacral area. Management issues specifically pertinent to pressure ulcers prevention in neurosurgery patients with hemodynamic instability are discussed. The chapter describes that a multidisciplinary institution-wide strategy is vital in the prevention of this "never event." Often it is perceived to be a nursing issue but the chapter clearly illustrates the vital importance of the involvement of multiple disciplines including physicians, nutritionists, and wound care nurses in the prevention of pressure ulcers.

**Chapter 15 on Diagnostic Error** discusses that the errors related to missed, delayed, or wrong diagnoses are common, costly, harmful, and a leading source of malpractice claims in the USA. However, these are a relatively ignored aspect of patient safety; the patient safety guru Robert Wachter wrote that diagnostic errors "don't get any respect" [27]. The chapter describes that diagnostic errors happen in all settings but are particularly common in the ED and ambulatory care and can be a source of significant morbidity and mortality. There is an in-depth discussion of the cognitive model of physician's clinical decision making as the basis for understanding various cognitive biases that may lead to diagnostic errors. A classification of diagnostic errors using DEER taxonomy is also described. The first case study involves the misdiagnosis of typhilitis in a 3-year old leading to delay in making the correct definitive diagnosis of appendicitis. The second case describes a potentially serious delay in diagnosis of congenital adrenal hyperplasia in an infant because the team incorrectly attributed the hyperkalemia to hemolysis. The chapter highlights that diagnostic errors are related to cognitive errors or system errors or most frequently due to a combination of both. The chapter concludes by providing practical tips for reducing diagnostic errors such as "metacognition" and "diagnostic pause."

**Chapter 16 on Patient Safety in Pediatrics** discusses the unique attributes of pediatric patient safety due to different physical characteristics, developmental issues, and the dependent/legal/vulnerable state of the children. The authors discuss the epidemiology of errors and patient harm in both outpatient and inpatient pediatric care. In the first case study, a 9-month-old infant presenting with scrotal pain ended up with orchiectomy as appropriate pediatric clinical and radiological expertise at the local community hospital was not available and there was a delay in transferring to a children's hospital. In the second case study, a 16-year-old boy was hospitalized for cellulitis and discharged but required readmission for a severe inflammatory bowel disease flare as this chronic disease was not recognized during the previous admission. Adolescents are a special challenge because they hesitate to complain, do not want to stay in the hospital, and may fail to advocate for themselves. In the third case study, a 5-year old with chronic lung disease suffered severe respiratory distress requiring intubation because the team forgot to order oral steroids after the taper of the IV form and multiple early warning signs were not recognized by the nursing and physician team due to poor communication. The chapter describes the PEWS (Pediatric Early Warning System) as a structured tool to improve care in a deteriorating patient.

**Chapter 17 on Patient Safety in Radiology** emphasizes the need for attention to patient safety in Radiology as a rapid growth in the use of imaging, particularly Computed Tomography (CT) scans, has nearly doubled the US population's exposure to ionizing radiation. In the first case study, a 70-year-old patient is found to have a 4 mm nodule in the right lower lobe of the lung (an "incidentaloma") during a CT chest for preoperative evaluation. This leads to a clinically unnecessary high-resolution CT chest three months later causing harmful radiation exposure. The analysis of the case describes the appropriate radiological follow up of incidentalomas, the need to review prior imaging studies, the application of the ALARA (As Low As Resonably Achievable) principle to minimize radiation exposure, the role of computerized decision support systems in proving real-time feedback to decrease inappropriate utilization of imaging tests, and the need for clear and direct communication between interpreting radiologist and ordering physician including the use of the critical test result management (CTRM) software. The second case study describes the frequent clinical dilemma of performing imaging in a pregnant patient with suspected appendicitis. The chapter describes the amount of radiation exposure incurred in various studies, the safety thresholds for the developing fetus, and provides practical recommendations for clinical use. The chapter stresses the importance of patient involvement in the decision-making process and of presenting information concerning the benefits and risks of proposed imaging studies clearly and honestly without creating unnecessary anxiety. Additional issues such as pediatric radiation safety (the Image Gently campaign), the MRI safety, and key role of communication are also discussed.

**Chapter 18 on Patient Safety in Anesthesia** describes that with the responsibility of caring for vulnerable patients in life-threatening situations, anesthesiologists must maintain a high level of vigilance and preparedness. The first case study describes the issue of substance abuse and physician impairment among anesthesiologists. These are concerns for all specialties but with ready access to narcotics and high stress levels, substance abuse poses a particularly strong risk for anesthesiologists. The solutions include restricting access to drugs, detailed accounting of drug usage, early detection of physician impairment, and educational programs to help identify impaired colleagues. In the second case, a patient suffers anoxic brain injury due to the failure in anticipating a difficult airway. The RCA illustrates that the success in airway management hinges on anticipation, planning, and preparedness, and that every case requires a preformed detailed rescue plan in the case of a difficult airway. The chapter also describes the American Society of Anesthesiologists (ASA) guidelines for difficult airway management and the emerging role of simulation in improving procedural skills, team communication, and emergency preparedness.

**Chapter 19 on Patient Safety in Behavioral Health** describes that behavioral health patients pose unique and complex safety challenges whether being treated in an emergency room, acute psychiatric unit, or general hospital. The typical harm risks encountered in behavioral health settings can be summarized using the SAFE MD mnemonic and include Suicide, Aggressive behavior, Falls, Elopement, Medical comorbidity, and Drug errors. Of note, in the USA, suicide ranks as the tenth leading cause of death and within the top four leading causes of death for persons from age 10 to 54. In the first case study, the escalating aggressive behavior of a patient in the ER leads to the application of wrist restraints, a worsening of agitation and eventually the adverse outcomes of two staff members being injured by the patient, and the patient sustaining a wrist fracture. In the second case, a young man with a past history of recurrent depression is admitted with worsening psychosis. The team underestimates his risk of self-harm and eventually he successfully commits suicide in his inpatient room. Detailed RCAs of both cases are described using the "fishbone model" leading to a discussion of the practical risk reduction strategies to mitigate safety risks in behavioral health patients. These strategies include the establishment of clear roles and responsibilities, work standards for communicating clinical information, clear guidelines for escalating safety concerns, ongoing environmental risk audits, and a culture of respect and sensitivity to potential "sanist" attitudes. The risk reduction strategies must be balanced against the patient's civil rights associated with least restrictive alternatives such as to be free of undue restraint.

**Chapter 20 on Patient Safety in Outpatient Care** proposes that the unique feature of ambulatory care safety is the central role of the patient and caregiver in ensuring safe delivery of care. While much of the patient safety movement has focused on inpatients, the chapter discusses the urgent need to recognize and implement solutions to prevent adverse events in outpatient care where most of the care is delivered. The chapter describes the epidemiology and the impact of adverse events in outpatient care. The first case study describes inadequate diuretic medication monitoring in a 66-year-old patient with diabetes, hypertension, and heart failure leading to the symptoms of hyponatremia due to a lack of coordination between his primary care physician, cardiologist, and endocrinologist. The case highlights the importance of multiple issues in outpatient safety such as treatment complexity, medication understanding, physician–patient communication, aggressive treatment goals, symptom recognition, and transition among multiple providers. In the second case study, a 77-year-old patient is referred from a rural area to a teaching hospital for knee replacement surgery where a chest x-ray reveals a suspicious lung mass leading to the cancelation of the surgery. However, there is no communication of the abnormality to the patient's primary care physician. This scenario will be all too familiar to most clinicians and raises various issues such as outpatient health system fragmentation and poor information availability, gaps in hospital documentation, poor notification of abnormal results, and the important role of patient awareness of abnormal test results. The authors have adopted the classic Wagner chronic disease model to provide a conceptual framework for patient safety and describe the underlying health system and community conditions, and patient and provider characteristics for safe provision of outpatient care with desired health outcomes.

**Chapter 21 on Error Disclosure** describes that the traditional ad hoc, legally oriented, "deny and defend, shut up and fight" adversarial model of disclosure of errors is ineffective in addressing and identifying key safety concerns in health delivery systems. The chapter presents an alternative, systems-focused approach to medical error disclosure and assessment. This system consists of standardized "error disclosure teams" and employs the "three Cs" throughout mediation and all error-related communication—Concern, Commitment, and Compassion. Some of the potential legal issues associated with apology and its use in disclosure systems are also reviewed. In the first case study, a patient suffers an intraoperative cardiac event and death due to inadvertent administration of a wrong medication caused by a syringe-swap error. In the second case study, there is a delay in the diagnosis of an eye infection leading to the need for the removal of the eye. The chapter describes the outcomes in the first case study for each stakeholder using the traditional system of error disclosure: the patient's family filed a lengthy and contentious lawsuit; the anesthesiologist settled the lawsuit with the family independently; the anesthesia resident quit her residency in distress; and the hospital settled with the family for an undisclosed sum. The chapter contrasts this with the outcome in the second case using the open system of error disclosure: the family settled the conflict in 8 months with the hospital assuming the cost of care; the family participated in the hospital's safety improvement efforts; the resident learned from the error and participated in educational activities; the hospital publicly thanked the family and patient for their help in improving patient safety; and the patient and family became advocates for the facility. Finally, practical implementable solutions are discussed to integrate these processes into the delivery system culture to promote patient safety.

**Chapter 22 on The Culture of Safety** describes that the culture is a function of the values, attitudes, perceptions, competencies, and patterns of behavior that influence the context in which care is delivered. There is emerging evidence that the culture of an organization has as much an impact on patient safety as the use of good clinical practices. The first case study illustrates the actions of an OR team, in the context of OR culture, after confronting a missing sponge with a negative intraoperative X-ray. The second case study describes the actions of an ED physician and nurse in the ordering and administration of an intravenous anticoagulant in the situation where the patient's weight is not readily available. Characteristics of a Culture of Safety include patient safety as an organizing principle, leadership engagement, teamwork, transparency, flexibility, and a learning environment. The authors discuss barriers to a culture of safety, surveys to measure the culture, and strategies to build and improve a safety culture.

**Chapter 23 on Second Victim** discusses that when a serious unanticipated adverse event occurs, while the patient as the recipient of the harm is clearly the "first victim," clinicians often also experience a harsh emotional response in the aftermath and may be described as "second victims." Without appropriate support and guidance, the distress experienced by healthcare providers may lead to long term consequences such as leaving their chosen fields prematurely or experiencing prolonged professional/personal suffering. In the first case study, an ED resident misses the diagnosis of an acute myocardial infarction and discharges the patient home. The patient returns later in critically ill condition requiring emergency intervention. In the second case study, the young daughter of an ED staff is brought by an ambulance in extremis. The ED team is unable to resuscitate the child of "one of their own" and she expires in the ED. In both cases, the clinicians involved in care suffer serious psychological distress and are "second victims" of adverse events. The authors, based on their research, describe a predictable recovery trajectory consisting of six distinct stages of the second victim phenomenon: (1) chaos and accident response, (2) intrusive reflections, (3) restoring personal integrity, (4) enduring the inquisition, (5) obtaining emotional first aid, and (6) moving on. The chapter concludes with the recommendation that health institutions should design a structured response plan that ensures ongoing surveillance for the identification of potential second victims as well as actions to mitigate emotional suffering immediately upon second victim identification.

# Who Is This Book for?

The book is written primarily for clinicians including physicians, nurses, and other healthcare professionals as well as those in training including medical students and house officers. The book should be useful for healthcare leaders and administrators at every level including the chief executive officer, chief medical officer, and chief nursing officer. Another group to benefit significantly from the book would be patient safety officers and quality and risk management professionals. They are often charged with conducting RCAs of adverse events in their institutions; they can use the analyses and solutions provided in the book as templates or examples for conducting RCAs.

Recognizing the importance of patient safety in training physicians and leaders of tomorrow, many medical schools are actively planning to formally incorporate patient safety courses in the medical school curriculum. In addition, the Accreditation Committee on Graduate Medical Education considers practice-based learning and systems-based practice as core competence for physicians in training providing further impetus for including patient safety in medical training. Medical schools and residency program educators should find the book a useful reference book for teaching patient safety using case-based learning method.

Hospitals—from small community hospitals to large academic medical centers—are faced with the challenge of disseminating the key principles of patient safety to all staff. Based on my own experience at a large urban academic hospital and communication with colleagues around the country, it is clear that the senior leadership at most hospitals has already committed to patient safety. However, the learning and commitment needs to disseminate from the boardroom to the bedside, from the administrators to the front line staff for it is the day to day practice of patient safety that will make care safer for patients. Hospital leaders should find this book a useful tool in educating and engaging clinical staff.

Finally, the book is written for a global audience. I recently spent 3 weeks in India as a Fulbright scholar focusing on patient safety at a large tertiary care medical center. The clinical stories and safety concerns of patients everywhere are the same globally. Although the patient safety solutions need to be customized according to the local environment of the hospital, the lessons learned from various cases in the book are generalizable and applicable to a global healthcare community.

Chicago, IL, USA                                               Abha Agrawal, M.D., F.A.C.P.

# References

1. Schimmel EM. The hazards of hospitalization. Ann Intern Med. 1964;60:100–10.
2. Steel K, Gertman PM, Crescenzi C, Anderson J. Iatrogenic illness on a general medical service at a university hospital. N Engl J Med. 1981;304(11):638–42.
3. Leape LL, Brennan TA, Laird N, Lawthers AG, Localio AR, Barnes BA, et al. The nature of adverse events in hospitalized patients. Results of the Harvard Medical Practice Study II. N Engl J Med. 1991;324(6):377–84.

4. Bedell SE, Deitz DC, Leeman D, Delbanco TL. Incidence and characteristics of preventable iatrogenic cardiac arrests. JAMA. 1991;265(21):2815–20.
5. Kohn LT, Corrigan J, Donaldson MS. To err is human: building a safer health system. Washington, DC: National Academy Press; 2000.
6. Levinson DR. Adverse events in hospitals: national incidence among medicare beneficiaries. Washington, DC: Department of Health and Human Services, Office of Inspector General; 2010.
7. Vincent C, Neale G, Woloshynowych M. Adverse events in British hospitals: preliminary retrospective record review. BMJ. 2001;322(7285):517–9.
8. Sari AB, Sheldon TA, Cracknell A, Turnbull A, Dobson Y, Grant C, et al. Extent, nature and consequences of adverse events: results of a retrospective casenote review in a large NHS hospital. Qual Saf Health Care. 2007;16(6):434–9.
9. Baker GR, Norton PG, Flintoft V, Blais R, Brown A, Cox J, et al. The Canadian adverse events study: the incidence of adverse events among hospital patients in Canada. CMAJ. 2004;170(11):1678–86.
10. Soop M, Fryksmark U, Koster M, Haglund B. The incidence of adverse events in Swedish hospitals: a retrospective medical record review study. Int J Qual Health Care. 2009;21(4):285–91.
11. Mendes W, Martins M, Rozenfeld S, Travassos C. The assessment of adverse events in hospitals in Brazil. Int J Qual Health Care. 2009;21(4):279–84.
12. Wilson RM, Runciman WB, Gibberd RW, Harrison BT, Newby L, Hamilton JD. The Quality in Australian Health Care Study. Med J Aust. 1995;163(9):458–71.
13. de Vries EN, Ramrattan MA, Smorenburg SM, Gouma DJ, Boermeester MA. The incidence and nature of in-hospital adverse events: a systematic review. Qual Saf Health Care. 2008;17(3): 216–23.
14. Donchin Y, Gopher D, Olin M, Badihi Y, Biesky M, Sprung CL, et al. A look into the nature and causes of human errors in the intensive care unit. Crit Care Med. 1995;23(2):294–300.
15. Wilson RM, Michel P, Olsen S, Gibberd RW, Vincent C, El-Assady R, et al. Patient safety in developing countries: retrospective estimation of scale and nature of harm to patients in hospital. BMJ. 2012;344:e832.
16. Henriksen K. Advances in patient safety: new directions and alternative approaches. Rockville, MD: Agency for Healthcare Research and Quality; 2008.
17. Berwick DM. Continuous improvement as an ideal in health care. N Engl J Med. 1989;320(1): 53–6.
18. Spath P, Minogue W. The soil, not the seed: Real problem with root cause analyses. 2008. Available from http://webmm.ahrq.gov/perspective.aspx?perspectiveID=62. Accessed 22 July 2012.
19. Leape LL. Error in medicine. JAMA. 1994;272(23):1851–7.
20. Mukherjee S. The emperor of all maladies: a biography of cancer. Large print ed. Waterville, ME: Thorndike Press; 2010.
21. Levinson DR. Hospital incident reporting systems do not capture most patient harm. Washington, DC: Department of Health and Human Services, Office of Inspector General; 2012.
22. Marx D. Just culture: a primer for health care executives 2001. Available from http://www.mers-tm.org/support/Marx_Primer.pdf. Accessed 22 Jul 2012.
23. Reason JT. Human error. Cambridge, England: Cambridge University Press; 1990.
24. Marx D. Whack-a-mole : the price we pay for expecting perfection. Plano, TX: By Your Side Studios; 2009.
25. Wachter RM, Pronovost PJ. Balancing "no blame" with accountability in patient safety. N Engl J Med. 2009;361(14):1401–6.
26. Wu AW, Lipshutz AK, Pronovost PJ. Effectiveness and efficiency of root cause analysis in medicine. JAMA. 2008;299(6):685–7.
27. Wachter RM. Why diagnostic errors don't get any respect- and what can be done about them. Health Aff (Millwood). 2010;29(9):1605–10.

# Acknowledgements

The seed for this book was planted during my deep involvement in conducting analyses of near misses and adverse events at Kings County Hospital in Brooklyn, NY. It rapidly became clear to me that there was an acute need to share the lessons learned from these analyses with other clinicians both inside and outside the organization. The seed germinated by sharing this experience with colleagues in the USA and around the world. We all had the same intense desire to share our insights with others in the healthcare community in the hope that this endeavor will make care safer for patients everywhere. Eventually, this small seed has culminated into a case-based comprehensive guide on Patient Safety.

I am immensely grateful to all the contributing authors. This book would not have been possible without their commitment to squeeze "yet another project" into their schedules and their unwavering faith in the project. Thank you for tolerating all those deadline reminders and still keeping a good cheer.

I had the good fortune of having a number of top leaders at the New York City Health and Hospitals Corporation during my time in New York who made Patient Safety as their number one priority for the largest municipal healthcare system in the country. They inspired me to aim high in the relentless pursuit of safer care and fostered an intellectual environment for this project to take root. Among them, I would like to acknowledge and thank Alan Aviles, Ramanathan Raju, M.D., Antonio Martin, Jean Leon, and Kathie Rones, M.D. Since my move to Chicago last year, I have found a superb leader and strong advocate of patient safety in José Sánchez at Norwegian American Hospital.

I thank Edmund Bourke, M.D., my mentor, not only for critiquing the multiple serial drafts of the manuscript, but also for being a constant source of motivation throughout my career. Vee, even though you were not with me physically, your spirit guided me throughout the project; for this, I am eternally grateful. Finally, Kanha, your addition to my life has made everything, including this project worthwhile; welcome to my world.

# Contents

# Contributors

**Abha Agrawal, M.D., F.A.C.P.** Norwegian American Hospital and Northwestern University Feinberg School of Medicine, Chicago, IL, USA

**Renuka Ananthamoorthy, M.D.** Department of Behavioral Health, Kings County Hospital Center, Brooklyn, NY, USA

**Vineet M. Arora, M.D., M.A.P.P.** Department of Medicine, University of Chicago, Chicago, IL, USA

**Robert J. Berding, J.D., M.S.** Regulatory Compliance, Risk Management and Patient Safety, Kings County Hospital Center, Brooklyn, NY,USA

**Grace M. Blaney, R.N., M.S.N.** CWOCN, Wound Care RN, Winthrop University Hospital, Nursing Administration, Mineola, NY, USA

**Cynthia J. Brown, M.D., M.S.P.H.** Geriatric Medicine Section, Department of Medicine/Gerontology, Geriatrics and Palliative Care, Birmingham VA Medical Center, University of Alabama at Birmingham, Birmingham, AL, USA

**Brian Bush, B.A** SUNY Downstate, College of Medicine, Brooklyn, NY, USA

**Enrico Coiera, M.B.B.S., Ph.D., F.A.C.M.I.** Centre for Health Informatics, University of New South Wales, Sydney, Australia

**Joseph Conigliaro, M.D., M.P.H., F.A.C.P.** Division of General Internal Medicine, Department of Internal Medicine, North Shore – LIJ Health Care System, Lake Success, NY, USA

**Regina Cregin, M.S., B.C.P.S., Pharm.D.** Department of Pharmacy, New York Hospital Queens, Flushing, NY, USA

**Erin A. Egan, M.D., J.D.** Neiswanger Institute for Bioethics and Health Policy, Loyola University, CO, USA

**Jeanne M. Farnan, M.D., M.H.P.E.** Department of Medicine, University of Chicago, Chicago, IL, USA

**Barbara Rabin Fastman, M.H.A., M.T.(A.S.C.P.)S.C., B.B.** Mount Sinai School of Medicine, Health Evidence and Policy, New York, NY, USA

**Erin Stucky Fisher, M.D., M.H.M.** University of California San Diego, San Diego, CA, USA

Rady Children's Hospital San Diego, MC, San Diego, CA, USA

**Ethan D. Fried, M.D., M.S., M.A.C.P.** Department of Internal Medicine, Columbia University College of Physicians and Surgeons, St. Luke's-Roosevelt Hospital, NY, USA

**Verna C. Gibbs, M.D.** A National Surgical Patient Safety Project to Prevent Retained Surgical Items, NoThing Left Behind, San Francisco, CA, USA

San Francisco VA Medical Center, San Francisco, CA, USA. University of California San Francisco Medical Center, San Francisco, CA, USA

**Kristin Hahn-Cover, M.D., F.A.C.P.** Department of Internal Medicine, Office of Clinical Effectiveness, University of Missouri Health System, Columbia, MO, USA

**Dea M. Hughes, M.P.H.** Department of Quality Management, VA New York Harbor Healthcare System, New York, NY, USA

**Alan Kantor, M.D.** Department of Radiology, Kings County Hospital Center, Brooklyn, NY, USA

**Harold S. Kaplan, M.D.** Mount Sinai School of Medicine, Health Evidence and Policy, New York, NY, USA

**Eric N. Klein, M.D.** Department of Surgery, SUNY Downstate Medical Center, Brooklyn, NY, USA

**Mei Kong, R.N., B.S., M.S.N.** Assistant Vice President, Patient Safety and Employee Safety, New York City Health and Hospitals Corporation, New York, NY, USA

**Bryan A. Liang, M.D., Ph.D., J.D.** Institute of Health Law Studies, California Western School of Law, San Diego, CA, USA

Department of Anesthesiology, San Diego Center for Patient Safety, University of California, San Diego School of Medicine, San Diego, CA, USA

**Kimberly M. Lovett, M.D.** Department of Family Medicine, Southern California Permanente Medical Group, El Cajon, CA, USA

**Rebecca S. (Suzie) Miltner, Ph.D., R.N.C.-O.B.** School of Nursing, Department of Community Health, Outcomes and Systems, VA Quality Scholars Program, Birmingham VA Medical Center, University of Alabama at Birmingham, Birmingham, AL, USA

**Susan Moffatt-Bruce, M.D., Ph.D.** Quality and Operations, Wexner Medical Center at The Ohio State University, Columbus, OH, USA

**Abdul Mondul, M.D.** Associate Medical Director and Patient Safety officer, Lincoln Medical Center, Weill Medical College at Cornell University, Bronx, NY, USA

**Ken Ong, M.D., M.P.H.** New York Hospital Queens, Clinical Informatics, Flushing, NY, USA

**Mei-Sing Ong, Ph.D.** Centre for Health Informatics, University of New South Wales, Sydney, Australia

**Patricia Ann O'Neill, R.N., M.D., F.A.C.S.** Department of Surgery, SUNY-Downstate/Kings County Hospital Center, Brooklyn, NY, USA

**Alberta T. Pedroja, Ph.D.** ATP Healthcare Services, LLC, Northridge, CA, USA

**Monica Santoro, M.S., B.S.N., C.P.H.Q.** Vice President and Chief Quality Officer, Patient Safety, Quality and Innovation, Winthrop University Hospital, Mineola, NY, USA

**Urmimala Sarkar, M.D., M.P.H.** Department of Medicine, University of California, San Francisco, San Francisco General Hospital and Trauma Center, San Francisco, CA, USA

**Susan D. Scott, R.N., M.S.N.** Sinclair School of Nursing, Office of Clinical Effectiveness – Patient Safety, University of Missouri Health System, Columbia, MO, USA

**Geeta Singhal, M.D., M.Ed.** Department of Pediatrics, Texas Children's Hospital, Houston, TX, USA

**Satid Thammasitboon, M.D., M.H.P.E.** Department of Pediatrics, Texas Children's Hospital, Baylor College of Medicine, Houston, TX, USA

**Supat Thammasitboon, M.D., M.S.C.R.** Pulmonary Diseases, Critical Care, and Environmental Medicine, Tulane University Health Sciences Center, New Orleans, LA, USA

**Rebecca S. Twersky, M.D., M.P.H.** Department of Anesthesia, SUNY Downstate Medical Center, Brooklyn, NY, USA

**Stephen Waite, M.D.** Department of Radiology, SUNY – Downstate Medical Center, Brooklyn, NY, USA

**Robert J. Weber, Pharm.D., M.S.** The Ohio State University Wexner Medical Center, Columbus, OH, USA

# Part I
# Concepts

# Chapter 1
# Patient Identification

Dea M. Hughes

> *"Give me a fruitful error any time, full of seeds, bursting with its own corrections."*
>
> Vilfredo Pareto

## Introduction

Patient misidentification is the failure to properly confirm the correct identity of a patient for whom clinical services are being provided [1]. Often during a misidentification, the correct identity of the patient, or vital details pertaining to the patient's care, are confused with that of another patient. Patient misidentifications are present during all types of care and result from a multitude of factors. If patient identification procedures are not the standard practice, then inpatients with roommates are vulnerable to misidentification, as are outpatients with common names. The severity of patient misidentifications varies greatly. Some events cause no harm (i.e., the patient almost received another patient's medication, but the error was detected before the medication was administered) and others are catastrophic in nature (i.e., the wrong patient was brought into the operating room and surgery commenced on the wrong patient).

The actual incidence of patient misidentifications in healthcare is unknown as the majority of these events go unreported. Over an 8-year period, the Joint Commission received 30 reports of invasive procedures being performed on the wrong patient [2]. Over a 1.5-year period, the United Kingdom (UK) National Health Service's, National Patient Safety Agency, received 236 reports of patient misidentifications

D.M. Hughes, M.P.H (✉)
Department of Quality Management, VA New York Harbor Healthcare System,
New York, NY 10009, USA
e-mail: dea.hughes@gmail.com

A. Agrawal (ed.), *Patient Safety: A Case-Based Comprehensive Guide*,
DOI 10.1007/978-1-4614-7419-7_1, © Springer Science+Business Media New York 2014

identifiers were not used. The nurse-to-nurse phone communication did not involve the use of two identifiers. As a result, staff thought they were communicating about the same patient, when in fact they were not. When the patient was called from the waiting room for her exam, two identifiers were not used. Finally, when the resident entered the room to examine the patient, two identifiers and *active identification* were not used during the greeting. Active identification involves asking the patient to state his or her name. For example, the provider should say, "Good morning. My name is Dr. Doctor. What is your name?" During passive identification, the provider would say, "Good morning Mr. Smith. I'm Dr. Doctor and I'm here to examine you." By stating the patient's name, the provider introduces the opportunity for misidentification by assuming the identity of the patient and eliminating the two-way communication engaging the patient in verifying his/her identity. Without such confirmation, the resident was unaware with whom he was communicating and that he was about to examine the wrong patient. Of note, patient misidentifications are more common when patients have similar or same names, surnames, dates of birth, or other demographic information [9]. Therefore, it becomes vitally important to use at least two identifiers to avoid similar name misidentifications.

(b) The team identified that during the nurse-to-nurse phone communication, write-down read-back was not used. Standard communication practices such as write-down read-back offer the opportunity to detect misidentification errors during verbal communications and are considered a best communication practice.

(c) When the patient was called into the exam room, the medical resident did not use active identification to confirm the correct identity of the patient. Due to the hierarchical nature of the physician–patient relationship, patients may be less likely to speak up and correct a physician if they are being addressed by the wrong name. The RCA team identified that the dermatology resident was not familiar with the process of active identification.

(d) The patient's wristband was never double-checked to confirm her identity. Patient wristbands are vigilantly placed on all patients prior to admission to facilitate identification processes. They are another vehicle by which a clinician can double-check the identity of a patient. Without checking the wristband, the final opportunity to correctly identify the patient was lost.

The RCA team identified the following root causes of this particular patient misidentification and the most relevant of the five rules of causation that applied, which are described in Table 1.2. Had at least one of these vulnerabilities been prevented, the patient would have been correctly identified.

After identifying the root causes, the team focused on implementing solutions to those procedural, cultural, communicative, and training-related vulnerabilities that led to the patient misidentification. The team agreed that the new procedures related to patient identification should be built into the current in situ simulation modules that were being conducted in the hospital. In situ simulation is an innovative approach to clinical education, which uses a realistic scenario to teach decision making within the complexity of interdisciplinary teamwork [10]. Additionally,

**Table 1.2**  Case 1: Root causes and five rules of causation

| Root cause | Category | Five rules of causation | | | | |
|---|---|---|---|---|---|---|
| | | 1 | 2 | 3 | 4 | 5 |
| The standard protocol for phone communications involving patient hand-offs did not involve the use of two identifiers, or write-down read-back, which contributed to the wrong patient being sent to Dermatology | Communication | ✓ | ✓ | | | ✓ |
| The organization lacked comprehensive education on how to properly identify patients, using active identification and two identifiers. As a result, staff were not familiar with the process of using two identifiers and the cultural norm was to only use one | Training | ✓ | ✓ | ✓ | | |

**Table 1.3**  Case 1: Action strength table

| Action | Type | Strength category | | |
|---|---|---|---|---|
| | | Strong | Intermediate | Weak |
| Write a standard protocol for staff to staff phone communications, which mandates the use of two patient identifiers and write-down read-back | Standardized process | ✓ | | |
| Use in situ simulation to train staff on how to appropriately communicate using these new standards (two identifiers and write-down read-back) | Education via simulation | | ✓ | |

clinical managers were tasked with developing standard processes that would incorporate the use of two identifiers and active patient identification when delivering care, treatment, services, and communicating critical patient information.

Finally, as a secondary recommendation to address efficiency, the team recommended that paper charts no longer travel with the patient since the hospital uses an EHR, which can be accessed anywhere in the hospital. Transport of the paper chart is a redundant process that does not contribute to the overall safety of the patient.

The RCA team crafted the following corrective actions highlighted in Table 1.3, to correct the systems issues that contributed to this patient misidentification. These actions are considered strong and intermediate fixes and therefore, address the root cause of the misidentification.

Perhaps the most important aspect of the RCA action plan is to ensure the actions are implemented and measured for effectiveness. The RCA team labored with writing the outcome measures and eventually agreed that multiple, quantifiable measures would best ascertain when these actions were implemented and how effective they were. Timelines and action completion dates were requested from clinical managers (i.e., ensure the protocol is written by x-date and confirm that the in situ simulations have occurred via attendance and training records by y-date).

**Fig. 1.2** Packed red blood cells ready for patient transfusion

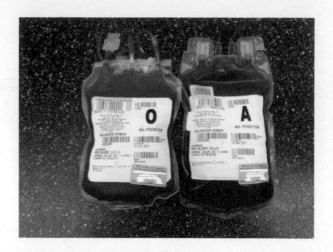

Additionally, employees who participated in the in situ simulation were evaluated based on if and how they used two identifiers during the clinical scenario. Simulations were repeated, if necessary, to ensure compliance with the new identification procedures.

The RCA team also recommended the use of "secret shoppers" to monitor adherence to the new patient identification protocols such as (a) using two identifiers and write-down read-back for all phone communications involving patient hand-offs and (b) using two identifiers, active identification, and checking the patient's wristband when available before examining the patient. The Centers for Medicare and Medicaid Services (CMS) has been successfully using "secret shoppers" for years to assess prescription drug programs for compliance with marketing requirements and the accuracy of information provided to customers [11].

Finally, the team tasked the Patient Safety department with tracking and trending all future patient misidentifications submitted via the institution's incident reporting system. The overall goals were a 100 % reduction in adverse misidentification events, monitoring of all close call misidentification events, and encouraging all staff to continue reporting these events through the Just Culture modalities.

## *Case Study 2 Analysis: Blood Drawn from Wrong Patient (Fig. 1.2)*

Transfusion errors related to patient misidentification are considered sentinel events, which are unexpected occurrences involving death or serious physical or psychological injury to patients [12]. Although the incident in Case 2 is a close call, it had the potential of becoming a sentinel event had the transfusion not been halted. As a result, the RCA process treats close-call sentinel events as if they were actual sentinel events and investigates them just as rigorously. Close-calls provide organizations with the opportunity to learn about an incident and correct system vulnerabilities.

During the analysis of this close-call sentinel event, the RCA team identified the following main breakdown points that contributed to the blood being drawn from the incorrect patient, which are highlighted on the yellow notes in Fig. 1.3.

When the blood was drawn for the transfusion, the RCA team identified four procedural vulnerabilities that contributed to the blood being drawn from the wrong patient: (1) the patient's specimen labels were not brought to the bedside so that they could be verified against the patient's wristband, (2) two identifiers were not used to properly identify the patient, (3) the tube was not labeled at the bedside after the blood was drawn, and (4) a second verification process (e.g., another person or technology) was not instituted. As previously stated, patients with similar or similar-sounding names are more likely to be misidentified, especially if two identifiers are not used. Additionally, blood specimens should always be labeled at the bedside or in front of the patient. This creates an environment of safety because it allows the patient to be involved in the identification process and creates patient confidence through transparency. Furthermore, a redundant safety system was not in place to ensure that this critical process went without error.

During the debriefing, the RCA team drilled down further with staff as to why the labels were not brought to the bedside. The team identified some misperceptions held by clinical staff that the labeling of blood tubes was considered an *administrative* duty and not a clinical duty. As a result, bringing the labels to the bedside was not perceived as an important part of the clinical process of drawing blood.

Additionally, the nurse agreed to label the tubes without having witnessed the blood draw. At some hospitals, two clinical staff members are involved in the process of drawing blood for a transfusion, especially if no other redundant identification system is in use. And, both members must be present at the patient's bedside when the blood is drawn and the tubes are labeled. Alternatively, at other hospitals, two blood specimens are required to ensure that the correct identity and blood type of the patient have been captured. In either case, redundant processes ensure a misidentification will be detected if it occurs. As discovered during this RCA, there was not a redundant system in place to ensure that this critical process went without error.

Finally, the RCA team found that an informal process was in place before the transfusion was initiated. Although two staff members were involved, there was not a standard checklist to review prior to the transfusion. Only by chance did nurse ask the patient about his blood type, which ultimately prevented the sentinel event from occurring. During surgical procedures, staff conducts a time-out prior to the commencement of the procedure to ensure (1) correct patient, (2) correct procedure, and (3) correct site. The World Health Organization summarizes this best safety practice in their comprehensive Surgical Safety Checklist, which outlines how the standard process, including the time-out, should occur before surgical procedures [13]. The RCA team identified that a lack of standardized process, including checklist, prior to initiating the transfusion created an unsafe environment and a lost opportunity for final verification of correct patient and correct blood type.

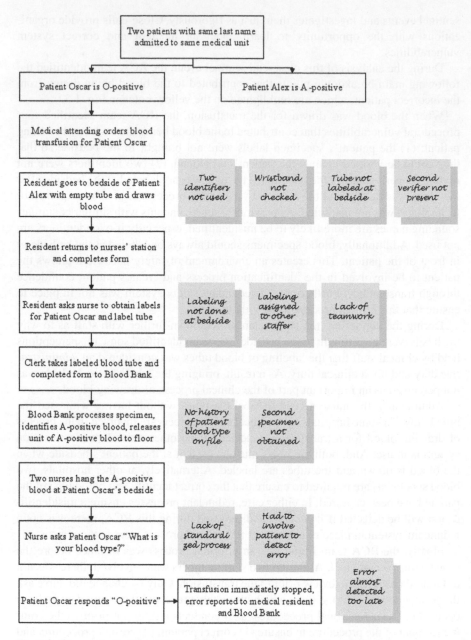

**Fig. 1.3** Case study 2—Blood drawn from wrong patient: flow chart analysis

The RCA team identified the following root causes of this close-call sentinel event, which involved blood that was drawn from the wrong patient and the most relevant of the five rules of causation that applied [Table 1.4].

**Table 1.4** Case 2: Root causes and five rules of causation

| Root cause | Category | Five rules of causation 1 | 2 | 3 | 4 | 5 |
|---|---|---|---|---|---|---|
| Due to cultural misperceptions, staff were not accustomed to bringing labels to the bedside when drawing blood for a transfusion. As a result, the opportunity to correctly identify the patient using the labels and the patient's wristband was lost | Procedures | ✓ | | | ✓ | |
| The recommended practice of using two identifiers and active identification at the bedside was not built into the standard process for drawing blood. As a result, the patient was not correctly identified | Procedures | ✓ | ✓ | | | |
| The organization did not have a standard process (i.e., checklist), such as a time-out, before the transfusion was initiated. As a result, the final opportunity to correctly identify the patient was almost lost | Communication | ✓ | | | | ✓ |

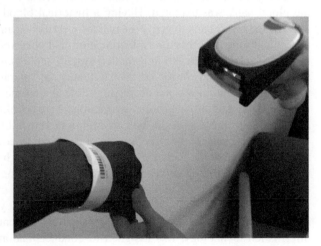

**Fig. 1.4** Scanning a patient's barcoded wristband

These root causes were validated against a literature review. In a study analyzing 227 RCAs conducted on patient misidentifications in laboratory medicine, it was found that the majority of misidentifications occurred during the pre-analytic phase of the process and that patient misidentifications accounted for 73 % of adverse events [14]. Furthermore, the study identified that during the pre-analytic phase, the majority of causal factors for those misidentifications involved printed labels, wristbands, two identifiers, and two-person verifications.

The RCA team felt the strongest fix for ensuring that both wristband and labels were used to identify the patient with the usage of two identifiers was by applying wireless barcode technology at the bedside (Fig. 1.4).

Barcode-based transfusion processes have been shown to be 15–20 times safer than current hospital practices [15]. Bar-coded transfusion verification systems confirm patient identity, display transfusions orders, track blood products, and maintain transfusion records. They eliminate opportunities for human error involving wristbands and patient labels and make the process safer for patients and more efficient. Additionally, they offer a redundant system to ensure patient safety and require that (1) the patient's wristband is used in the identification process and (2) that it is checked against the labels, which are applied to the blood tubes at the bedside after the blood is drawn. The usage of barcode technology with this standardized process would eliminate the need to involve another staff member during the pre-analytic blood draw process. As the hospital learned from this incident, adding that second person to the process does not necessarily make it safer.

In addition to barcode technology, a formal process including the usage of a checklist, much like the surgical time-out, should be instituted using two staff members at the patient's bedside before the transfusion is initiated. The time-out is considered a best safety practice and now widely accepted among staff who perform invasive and surgical procedures. Therefore, the process of blood transfusion could also benefit from this safety feature.

Finally, the possibility of maintaining a historical blood type for all patients in the blood bank was explored. Having a historical blood type on file would have allowed the blood bank to verify the patient specimen against an accurate blood type and quickly identify that a misidentification had occurred. Unfortunately, without having an integrated health information system and a patient population for which a blood type is already on record, such as that of the Veterans Health Administration, this hard-fix solution was not deemed feasible at the time.

Therefore, the following corrective actions were developed to address the identified systems vulnerabilities, which are highlighted in Table 1.5.

In order to measure the effectiveness of these proposed strategies, the RCA team recommended that the Patient Safety department in the hospital work closely with the information technology team responsible for installing and maintaining the barcode technology to track and trend data associated with the new system. All usage and scanning discrepancies were to be tracked for the first year post-implementation. Additionally, an implementation team consisting of patient safety, clinical staff, and information technology, was assigned to conduct random rounds on the units to ensure that barcode technology is being utilized accurately and to resolve any technical issues that staff may encounter. Furthermore, the implementation team was charged with monitoring how staff are interacting with each other and the technology. Finally, a separate Patient Safety team would monitor when transfusions are taking place in the hospital and round on the units during those times to observe and ensure that a time-out, checklist, and two engaged staff members are involved in the transfusion initiation process. Due to the sentinel nature of mistakes made in this context, staff must have 100 % confidence that they are drawing blood from or transfusing the correct patient with a unit of blood.

**Table 1.5** Case 2: Blood drawn from wrong patient: action strength table

| Action | Type | Strong | Intermediate | Weak |
|---|---|---|---|---|
| | | Strength category | | |
| Implement the usage of wireless barcode technology at the bedside to confirm accurate patient identity using two identifiers during the specimen collection process | New nonmedical device | ✓ | | |
| Educate all staff to use two patient identifiers when drawing blood and during all aspects of the transfusion administration process | Training | | | ✓ |
| Implement a time-out with checklist, that involves two staff members who are actively involved and present, before the initiation of the blood transfusion, to confirm correct patient and correct blood type | Standardized process | ✓ | | |

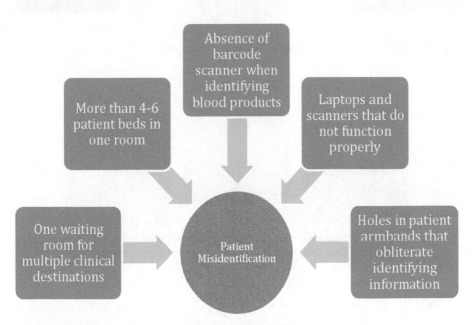

**Fig. 1.5** Environmental factors that contribute to patient misidentification

## Discussion

During the RCA of both these cases of patient misidentification, several key lessons were learned. Patient misidentifications are common occurrences within hospitals and have the potential for having devastating consequences. Additionally, many factors contribute to patient misidentification, which are highlighted in Figs. 1.5, 1.6, and 1.7.

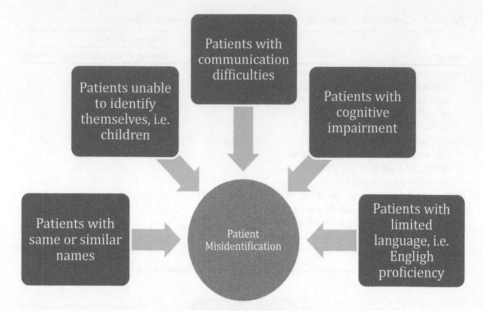

**Fig. 1.6** Patient factors that contribute to patient misidentification

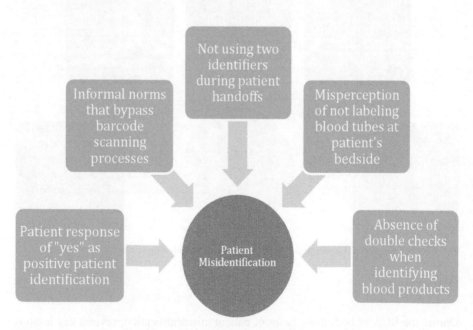

**Fig. 1.7** Cultural factors that contribute to patient misidentification

These factors can occur at any stage in the healthcare delivery process. A lack of redundant nonhuman methods for identification, such as barcode scanning technology, increases the likelihood of patient misidentifications. Patient factors, such as patients with same or similar names, introduce the possibility of misidentification if more than one identifier is not used to actively identify the patient. Finally, cultural factors and deviations from standard practices continue to put patients at risk for misidentification.

## Key Lessons Learned

As presented through the analyses of the two case studies in this chapter, there are many factors that contribute to patient misidentification. Below are some key takeaways that will help to ensure accurate patient identification and hopefully eliminate the occurrence of these preventable and distressing events.

- Two identifiers must be used during all aspects of patient care.
  Wristbands are a second method for identifying the patient and should be read or scanned.
- Write-down and read-back of the patient's identity should take place during phone communication about a patient.
- *Active identification* (asking patient to state his or her name) should be used during all verbal communications with the patient; passive identification should be avoided.
- Redundant systems that are technologically based (e.g., bar-coded technology) are hard-fixes to ensure the correct identity of patients.
- Cultural misperceptions about the importance of patient identification, labeling tubes at the bedside, and other practices can be addressed through simulation type training.
- Best practices, such as time-outs, should be adopted when appropriate.

## References

1. Bittle M, Charache P, Wassilchalk D. Registration-associated patient misidentification in an academic medical center: causes and corrections. Jt Comm J Qual Patient Saf. 2007; 33(1):25–33.
2. Pennsylvania Patient Safety Authority. Patient identification. Patient Saf Advis. 2004;1(2): 8–10.
3. National Health Service, United Kingdom, National Patient Safety Agency. Wristbands for hospital inpatients improves safety. Safe Practice Notice, 22 Nov 2005 (Ref: NPSA/2005/11).
4. National Health Service, United Kingdom, National Patient Safety Agency, NRLS quarterly data workbook up to September 2011. Available at http://www.nrls.npsa.nhs.uk/resources/collections/quarterly-data-summaries/?entryid45=133438. Last accessed 28 Jun 2012.
5. Mannos D. NCPS patient misidentification study: a summary of root cause analysis. Top Patient Saf. 2003;3(1):14.

6. Bagian J, Gosbee J, Lee C, et al. The veterans affairs root cause analysis system. J Qual Improv. 2002;28(10):531–45.
7. United States Department of Veterans Affairs, National Center for Patient Safety. Triage and triggering questions: five rules [homepage on the internet]. [Updated 30 Jun 2011]. Available at http://www.patientsafety.gov/CogAids/Triage/index.html#page=page-9. Last accessed 28 Jun 2012.
8. Ball M, Garets D, Handler T. Leveraging information technology towards enhancing patient care and a culture of safety in the US. Methods Inf Med. 2003;42(5):503–8.
9. National Health Service, United Kingdom, National Patient Safety Agency. Patient identification errors from failure to use or check ID numbers correctly [homepage on the internet]. [Updated Nov 2008]. Available at http://www.nrls.npsa.nhs.uk/resources/?EntryId45=59855. Last accessed 28 Jun 2012.
10. Miller K, Riley W, Davis S, Hansen H. In situ simulation, a method of experiential learning to promote safety and team behavior. J Perinat Neonatal Nurs. 2008;22(2):105–13.
11. Mundy A. Uncle Sam, secret Medicare shopper. 2008. Available at http://blogs.wsj.com/health/2008/12/05/uncle-sam-secret-medicare-snooper/. Last accessed 28 Jun 2012.
12. The Joint Commission, Sentinel events [homepage on the internet]. [Updated Jan 2011]. Available at http://www.jointcommission.org/assets/1/6/2011_CAMH_SE.pdf. Last accessed 1 Jul 2012.
13. World Health Organization, WHO surgical safety checklist and implementation manual [homepage on the internet]. [Updated 2011]. Available at http://www.who.int/patientsafety/safesurgery/tools_resources/SSSL_Checklist_finalJun08.pdf. Last accessed 1 Jul 2012.
14. Dunn E, Moga P. Patient misidentification in laboratory medicine: a qualitative analysis of 227 root cause analysis reports in the Veterans Health Administration. Arch Pathol Lab Med. 2010;134:244–55.
15. Stone E, Keenan P. Barcode technology for positive patient identification prior to transfusion. Patient Saf Qual Healthc. 2011;8(4):26–32.

# Chapter 2
# Teamwork and Communication

Joseph Conigliaro

*"Seek first to understand, then to be understood."*

Stephen R. Covey

## Introduction

Healthcare delivery has become a complex endeavor whereby differing members of the healthcare team bring important skills, experience, and expertise to bear in the care of the patient. The complexity of the environment is characterized by rapidly evolving, ambiguous situations, complex, multicomponent decisions, informational overload, severe time pressure, severe consequences for error, and performance/command pressure [1].

This is most pronounced in the operating room (OR) and in the intensive care unit (ICU) but is present in all settings of health care where physicians, nurses, and other healthcare providers work together. This interprofessional and interdisciplinary approach to care has tremendous positive advantages; yet, there are risks that are created through inadequate or ineffective communication among healthcare workers. In addition, when respective roles of healthcare workers are poorly understood or are in conflict with each other, the potential for error is magnified. Finally, the traditional hierarchy or authority gradient between nurses and physicians as well as other members of the healthcare team can undermine teamwork, creating an inability to communicate honestly and effectively.

The goals of this chapter are to define the concept and key attributes of teams and to highlight the importance of communication in the healthcare setting through two

J. Conigliaro, M.D., M.P.H., F.A.C.P. (✉)
Division of General Internal Medicine, Department of Internal Medicine, North Shore – LIJ
Health Care System, 2001 Marcus Avenue, Suite S160, Lake Success, NY 11042, USA
e-mail: jconigliaro@nshs.edu

A. Agrawal (ed.), *Patient Safety: A Case-Based Comprehensive Guide*,
DOI 10.1007/978-1-4614-7419-7_2, © Springer Science+Business Media New York 2014

clinical case studies. The chapter will describe various approaches to improve team-work and communication in health care including crew resource management (CRM) and TeamSTEPPS™ and describe some practical strategies such as "SBAR" and "CUS" for application in various healthcare settings.

## Benefits of Team-Based Approach

Teamwork plays an important role in the delivery of safe care. The Joint Commission on the Accreditation of Healthcare Organizations (The Joint Commission) notes that a majority of sentinel events include the failure of team work and communication as a contributing factor to adverse events [2]. This relationship is linked both to the perception of a cohesive and well-functioning team on patient safety as well as its effect on staff well-being and satisfaction [3]. Another study found that the lack of teamwork or poor coordination among providers in an organization is associated with delays in testing and in the communication of conflicting information to patients [4].

The benefits of improved teamwork are also well documented in the literature. Both perceived and measured high levels of teamwork result in enhanced effectiveness [5], fewer and shorter delays in patient care, improved staff morale and job satisfaction, increased efficiency of care, lower staff stress, and improved patient satisfaction [6–8]. One study reported that the implementation of teamwork initiatives reduced their clinical error rate from 30.9 % to 4.4 % [9]. Another study reported a 50 % reduction in adverse outcomes after team training [10].

Despite mounting evidence of the benefits of improved teamwork and communication in literature and increasing support by policy makers and professional bodies, teamwork occurs infrequently and is often misunderstood by healthcare professionals. For example, in one report of emergency department, clinicians had the perception of teamwork as a program that administrators implement so everyone will "like each other [9]." Obviously, this overly simplistic assumption must be overcome as effective teamwork is much more than "liking each other."

## Defining Team and Teamwork

The *team,* as commonly described in the literature, consists of two or more individuals, who have specific roles, who perform interdependent tasks, are adaptable, and share a common goal [11]. As stated in a report by the Agency for Healthcare Research and Quality (AHRQ) on teamwork (http://www.ahrq.gov/teamsteppstools/instructor/fundamentals/module2/slteamstruct.pdf), "the ratio of We's to I's the best indicator of the development of a team."

**Fig. 2.1** Paradigm shift needed to occur from thinking as an individual to thinking as a team. Reprinted with permission from Heidi King et al., in the 2008 *Advances in Patient Safety: New Directions and Alternative Approaches* Vol. 3. Performance and Tools. [AHRQ Publications No. 08-0034-3] pp 11

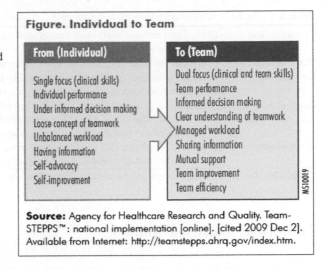

It should be clear that teamwork is not simply "feeling close" to your team members. Instead, it is a set of interrelated behaviors, cognitions, and attitudes that combine to facilitate coordinated adaptive performance. A key attribute of high-performing teams is a recognition of their interdependence, an awareness of their collective efficacy, and an intuitive sense of their "teamness." Figure 2.1 describes a fundamental paradigm shift that needs to occur from thinking as an individual to thinking as a team [12].

Teamwork is also not an automatic result of placing people together in one place. Indeed, teamwork does not require that you work with your team members on a permanent basis from day-to-day; rather teamwork is sustained by a shared set of teamwork skills with shared goal that can transfer across teams and situations [9].

## The Five "C"s of Effective Teamwork

Wenger has proposed that effective teamwork in the healthcare setting requires the presence of the "5 Cs" as outlined in Table 2.1 [14]. The "5 Cs" refer to key properties of a true team that includes a common goal, assigned with collaborative tasks. As a group, individuals are committed to attaining those goals. High functioning teams also possess a variety of competencies including individual knowledge, skills, behaviors, and attitudes necessary to accomplish their assigned role in the team's activities. Communication and coordination are also essential processes for the successful attainment of team-based goals.

**Table 2.1** The 5 C's of effective teamwork in health care

| | |
|---|---|
| Common Goal | Every Team member shares and understands the short and long term goals of the team and the organization |
| Commitment | Every team member is committed to attaining the team's goals |
| Competence | Every team member has the knowledge, skills, behaviors and attitudes necessary to accomplish successfully their role in the team's activities |
| Communication | Team members communicate effectively and efficiently with each other, with their patients and with other parties (whether animate or inanimate) through whatever means are required to accomplish the desired goals |
| Coordination | Team members efficiently and effectively work together and with other needed technology, people, and resources to accomplish desired goals |

Refined by Wenger for health care based on work by Katzenbach and Smith (13).

## Knowledge, Skills, and Attitude

To work effectively, team members must possess specific knowledge, skills, and attitudes both individually and collectively. This could also be translated into understanding what team members think (knowledge), do (skills), and feel (attitude) [15].

Team members need to *know* the task ahead and the goals of the team. They must be knowledgeable of their own and teammate's task responsibilities and be aware of each other's strengths and weaknesses so that challenges can be anticipated and overcome quickly. Effective teams must *do* certain tasks at a specified time in a specified sequence and must communicate effectively. They must also monitor each team member's performance. Finally, team members must *feel* motivated, have mutual trust and cohesion, and have a positive disposition toward working together in the team [15, 16].

## Nurses and Physicians

Traditionally the healthcare industry has focused on the individual technical training of physicians and nurses separately where physicians individually or as a group are taught knowledge and clinical skills and then practice those knowledge and skills in silos and in isolation from other healthcare providers. Similarly nurses are trained alone and in a graded manner that does not include other healthcare trainees early on. After completing their training, the work of nurses and physicians has traditionally been and for the most part continues to be independent. This professional disconnect has been associated with differences in the perception and conceptualization about how each profession views a clinical situation and how nurses and physicians communicate that clinical situation to each other. A recent cross-sectional survey on interdisciplinary collaboration of 136 ICU nurses and 48 physicians in the UK found that nurses and physicians have differing perceptions of interdisciplinary communication, with nurses reporting lower levels of communication openness between nurses and physicians [17]. In the same study, medical students and physicians in training, when compared with senior physicians, also reported lower levels of communication openness between doctors. Communication openness among ICU team members predicted the degree to which individuals reported an understanding of patient care goals and perceptions of the quality of unit leadership predicted open communication [17]. The current efforts around interprofessional education and team-based learning are designed, in part, to mitigate the results of this "communication chasm" at an early point in the training of both physicians and nurses.

Interprofessional training at an early stage in the training of physicians and nurses can have a lasting impact later on professional career, on patient outcomes, and on error rates. Favorable attitudes about teamwork have been associated with an increase in error reduction behaviors in the aviation industry [18] and improved patient outcomes in intensive care units [17, 19, 20]. In addition improved communication is associated with decreased nurse turnover in the operating room [21]. Effective teams and teamwork is associated with better job satisfaction [22] and less sick time taken off from work [23]. Alternatively, discrepant attitudes about teamwork have been suggested as a considerable source of nurse dissatisfaction with their profession [24] that has been implicated as a contributing factor to the current nursing shortage [25].

In studies of Medical Team Training at the Department of Veterans Affairs Medical Centers, Mills et al. found a pattern of disagreement among physicians and nurses whereby surgeons usually perceived a stronger organizational culture of safety, better communication, and better teamwork compared to either nurses or anesthesiologists [26]. These differences can undermine the "Common Goal" factor associated with an effective team and can represent barriers to effective teams and teamwork that need to be recognized and addressed if safe clinical care is to result from team-based care.

## Disruptive Behavior

Coordination refers to the ability of team members to efficiently and effectively work together and with other needed technology, people, and resources to accomplish the desired goals. A potential barrier or factor that can undermine coordination in team-based care is disruptive behavior among healthcare workers. Disruptive behavior is often a manifestation of the increased stress that the healthcare environment places on members of the healthcare work force. Disruptive behavior may also stem from the traditional hierarchical authority structure among physicians, nurses, and other providers. Disruptive behavior, however, is not solely a behavior seen in physicians but is noted in other healthcare providers as well. As an indication of the diminishing tolerance for such behavior, The Joint Commission now requires that "all accredited hospitals have a code of conduct that defines acceptable, disruptive, and inappropriate behaviors (Element of Performance [EP] 4) and that healthcare facility leaders create and implement a process for managing disruptive and inappropriate behaviors" [27].

Disruptive behavior among team members has been found to be associated with adverse events. A recent survey of voluntary hospitals was conducted to assess the significance of disruptive behaviors and their effect on communication and coordination and on the impact on patient care [28]. The study included 4,530 participants, (2,846 nurses, 944 physicians, 40 administrative executives, and 700 "other"). Seventy seven percent of the total respondents reported that they had witnessed disruptive behavior in physicians (88 % of the nurses and 51 % of the physicians). Sixty-five percent of the respondents reported witnessing disruptive behavior in nurses (73 % of the nurses and 48 % of the physicians). Sixty-seven percent of the respondents agreed that disruptive behaviors were linked with adverse events with 71 % stating a link to medical errors and 27 % with patient mortality [28].

## Communication

Communication is the fundamental underpinning of good teamwork. Teamwork behaviors related to high clinical performance have identifiable patterns of communication, management, and leadership that support effective teamwork. In a study measuring the impact of organizational climate safety factors (OCSFs) and coworkers' communication and collaboration on risk-adjusted surgical morbidity and mortality, the staff of 52 general and vascular surgery services at 44 Veterans Affairs Medical Centers and 8 academic medical centers were surveyed [29]. Data from The National Surgical Quality Improvement Program was used to assess risk-adjusted morbidity and mortality. Correlations between outcomes and OCSFs and between outcomes and communication and collaboration were calculated with attending and resident physicians, nurses, and other providers. OCSF measures of teamwork climate, safety climate, working conditions, recognition of stress effects, job satisfaction, and burnout did not correlate with risk-adjusted outcomes but reported levels of communication and collaboration between attending and resident residents correlated with risk-adjusted morbidity suggesting the importance of physicians' coordination and decision-making roles on surgical teams in providing high-quality and safe care [29].

## Case Studies

### Case 1: Respiratory Arrest Due to Inadvertent Administration of Paralytic Agent

*A 64-year-old male is scheduled for surgical repair of a hallux varus of the left first metatarsal. In the preoperative area the anesthesiologist asks the nurse anesthetist to draw up to two milliliters of a paralytic agent as well as a saline flush. The syringes are left on the table near the intubation tray. The pre-op nurse takes the syringe containing the paralytic agent and injects into the right deltoid of the patient. Three minutes later the patient's oxygen monitor alarms for oxygen desaturation and he is observed to have no respirations. An emergency intubation is performed and the patient is resuscitated.*

A root cause analysis of the incident notes the lack of a debriefing of the surgical team before the start of the procedure including no communication between the nurse anesthetist and the pre-op nurse. In response to the event, mandatory briefing and debriefings using a structured communication format are ordered before and after each surgical procedure.

This case illustrates the common deficiencies of interpersonal communication including a lack of formal interaction prior to the procedure- or team-based activity. Using the 5 C's model adapted by Wenger, one would note the lack of communication and coordination. Use of a debriefing meeting would formalize the components of the 5 Cs into the preoperative activities of the patient's care.

## Case 2: Pulmonary Embolism Due to Delay in Heparin Ordering and Administration

*The medical team on the acute care inpatient service is rounding when the attending is presented an 84-year-old female patient with dementia, hypertension, and obesity, who was admitted after a fall and a fractured hip. The attending notes that there are no lower extremity compression devices on the patient and that no order for prophylactic heparin has been written. Given his concern for a deep venous thrombosis in this patient, he asks the intern to place an order for both. The intern decides to wait to place the order at the end of rounds 90 min later. By the time the nurse carries out the order, it is 4 h after the attending has seen the patient. Two hours later a cardiac arrest is called and the patient is found unresponsive. Autopsy reveals a large pulmonary embolus.*

Root cause analysis of the case notes the delay in communication between the nurse and the physician as a contributing factor to the event. The hospital addresses this deficiency by establishing regular interprofessional rounds whereby a nurse on the ward participates in all morning rounds that include the attending.

## Discussion

Healthcare's focus on individual training of physicians and nurses is similar to the aviation industry's approach more than 25 years ago where the task of flying was the responsibility primarily of the pilot. In reality, both the pilot and the physician accomplish their work as part of a larger team [30]. Aviation's approach to expand from the performance of the solo pilot was in recognition that the cockpit crew should also, with their combined expertise and roles, be responsible for the safety of the passengers. In response, Crew Resource Management (CRM) and similar standardized teamwork models that have been uniformly adopted in the aviation industry to reduce the risk of errors [31] are finding new uses in medicine. In the case of medical care, the care of the patient is recognized as being not only the responsibility of the physician but also of the nurse caring for the patient, the pharmacist preparing the patient's medications, and other providers of patient care including the social worker and the physical therapist. The Institute of Medicine (IOM) report, "Crossing the Quality Chasm" recommends, among other things, that healthcare organizations implement patient safety programs that "promote team functioning" and that healthcare systems should "train in teams those who are expected to work in teams" [32]. Other calls for interdisciplinary teamwork come from The Joint Commission and the report of the President's Advisory Committee on Consumer Protection and Quality in the Health Care Industry [33].

## Crew Resource Management

Cockpit or Crew Resource Management (CRM) is a system of team building, communication, and task management used for many years in military and civilian aviation. CRM or similar programs can be defined as the effective use of all available resources, people, processes, facilities, equipment, and environment (the cockpit or healthcare environment), by teams (the crew or healthcare team) or individuals to safely and efficiently accomplish an assigned mission or task. Since the majority of commercial flight accidents were caused in part from communication failures among crew members, CRM was designed to standardize communication and teamwork to eliminate critical and fatal errors by the flight team. Currently, CRM is required worldwide throughout the aviation industry and is a required aspect of the training of pilots.

The CRM process has been exported to a variety of other industrial settings which share the same characteristics of aviation; settings with high complexity that use advanced technology, work in high stress environments, with an element of danger, have a hierarchal structure, and with major penalties or adverse outcomes for failure. These include diverse industries such as nuclear technology, chemical manufacturing, and health care. The goal of adapting CRM to health care was to increase patient safety and improve clinical outcomes through better communication and teamwork. In the healthcare setting, training in CRM has been associated with major reductions in both observed errors and adverse outcomes. For example, in a study evaluating the effect of implementing the Veterans Health Administration

(VHA) version of CRM, known as Medical Team Training, on patient outcomes, facilities who underwent training had a greater than 50 % decline in surgical mortality compared to facilities that did not undergo training. This reduction in mortality was correlated with a dose–response relationship to the amount of Medical Team Training received [34].

CRM emphasizes team building, briefings, situational awareness, stress management, and decision-making strategies. Human factor issues that extend to other team members in joint training also helps to decrease team errors [9]. CRM Training has been shown to improve attitudes toward fatigue management, team building, communication, recognizing adverse events, team decision making, and performance feedback. CRM-trained participants also felt that such training would reduce errors and improve patient safety [35]. Studies identify several key concepts in CRM that relate to team building. These include managing fatigue and workload; stress management, creating and managing the team, recognizing adverse situations, cross-checking, and communicating; and assertiveness, developing and applying shared mental models for decision making, situational awareness, and giving and receiving performance feedback [36].

## The Team STEPPS approach

Team Strategies and Tools to Enhance Performance and Patient Safety (TeamSTEPPS) is a systematic approach developed by the Department of Defense (DoD) and the Agency for Healthcare Research and Quality (AHRQ) to integrate teamwork into practice. It is designed to improve the quality and safety as well as the efficiency of health care. TeamSTEPPS is based on research related to teamwork, team training, and culture change. The components of TeamSTEPPS are based on a list of competencies developed by Baker as important elements of teamwork in the professional education of physicians [37].

Figure 2.2 illustrates the critical concepts related to the teamwork training approach in TeamSTEPPS [38]. Individuals earn four primary trainable teamwork skills: leadership, communication, situation monitoring, and mutual support. By building a fundamental level of competency in each of those skills, research has shown that the team can enhance three types of teamwork outcomes: performance, knowledge, attitudes. Table 2.2 lists the components of the TeamSTEPPS curriculum. These include the barriers that undermine the team and its effectiveness as well the tools and strategies to overcome them. Outcomes from TeamSTEPPS include intermediate process-level outcomes such as a shared mental model and mutual trust as well as improved patient outcomes and patient safety.

**Fig. 2.2** The TeamSTEPPS
conceptual framework.
Adapted with permission
from "Barriers to Team
Effectiveness" in
*TeamSTEPPS® Fundamentals
Course*: Module 7.
Summary—Putting It All
Together (Instructor's
Materials)

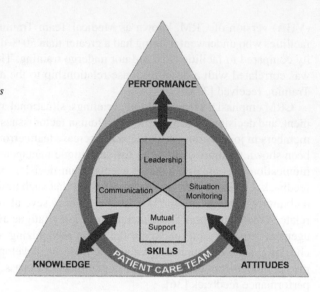

**Table 2.2** Components of the TeamSTEPPS curriculum.

| Barriers | Tools and Strategies | Outcomes |
| --- | --- | --- |
| Inconsistency in team membership | Brief | Shared Mental Model |
| Lack of time | Huddle | Adaptability |
| Lack of information sharing | Debrief | Team Orientation |
| Hierarchy | STEP[1] | Mutual Trust |
| Defensiveness | Cross Monitoring | Team Performance |
| Conventional thinking | Feedback | Patient Safety |
| Complacency | Advocacy and Assertion | |
| Varying communication Styles | Two Challenge Rule | |
| Conflict | CUS[2] | |
| Lack of coordination and follow-up | DESC Script[3] | |
| with co-workers | Collaboration | |
| Distractions | SBAR[4] | |
| Fatigue | Call-Out | |
| Workload | Check-Back | |
| Misinterpretation of cues | Handoff | |
| Lack of role clarity | | |

[1]STEP: Status of the Patient, Team Members, Environment, Status toward goal
[2]CUS stands for "I'm concerned, I'm uncomfortable, this is unsafe, or I'm scared"
[3]DESC Script Model:
1. **Describe** the actions or behavior that you see as taking place;
2. **Express** why that behavior is an issue?
3. **Specify** the resulting actions or change of behavior you would like to effect;
4. **Clarify** the consequences for failing to change behavior or meet demands.
[4]SBAR: Situation, Background, Assessment, and Recommendation

## SBAR

One of the many barriers that can potentially contribute to communication difficulties between clinicians is a lack of structure and standardization for communications. As mentioned earlier in this chapter, differences in the communication styles between nurses and physicians are also a major contributing factor. These variations in communication style can cause misinterpretation, lack of clarity, and frustration among team members.

Therefore, as part of the TeamSTEPPS model, structured and explicitly designed forms of communication have been recommended that reduce ambiguity, enhance clarity, and send an unequivocal signal, when needed, that a different action is required. Read-backs or check backs, Situation–Background–Assessment–Recommendation (SBAR), critical assertions (CUS), briefings, and debriefings are the tools and strategies that have been increasingly used in health care and in particular as a hand-off mechanism between team members and between teams.

SBAR (**S**ituation, **B**ackground, **A**ssessment, and **R**ecommendation) is the most common shared mental model of communication that can be incorporated in all manner of communication during hand-off and shift changes (Fig. 2.3) [39].

Using the first case example described earlier in the chapter, the nurse anesthetist to pre-op nurse communication using the SBAR format could look like this.

Situation: "Hi Rick, this is Darlene the nurse anesthetist for the case. Mr. Jones is being prepped for a hallux valgus repair."
Background: "We will be doing Mr. Jones with general anesthesia. He has a history of mild COPD."
Assessment: "I want to use a paralytic agent prior to intubation to ease induction."
Recommendation: "I've drawn up the agent with a saline flush. I'll let you know when I am ready to intubate the patient. At that point you can administer the agent."

Likewise an SBAR-based communication between the intern and the nurse described in case two could look like this.

Situation: "Hi Marcie, this is John the intern taking care of Mrs. Simmons in Room 347. I need to update you on her care."
Background: "The attending, Dr. Collins, is concerned that she is at high risk for a deep venous thrombosis (DVT) and a pulmonary embolism (PE). She has a history of mild dementia, hypertension and obesity."
Assessment: "We need to start appropriate DVT prophylaxis now."
Recommendation: "I placed an order in the electronic medical record (EMR) for lower extremity compression devices and heparin to be started now. Let me know if you have any questions."

## Critical Language

Medicine's complex and hierarchical environment can make it difficult for people to speak up with concerns. As already mentioned, power and authority differences,

| SBAR Communication | |
|---|---|
| Use the following SBAR steps to communicate issues, problems or opportunities for improvement to coworkers or supervisors. SBAR can be applied to both written and verbal communications. | |
| **SITUATION** **State what is happening at the present time that has warranted the SBAR communication.** | *Example:* Patients and visitors are entering the medical center through the wrong doors and getting lost trying to find their destination. |
| **BACKGROUND** **Explain circumstances leading up to this situation. Put the situation into context for the reader/listener** | *Example:* The campus has many buildings and is accessible from both E. Washington St. and Eastland Dr. Other entrances are more noticeable than the hospital's main entrance. MD offices do not have good maps to mark and hand to patients when sending them to our campus, and they often misdirect patients. |
| **Assessment** **What do you think the problem is?** | *Example:* People need something that they can carry with them when they are coming to the hospital so they park outside the appropriate entrance. |
| **RECOMMENDATION** **What would you do to correct the problem?** | *Example:* Create a campus visitor guide that includes an "aerial" map of the campus as well as a community map and floor by floor maps. Distribute widely, including to physician offices. Make them available to visitors in admission packets and at all entrances. |

**Fig. 2.3** SBAR Communication. ©Joint Commission Resources: *Joint Commission Journal on Quality & Patient Safety*. Oakbrook Terrace, IL: Joint Commission on Accreditation of Healthcare Organizations, (2006) [39]. Reprinted with permission

lack of psychological safety, cultural norms, stress and fast paced situations, and ambiguity regarding the plan of action further complicate the situation. The adoption of critical language, derived from the CUS program at United Airlines, can be very effective in overcoming these challenges [40]. CUS stands for "I'm concerned, I'm uncomfortable, this is unsafe, or I'm scared" used in an escalating sequence, and is adopted within the culture as meaning: "we have a serious problem, stop and listen to me." Critical language such as CUS creates a clearly agreed upon communication model, that helps avoid the natural tendency in health care to speak indirectly and deferentially. Critical language can also be used in conjunction with standardized communication protocols such as SBAR.

Using the example of case two, an SBAR structured communication incorporating critical language could look like this:

Situation: Dr. Ross, this is Peter the nurse from 3 south. Mrs. Simmons in 347 is in respiratory distress. **I'm really concerned about her!**

**Background:** She has dementia and hypertension and has been more agitated this afternoon.

**Assessment:** Her breath sounds are diminished and she looks to be tachypneic. She's not moving much air. I think she needs a treatment before she stops breathing.

**Recommendation: I'm uncomfortable about her condition.** I'd like you to come and evaluate her immediately.

## Conclusion and Key Lessons Learned

- Lack of communication has been implicated as a major contributing factor for adverse and sentinel events.
- Effective teamwork and communication decreases adverse events and improves staff satisfaction and effectiveness.
- Components of an effective team include the 5 "Cs": a common goal; commitment; competence; communication; and coordination.
- Effective teams share: knowledge, skills, and attitudes.
- Physicians and nurses have differing communication styles that can lead to miscommunication and frustration.
- Disruptive behavior on the part of any healthcare worker can undermine communication and team effectiveness and represents a significant barrier to effective teams.
- Shared and standardized communication models taken from the aviation industry such as SBAR and CUS are effective to overcome traditional communication barriers and should be taught to physicians and nurses in training in a combined manner.
- MedTeams and TeamSTEPPS are standardized and effective teaching curricula to develop effective teams.

## References

1. Diagnostic monitoring of skill and knowledge acquisition. Hillsdale, NJ: Lawrence Erlbaum Associates; 1990.
2. Joint commission on accreditation of healthcare organizations: sentinel event data summary. http://www.jointcommission.org/sentinel_event_statistics_quarterly/. Last accessed 27 Jun 2012.
3. Manser T. Teamwork and patient safety in dynamic domains of healthcare: a review of the literature. Acta Anaesthesiol Scand. 2009;53:143–51.
4. Baggs JG, Schmitt MH, Mushlin AI, et al. Association between nurse-physician collaboration and patient outcomes in three intensive care units. Crit Care Med. 1999;27:1991–8.
5. Risser DT, Rice MM, Salisbury ML, Simon R, Gregory J, Berns SD. The potential for improved teamwork to reduce medical errors in the emergency department. Ann Emerg Med. 1999;34(3):373–83.
6. Firth-Cozens J. Multidisciplinary teamwork: the good, bad, and everything in between. Qual Health Care. 2001;10(2):65–6.

7. Majzun R. The role of teamwork in improving patient satisfaction. Group Prac J. 1998;47(4):12–6.
8. Sexton JB, Thomas EJ, Helmreich RL. Error, stress, and teamwork in medicine and aviation: cross sectional surveys. Br Med J. 2000;320:745–9.
9. Morey JC, Simon R, Jay GD, Wears RL, Salisbury M, Dukes KA, et al. Error reduction and performance improvement in the emergency department through formal teamwork training: evaluation results of the MedTeams project. Health Serv Res. 2002;37(6):1553–81.
10. Mann S, Marcus R, Sachs B. Lessons from the cockpit: how team training can reduce errors from L&D. Contemp Ob Gyn. 2006;51:34–45.
11. Salas E, Dickinson TL, Converse SA, Tannenbaum S. Toward an understanding of team performance and training. In: Swezey RW, Salas E, editors. Teams: their training and performance. Norwood, NJ: Ablex; 1992.
12. Agency for Healthcare Research and Quality. Team-STEPPS: national implementation [online]. http://teamstepps.ahrq.goc/index.htm. Last accessed 27 Jun 2012.
13. Katzenbach JR, Smith DK. The wisdom of teams: creating the high performance organization. Cambridge, MA: Harvard Business School Press; 1993.
14. Weinger MB, Blike GT. Intubation Mishap. AHRQ WebM&M: Case & commentary. http://Webmm.ahrq.gov/printviewCase.aspx?caseID=29
15. Cannon-Bowers JA, Tannenbaum SI, Salas E, Volpe CE. Defining competencies and establishing team training requirements. In: Guzzo RA, Salas E, editors. Team effectiveness and decision-making in organizations. San Francisco, CA: Jossey-Bass Associates; 1995. p. 333.
16. Sims DE, Salas E, Burke SC. Is there a 'Big Five' in teamwork? 19th Annual meeting of the society for industrial and organizational psychology, Chicago, IL; 2004, p. 4.
17. Reader TW, Flin R, Mearns K, Cuthbertson BH. Interdisciplinary communication in the intensive care unit. J Anaesth. 2007;98:347–52.
18. Helmreich RL, Foushee HC, Benson R, Russini W. Cockpit resource management: exploring the attitude-performance linkage. Aviat Space Environ Med. 1986;57:1198–200.
19. Shortell SM, Zimmerman JE, Rousseau DM, et al. The performance of intensive care units: does good management make a difference? Med Care. 1994;32:508–25.
20. Knaus WA, Draper EA, Wagner DP, Zimmerman JE. An evaluation of outcome from intensive care in major medical centers. Ann Intern Med. 1986;104:410–8.
21. DeFontes J, Subida S. Preoperative safety briefing project. Perm J. 2004;8:21–7.
22. Posner B, Randolph W. Perceived situation moderators of the relationship between role ambiguity, job satisfaction, and effectiveness. J Soc Psychol. 1979;109:237–44.
23. Kivimaki M, Sutinen R, Elovainio M, et al. Sickness absence in hospital physicians: 2 year follow up study on determinants. Occup Environ Med. 2001;58:361–6.
24. Aiken LH, Clarke SP, Sloane DM, et al. Nurses' reports on hospital care in five countries. Health Aff (Millwood). 2001;20:43–53.
25. Bednash G. The decreasing supply of registered nurses: inevitable future or call to action? JAMA. 2000;283:2985–7.
26. Mills P, Neily J, Dunn E. Teamwork and communication in surgical teams: implications for patient safety. J Am Coll Surg. 2008;206:107–12.
27. Joint Commission on Accreditation of Healthcare Organizations. Leadership standard clarified to address behaviors that undermine a safety culture joint commission perspectives®. 2012; 32:1.
28. Rosenstein AH, Daniel MO. A survey of the impact of disruptive behaviors and communication defects on patient safety. Jt Comm J Qual Patient Saf. 2008;34(8):462–71.
29. Davenport DL, Henderson WG, Mosca CL, et al. Risk-adjusted morbidity in teaching hospitals correlates with reported levels of communication and collaboration on surgical teams but not with scale measures of teamwork climate, safety climate, or working conditions. J Am Coll Surg. 2007;205:778–84.
30. Prince C, Salas E. Training and research for teamwork in the military aircrew. In: Wiener EL, Kanki BG, Helmreich RL, editors. Cockpit resource management. Orlando, FL: Academic; 1993. p. 337–66.

31. Povenmire HK, Rockway MR, Bunecke JL, Patton MW. Cockpit resource management skills enhance combat mission performance in a B-52 simulator. In: Jensen RS, editor. Proceedings of the fifth international symposium on aviation psychology. Columbus, OH: Association of Aviation Psychologists; 1989. p. 489–94.
32. Institute of Medicine (U.S.). Committee on quality of health care in America. Crossing the quality chasm: a new health system for the 21st century. Washington, DC: National Academy Press; 2001.
33. Drinka TJK, Clark PG. Health care teamwork: interdisciplinary practice and teaching. Westport, CT: Auburn House/Greenwood; 2000.
34. Neily J, Mills PD, Young-Xu Y, Carney BT, West P, Berger DH, et al. Association between implementation of a medical team training program and surgical mortality. JAMA. 2010;304(15):1693–700.
35. Grogan EL, Stiles RA, France DJ, et al. The impact of aviation-based teamwork training on the attitudes of health-care professionals. J Am Coll Surg. 2004;199:843–8.
36. McConaughey E. Crew resource management in healthcare: the evolution of teamwork training and MedTeams®. J Perinat Neonatal Nurs. 2008;22:96–104.
37. Baker DP, Gustafson S, Beaubien J, Salas E, Barach P. Medical teamwork and patient safety: the evidence-based relation. AHRQ Publication No. 05–0053. Rockville, MD; 2005.
38. King HB, Battles J, Baker DP, Alonso A, Salas E, Webster J, Toomey L, Salisbury M. TeamSTEPPS™: team strategies and tools to enhance performance and patient safety. http://www.ahrq.gov/downloads/pub/advances2/vol3/Advances-King_1.pdf. Last accessed 27 Jun 2012.
39. Haig KM, Sutton S, Whittington J. National Patient Safety Goals. SBAR: a shared mental model for improving communication between clinicians. Jt Comm J Qual Patient Saf. 2006;32:167–75.
40. Leonard M, Graham S, Bonacum D. The human factor: the critical importance of effective teamwork and communication in providing safe care. Quality and Safety in Health Care 2004;13(suppl 1):i85–i90.

# Chapter 3
# Handoff and Care Transitions

Mei-Sing Ong and Enrico Coiera

> *"The single biggest problem in communication is the illusion it has taken place."*
>
> George Bernard Shaw

## Introduction

A *handoff* is a transfer of responsibility and accountability. Clinical handoffs transfer the responsibility for some or all aspects of patient care to another clinician or team, either on a temporary or a permanent basis [1]. Handoffs occur at transitions of care when patients move from one institution to the other, from one care setting to another (e.g., intensive care unit to floor) or at shift changes within a hospital. During handoff, clinicians exchange patient information and may jointly plan the next steps in care. Effective handoff is critical in ensuring care continuity and patient safety, as failure to communicate critical information during handoff can lead to uncertainty in decisions about patient care and result in suboptimal care.

Numerous published reports demonstrate that ineffective handoff and communication is a major contributor to adverse clinical events and outcomes. A 1994 report found that cross-coverage of medical inpatients among more than one clinician is associated with a fivefold increase in the risk of an adverse event [2]. In a review of 122 malpractice claims in which patients had alleged a missed or delayed diagnosis in the Emergency Department (ED), inadequate handoffs contributed to 24 % of the cases [3]. Another study of 889 malpractice claims found that communication failures during handoff were implicated as the cause in 19 % of cases involving medical trainees and 13 % of other cases [4].

M.-S. Ong, Ph.D. (✉) • E. Coiera, M.B.B.S., Ph.D., F.A.C.M.I.
Centre for Health Informatics, University of New South Wales, Sydney,
NSW 2052, Australia
e-mail: m.ong@unsw.edu.au; e.coiera@unsw.edu.au

A. Agrawal (ed.), *Patient Safety: A Case-Based Comprehensive Guide*,
DOI 10.1007/978-1-4614-7419-7_3, © Springer Science+Business Media New York 2014

In the current landscape of decentralized and increasingly specialized and fragmented healthcare services, managing the quality of the handoff process is thus a critical component of any safety and quality initiative. The Joint Commission introduced "Effective handoff and communication" among staff as a National Patient Safety Goal in 2009 (http://www.jointcommission.org) and then subsequently, recognizing the vital importance of the issue, this goal was moved into a Joint Commission "standard" that all accredited hospitals must achieve (*Chapter: Provision of care, treatment, and services; Element of Performance 2*). However, getting the policy balance right is challenging. For example, recent attempts by the US Accreditation Council for Graduate Medical Education (AGCME) to cut extended shifts and reduce work hours for medical residents has a side-effect of increasing handoff rates [5].

In recent years, much effort and research have been directed at improving handoff practices. Still, reports of handoff failures continue primarily because handoff is a complex process, with many different potential failure points [6]. In this chapter, we present two case studies of patient harm events caused by ineffective handoff. Our goal is to examine some of the challenges faced by clinicians at handoff and to provide some insights into how these obstacles can be overcome.

# Case Studies

## Case 1: Poor Management of Postpartum Hemorrhage

### Clinical Summary

*An emergency cesarean was performed on a 29-year-old patient, Ms. J, during which a healthy baby was delivered. A uterine tear occurred during the delivery resulting in excessive blood loss. The tear was repaired, and the wound was observed for a period of time to ensure there was no further bleeding. Ms. J was then transferred to the recovery room. In the recovery room, the care of the patient was delegated to a recovery nurse, who was also assisting with the cesarean procedure. At handoff, the recovery nurse was not informed of the uterine tear. Details of the surgery and postoperative care instructions were documented and given to the nurse. The nurse read the report, but not in its entirety. Within a short time after the admission to the recovery room, Ms. J started to bleed internally and her blood pressure progressively declined. Unaware that the patient had experienced postpartum hemorrhage during the cesarean section, the nurse failed to recognize the significance of the blood loss and changes in blood pressure. She felt that the priority was to clean up the blood and the patient. She was also unaware of the need to observe the patient's fundal height and to perform fundal massage. Three hours after admission to the recovery room, an experienced nurse noticed the blood loss and notified the medical team immediately. Ms. J was transferred to the operating theater where she underwent further surgery. A large blood clot was removed from her uterus, which was found to be atonic. Ms. J had a cardiac arrest on the operating theater table. She was resuscitated, but died shortly after in the intensive care unit. The cause of her death was multisystem organ failure following postpartum hemorrhage.*

## Case 2: Opioid-Induced Respiratory Depression in a Head Injury Patient

### Clinical Summary

*A 16-year-old girl, Ms. A, was admitted to a neurosurgical unit after sustaining a closed head injury. On admission, she was examined by a neurosurgical fellow and diagnosed with a mild head injury. The fellow did not inform the on-call-attending physician of the admission. As a result, Ms. A's case was not reviewed by the attending physician on the day of admission. The following day, the attending physician attended the ward with a senior resident. On reviewing the patient, the attending physician formed the view that she had dural lacerations with bone fragments within the brain. Ms. A was scheduled for a surgery to elevate her skull the following morning. The attending physician further stated that he was constrained regarding the amount of analgesia that could be given to Ms. A and gave a verbal order that analgesia was to be determined by the attending physician or fellow. No medical notes were taken at this ward round, and the attending physician's instructions were not documented. After the ward round, the patient was left in the care of the neurology resident. Early that afternoon, in response to Ms. A's ongoing pain, the resident decided to alter the pain management regime. She did not discuss her decision with a senior member of the team. Later that evening, an anesthetic fellow reviewed Ms. A for a preoperative anesthetic consultation. In response to the patient's severe pain, the anesthetic fellow further increased the dose and frequency of pain relief, without consulting with the primary care team. Neurological observations of the patient through the early night remained stable. When observations were due again later in the night, the responsible nurse decided that it was not necessary as patient was "sleeping comfortably." Early next morning, Ms. A was found to be unresponsive and died shortly after unsuccessful resuscitation attempts. Coroner's inquest into the case indicated that the patient died from a respiratory arrest due to the depressive effect of opiate medication.*

## Root Cause Analysis

### Case 1

**What Happened?**

Ms. J's death resulted from the delay in the recognition and treatment of postpartum hemorrhage (PPH). PPH is a leading cause of preventable maternal death. Patients suffering from PPH can deteriorate quickly, unless immediate medical care is provided. Common causes of PPH include failure of the uterus to contract adequately after birth (atonic PPH), and trauma to the genital tract (traumatic PPH) [7]. After experiencing a uterine tear during an emergency cesarean, Ms. J was at risk of developing PPH. Appropriate postoperative care should have included the careful monitoring of fundal

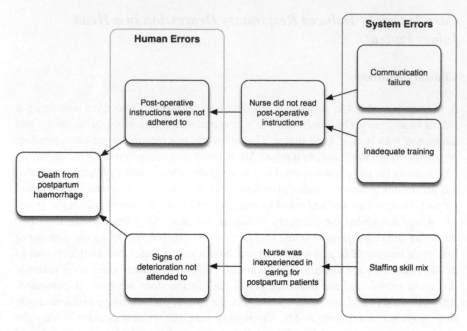

**Fig. 3.1** Case 1: Poor management of postpartum hemorrhage—cause and effect diagram

height, and the massaging of the uterus to expel blood and blood clots. These were clearly specified in the postoperative notes. However, neither having read the notes nor having been alerted at handoff, the nurse failed to comply with these care instructions. Ms. J continued to bleed throughout her stay at the recovery room. Early signs of deterioration were not recognized, and the delay in treatment ultimately led to her death.

## Why Did It Happen?

Multiple human and systemic errors contributed to this unfortunate event (Fig. 3.1). On the outset, the death of Ms. J was caused by the nurse's failure in adhering to postoperative instructions and in recognizing vital signs of deterioration. The underlying systemic causes of these errors were much more complex.

Communication Failure

Communication breakdown between the operating theater and recovery room was a major contributor to the incident. While details of the cesarean procedure and postoperative care instructions were documented, the recovery room nurse was not verbally briefed on the patient's condition and the care she required. Assumptions were made that since the nurse was present at the operation, the patient's condition should have been obvious and being a nurse looking after a postoperative obstetric patient, the nurse would have been trained to provide appropriate care. Unfortunately, the nurse had no prior experience in caring for a postoperative patient who had suffered a PPH. While she was present at the surgery, she was unaware of the amount of blood the patient had lost and appeared to have no specific recollection of the procedure to repair the uterine tear. Had a proper handoff been given, it may have been ascertained at an early stage that the nurse did not have the requisite skills to care for the patient. Further, had she been informed of the blood loss during the cesarean section, she might possibly have been more acutely aware of the need to closely monitor vital signs. The blood loss and the declining blood pressure that she observed may have resulted in her alerting the medical team straight away.

## Inadequate Training

If a staff member does not have the skills to deal with a particular crisis, they should at least be trained to identify it and seek assistance. The hospital had a formal protocol for treating PPH. However, it appeared that the nurse had never seen the protocol nor was she trained in the identification of symptoms of PPH. Her observations of the patient's declining blood pressure and increased blood loss should have resulted in an immediate call for assistance. However, not appreciating the gravity of the situation, her priority was to seek for assistance to clean the blood. Another nurse who came to assist recognized the urgency of the situation immediately, and the medical team was notified. However, by then, it was already too late.

## Poor Staff Allocation

An important contributor to this incident is the failure of hospital administrators in ensuring that rostered staff have the skills to identify and deal with a particular medical condition. Ms. J was entrusted to a nurse who was unskilled to provide the care required. On the day of the incident, resource constraint was not an issue, as there were other more experienced nurses in the hospital who could have assisted with caring for Ms. J. The recovery room was relatively quiet, and Ms. J was the only patient. Poor organization on the part of the hospital administrators meant that the patient was denied the best care that she could have received.

## How Can It Be Prevented?

The incident could have been prevented by adequate handoff communication during the transfer of care. The transferring team should not rely on written documentation alone to communicate patient information, since written notes are easily overlooked. Verbal handoff of critical information is vital to ensure that the receiving team understands the care required and is capable of providing it.

The need for better coordination of available resources is also evident. Hospital administrators have the responsibility to ensure that the rostered staff has the skills to provide the care required. Further, it is imperative that new staff members are introduced to the relevant hospital protocols. Deficits in skills or experience should be identified early in the induction stage, so that appropriate training can be provided.

## Case 2

### What Happened?

Ms. A's death was most likely a result of opioid-induced respiratory depression. Respiratory depression is recognized as a serious complication of opioid analgesic therapy. Opioids can impair central nervous system respiratory drive, resulting in alveolar hypoventilation and inability to adequately eliminate carbon dioxide, and eventually to adequately exchange oxygen [8, 9]. The respiratory depressant effects of opioids may be markedly exaggerated in the presence of head injury, due to the increased intracranial tension. Further, opioid-naive patients (individuals who are not chronically receiving opioid analgesics on a daily basis) are far more susceptible to respiratory depression. In the case of Ms. A, both these risk factors were present. Being opioid-naive and having sustained a head injury, Ms. A was at risk of developing opioid-induced respiratory depression. The amount of analgesia prescribed to Ms. A exceeded the usual dosage given to a head injury patient and was likely the cause of her sudden decline and eventual death.

### Why Did It Happen?

A confluence of human and system errors resulted in this unfortunate event (Fig. 3.2). At the human level, poor clinical judgments were made by the neurosurgical resident and the anesthetist fellow in the prescription of opioids. Additionally, the nurse's failure to perform routine neurological observations meant that the patient's decline was left undetected. The neurosurgical fellow's failure in notifying the on-call-attending physician of the patient's admission, while not directly linked to the cause of death, had contributed to the unfolding of the events. Had the patient been reviewed by the attending physician on the day of admission, the surgery would have been scheduled earlier, and this unfortunate incident could arguably have been averted. Multiple systemic failures facilitated these human errors.

Poor Staffing Level and Inadequate Supervision

Poor management of staff resources played a major role in the incident. On the day of Ms. A's admission, the neurosurgical unit was understaffed, as two fellows were on a training program. The remaining fellow was overburdened with heavy workload, which contributed to his failure to inform the on-call-attending physician of the admission. On the following day, due to lack of staff, a resident with little experience was placed in charge of the ward for the first time, merely 2

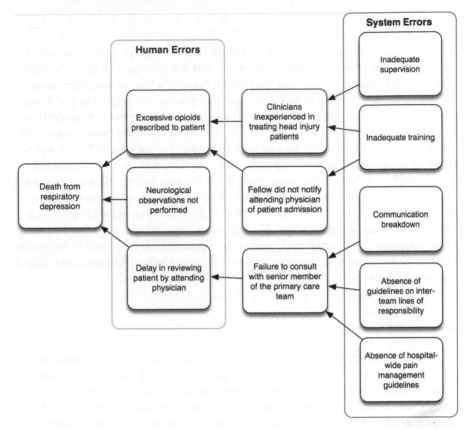

**Fig. 3.2** Case 2: Opioid-induced respiratory depression in a head injury patient—cause and effect diagram

weeks after she had commenced rotation there. The resident had no prior experience in managing patients with head injury. Based on her previous training at an orthopedic ward, she believed it was her responsibility to prescribe analgesic drugs without consulting with a senior member of the team. The general danger of narcotics in head injury patients was not appreciated. Poor staffing level left the inexperienced resident with little support from senior members of the team. The responsibility imposed on the resident was disproportionate to her level of knowledge and experience. Consequently, poor clinical decisions were made leading to adverse patient outcome.

Communication Failures

A chain of communication failures resulted in the worst possible outcome for Ms. A. First, the on-call-attending physician was not informed of Ms. A's admission, leading to delay in reviewing the patient. Second, while the resident was present during the ward round with the attending physician, the attending physician's instruction not to alter the pain management regime did not appear to be understood. It was also the resident's responsibility to document the ward round, which she failed to do. As a result of the lack of documentation, the anesthetist fellow was not aware of the attending physician's order that analgesia was to be determined by a senior member of the neurosurgical team. When the anesthetist fellow discussed his decision to increase the doses of oxycodone hydrochloride with an attending physician anesthetist, he was advised to have the patient reviewed and the changes authorized by the neurosurgical team. Unfortunately, this advice was not followed. Thus, communication breakdown within the neurosurgical team and between the anesthetic and neurosurgical teams resulted in poor clinical decisions and a fatal outcome for the patient.

Lack of Guidelines

At the organizational level, there was a lack of hospital-wide pain management guidelines. As a result, new staff members were unaware of the need to escalate to senior medical staff changes in pain management for patients with head injury. Further, the absence of inter-team lines of responsibility for treating pain and prescribing analgesia resulted in multiple team involvement in pain management beyond the primary care team.

**How Can It Be Prevented?**

Following Ms. A's tragic death, several measures were implemented by the hospital to prevent similar incidents from occurring. These included the development of an acute pain management policy and procedure for use in the neurosurgery department, establishing that decisions regarding the prescription of analgesia outside the terms of the guidelines can only be made by a neurosurgical fellow or attending physician. Tutorial and orientation program were implemented to ensure that junior practitioners and new staff members were aware of these guidelines. In-house education for medical and nursing staff regarding pain management treatment was also introduced. Additionally, the hospital implemented a system for dealing with periods where there is reduced fellow coverage due to training requirement, pursuant to which the head of department is responsible for ensuring that adequate cover is documented.

# Discussion

The case studies presented in this chapter are classic illustrations of James Reason's Swiss cheese model, where multiple system and human errors cumulatively cascaded into an adverse event. In both cases, the trajectory of error began during the transition of patient care between providers. The first involves the transfer of care from the operating theater to the recovery room, and the second involves the transfer of care between a neurosurgical attending physician and a resident.

## *Transition of Care: A Point of Vulnerability*

It is widely recognized that transition of care is a point of vulnerability in patient safety. There are five main types of transition (Fig. 3.3) (1) interhospital—the transfer of care when a patient is transferred from one facility to another; (2) interdepartmental—the transfer of care during an inpatient transfer; (3) inter-shift—the transfer of care during shift changes; (4) interprofessional—the transfer of care between medical teams; (5) intra-team—the transfer of care between members of the same team. During these transitions, there is a handoff of responsibility from one clinician to another that involves the transfer of rights, duties, and obligations for the care of patients [10]. Existing studies show that current handoff practices are deficient; handoffs are typically unstructured and highly variable in content and process. Thus, handoff failures during transitions of care are common, leading to poor clinical decisions and suboptimal patient care [6, 11].

## *Barriers to Effective Handoff Communication*

### The Diversity of Teams

The prevalence of handoff failures is partially a result of the complexity of clinical interactions. Patient handoff often involves multiple teams, with differing expertise, work processes, and culture. Even within the same team, the level of knowledge and experience between team members can vary greatly. These differences can impede effective communication.

Information is interpreted differently by different individuals. The amount of information required to communicate a particular message depends on the degree to which the sender and receiver share mental models of the world or common ground [12]. The greater the common ground between the sender and receiver of a message, the less the message needs to contain, and the more that can be assumed. A message that contains more information than is required by an expert might be insufficiently

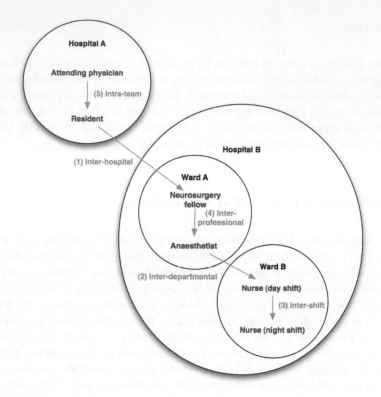

**Fig. 3.3** Transitions of care

informative for a novice. Failure to recognize or appreciate that others do not share a mental model or *common ground* is a major barrier to effective clinical communication [12, 13].

Communication between senior clinicians and their junior counterparts is often plagued by this problem. More experienced clinicians tend to assume too much about the knowledge and skill level of their junior counterparts and fail to provide sufficient information during handoff of patient care [14, 15]. This is evident in both the case studies presented. In the first case, despite being a witness to Ms. J's cesarean section, the recovery nurse was oblivious to the events that occurred during the surgery. The obstetrician wrongly equated her presence at the surgery to an understanding of the patient's condition and therefore did not consider that a verbal briefing of surgical events was necessary. In addition, the obstetrician assumed that since the nurse assisted with the cesarean section, she was experienced in caring for postpartum patients. As a result, postoperative care instructions were not verbally communicated during the transfer of patient to the recovery room. Thus, wrong assumptions of the nurse's level of knowledge led to poor handoff communication.

In the second case, the neurosurgery-attending physician clearly instructed the resident not to change the patient's pain management regime without consulting a senior member of the team. Similarly, the anesthetist fellow was advised by her supervisor to discuss with the neurosurgery team her decision to increase the patient's analgesia. In both instances, the instructions were not complied with. The message was somehow overlooked due to their lack of appreciation for the risk of analgesia on head injury patients.

Communication difficulty between teams during care transitions can further be exacerbated by the ambiguity in roles and differences in work processes. Studies have shown that clinicians often report not knowing when the transfer of care takes place and to whom handoff should be given [16, 17]. Even within a team, poorly defined boundaries of responsibility are not uncommon [14]. Under such circumstances, tasks that are not explicitly assigned to an identified provider can easily get lost [18]. Problems can also arise when multiple clinicians assume responsibility for a task in the absence of well-defined inter-team lines of responsibility. This is evident in the second case study, where multiple team involvement in pain management of the patient resulted in the overprescription of analgesia.

**Time and Resource Constraints**

Time and resource constraints compound the communication challenges. Clinicians are often expected to operate under limited resources. When workload is high, clinical communication becomes less interactive and rushed [19]. Communication failures also abound when clinicians are fatigued. This is evident in the second case study. Overburdened with the heavy workload, the neurosurgical fellow failed to inform the on-call-attending physician of Ms. A's admission, resulting in delay in reviewing the patient. Existing literature contains many examples of communication breakdown caused by time and resource constraints. For example, studies on handoff in the ED showed that patients are commonly transferred to an inpatient ward without adequate handoff, due to the urgency of treating emergency patients [19]. Even when handoff is provided, the information given is often outdated, as emergency physicians may not have time to review the patient again before the transfer, and are therefore unaware of new developments or current vital signs.

## Delegating Care: The Importance of Supervision

A major contributor to these incidents was the lack of experience of the care providers. In both cases, patient care was delegated to an inexperienced practitioner, who was expected to perform beyond their level of competencies without adequate supervision. Consequently, poor clinical decisions were made, resulting in harm to the patient.

The healthcare system is often heavily reliant on physicians-in-training for the day-to-day provision of medical care. Balancing the need to provide medical training to junior practitioners and patient safety is a challenge. Ideally, junior practitioners should only carry out tasks within their competency and have a responsibility to contact senior staff if they get out of their depth. Unfortunately, due to their lack of experience, junior doctors may fail to recognize when they need assistance. As a result, they may take on more responsibility than is appropriate, involving senior staff too late, or failing to contact them at all. This is evident in both case studies. In the first incident, the recovery room nurse failed to recognize that the patient was deteriorating. And in the second case, the neurosurgery resident and anesthetist fellow were unaware of the danger of analgesia in head injury patients. Thus, they failed to seek advice from a senior staff member. Indeed, several studies have shown that junior doctors often have difficulty in identifying their own clinical limitations [20–22]. A detailed discussion of the issue around graduate medical education and patient safety can be found in Chap. 4.

## Improvement Strategies

### Standardization

The need for strategies that support safe and reliable patient handoff is evident. A common mechanism for minimizing breakdowns in communication is to develop standard communication protocols. Standardization defines best practices and helps set normative standards for what is expected in a communication event. Message standardization leads to consistency in the message structure, reduces the opportunity for misunderstanding between medical teams, and assists in the detection of errors of omission. For example, ambiguity in roles and responsibilities can be managed by defining expectations for each team member [6]. Communication breakdown caused by differences in the level of experience and knowledge can potentially be diminished by standardizing the handoff protocol between senior and junior clinicians and providing guidelines for delegating care to junior clinicians. Several methods for standardization are summarized in Table 3.1.

Handoff protocols should cover both verbal and written communication of patient information. Verbal handoff facilitates interactive questionings between providers, during which patient care plans can be clarified, and the ability of the receiving team to manage the patient can be assessed. Written handoff ensures there is a persistent copy of critical information, which is not "lost in translation," and can be time effective, as there may be limited opportunity for communication between clinicians after a shift change or transfer.

### The Role of Information Technology

Time and resource constraints often preclude adequate handoff between clinicians [6]. Information technology such as electronic health records (EHR) can facilitate the access to patient information in a distributed manner. Using an EHR, patient information can be consolidated into a single system that can be accessed

**Table 3.1** Methods for standardizing handoff communication

*Read-backs*: Read-back requires the recipient of a message to repeat back the information to the communicator. By ensuring closed loop communication, the method can ensure critical information is not missed or heard incorrectly [23]. The use of standard read-back protocols can minimize the misinterpretation of communicated information between two parties [24]. In one study, read-back was implemented for telephone reports of critical laboratory results and detected and corrected errors in 3.5 % telephone exchanges [25]

*Standardized sign-out templates*: Written sign-out information can be presented in a predefined structure. This might include critical fields that need to be filled out, such as allergy status, medication history, and preference for treatment. Simple sign-out templates have been shown to be effective in ensuring critical information is communicated during care transitions [26, 27]

*Structured goals*: The use of a structured daily goals form in the intensive care unit produced a significant improvement in the percentage of residents and nurses who understood the goals of care for the day and reduced ICU length of stay [28]. At baseline, less than 10 % of residents and nurses in the study understood the goals of care for the day. After implementing the daily goals form, greater than 95 % of nurses and residents understood the goals of care for the day. The ICU length of stay decreased from a mean of 2.2 days to 1.1 days

*SBAR (Situation, Background, Assessment, Recommendation)*: Communication can be improved by imposing a standardized structure, such as SBAR. The structure of SBAR consists of a brief description of the situation, followed by the background and the clinician's specific assessment and complete recommendation [29]. By providing a common framework for information sharing, ambiguity in handoff communication can be minimized [30, 31]

anytime, in different localities, and by different team members. Improving the electronic availability of critical information can decrease misinformation, facilitate recognition of clinical changes, and increase the transparency of responsibility changes to other specialties [19]. There is an increasing body of work demonstrating the benefits of information technology in facilitating information exchange. In one study, the implementation of a computerized handoff system reduced the overall number of patients missed on resident rounds by half [32]. In another study, computerized handoffs reduced the rate of preventable adverse events from 1.7 to 1.2 % [5].

Another advantage of gathering information through information technology is the ability to standardize information to ensure completeness and legibility. For example, computerized physician order entry (CPOE) can be structured so that each medication order includes a dose, route, and frequency [33]. Additionally, *forcing functions* (features that restrict how a task may be performed [34]) can be implemented to ensure that critical information is provided by clinicians (Fig. 3.4).

## The Role of Supervision During Handoff

There is much room for improved trainee supervision. Currently, medical training often involves throwing trainees into the deep end. Supervision is largely "reactive," where assistance is provided when requested. This approach is inadequate, as junior

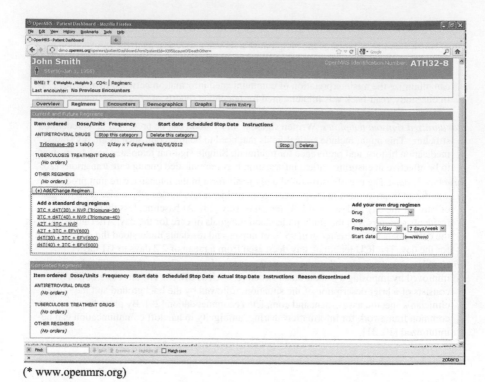

(* www.openmrs.org)

**Fig. 3.4** Screenshot of an open source EHR system, known as the OpenMRS (http://www. openmrs.org). Common drug regimens are listed to facilitate correct prescription based on the recommended practice

practitioners often do not have a realistic understanding of their own clinical limitations. It is imperative for supervisors to know the competencies of their trainees when handing patient care. Junior practitioners working in a new specialty should be provided with close supervision with regular checking. As they gain experience, more responsibilities can be given with less supervision.

There is also a need to provide support to senior practitioners in their supervisory roles. Senior practitioners are under ever-increasing pressure and are often not supported to pass on their skills to junior colleagues. Despite the implications of poor supervision on patient safety, the supervisory responsibilities of attending physicians are poorly defined. The skills necessary to supervise junior practitioners have either never been taught or taught suboptimally. An audit carried out by The Royal College of Anaesthetists found that fewer than half of department provided written guidance on attending physician supervision for trainees [35]. Further, most attending physicians found conflicting demands of service and supervision difficult. Unless these systemic issues are addressed, the risk posed by inexperienced practitioners will continue to persist.

## Conclusion and Key Lessons Learned

In this chapter, we explore some of the challenges with patient handoff through two case studies. Several organizational issues contributed to the adverse outcome in these case studies. We have addressed the problems with communication failures and inadequate supervision during transition of care. Other systemic issues featured in the case studies include poor resource coordination, which resulted in inexperienced practitioners being imposed responsibilities that were beyond their level of competencies, and the lack of training and induction program provided to new staff members. Some strategies for addressing these issues are summarized in Table 3.2.

Unfortunately the problems identified in our case studies have existed for a number of years and regrettably the same errors are likely to recur. Many strategies to improve handoff failed to translate into safety for patients, due to lack of compliance on the part of the clinicians. Clinicians can become desensitized to risky practices. Daily violations become routine, and since everyone is doing the wrong thing, no one can be held responsible. This phenomenon is known as *normalization of deviance* [36]. Ultimately, safe patient handoff can only be achieved when there is an unwavering commitment and dedication from all levels in the organization to create a culture of safety and collaboration.

**Table 3.2** Key issues identified and recommended improvement strategies

| Key issues | Improvement strategies |
| --- | --- |
| *Policy standards* | |
| Absence of guidelines and inter-team lines of responsibility | Standardize critical clinical processes (e.g., pain management) and inter-team lines of responsibility |
| Staff members unfamiliar with hospital protocols and escalation process | Induction program to ensure all new staff members are familiar with relevant protocols |
| *Work environment* | |
| Poor staffing levels and mix of skills | Provide adequate supervision for junior staff members |
| Workload and resource constraints | Better coordination of available resources, including early identification of deficits in knowledge and skills |
| *Teamwork* | |
| Communication failure caused by diversity in expertise and expectations | Standardize handoff communication, including both verbal and written handoff |
| Poor availability of information | Provide digital access to patient information so as to facilitate distributed information transfer |

# References

1. National Patient Safety Agency / British Medical Association. Safe handover: safe patients. Guidance on clinical handover for clinicians and managers. Available on bma.org.uk/-/media/ Files/.../safe%20handover%20safe%20patients.pdf. Accessed 11 Jul 2013.
2. Petersen LA, Brennan TA, O'Neil AC, Cook EF, Lee TH. Does housestaff discontinuity of care increase the risk for preventable adverse events? Ann Intern Med. 1994;121(11):866–72.
3. Kachalia A, Gandhi TK, Puopolo AL, Yoon C, Thomas EJ, Griffey R, et al. Missed and delayed diagnoses in the emergency department: a study of closed malpractice claims from 4 liability insurers. Ann Emerg Med. 2007;49(2):196–205.
4. Singh H, Thomas EJ, Petersen LA, Studdert DM. Medical errors involving trainees: a study of closed malpractice claims from 5 insurers. Arch Intern Med. 2007;167(19):2030–6.
5. Institute of Medicine. Resident duty hours: enhancing sleep, supervision and safety. Washington, DC: The National Academies Press; 2009.
6. Ong MS, Coiera E. A systematic review of failures in handoff communication during intrahospital transfers. Jt Comm J Qual Patient Saf. 2011;37(6):274–84.
7. Carroli G, Cuesta C, Abalos E, Gulmezoglu AM. Epidemiology of postpartum haemorrhage: a systematic review. Best Pract Res Clin Obstet Gynaecol. 2008;22(6):999–1012.
8. Gallager R. Killing the symptom without killing the patient. Can Farm Physician. 2010;56(6):544–6.
9. Roussos C, Koutsoukou A. Respiratory failure. Eur Respir J Suppl. 2003;47:3s–14.
10. Solet DJ, Norvell JM, Rutan GH, Frankel RM. Lost in translation: challenges and opportunities in physician-to-physician communication during patient handoffs. Acad Med. 2005;80(12):1094–9.
11. Arora V, Johnson J, Lovinger D, Humphrey HJ, Meltzer DO. Communication failures in patient sign-out and suggestions for improvement: a critical incident analysis. Qual Saf Health Care. 2005;14(6):401–7.
12. Coiera E. Guide to health informatics. 2nd ed. London: Hodder Arnold; 2003. p. 39.
13. Coiera E. When communication is better than computation. J Am Med Inform Assoc. 2000;7(3):277–86.
14. Williams RG, Silverman R, Schwind C, Fortune JB, Sutyak J, Horvath KD, et al. Surgeon information transfer and communication: factors affecting quality and efficiency of inpatient care. Ann Surg. 2007;245(2):159–69.
15. Sutcliffe KM, Lewton E, Rosenthal MM. Communication failures: an insidious contributor to medical mishaps. Acad Med. 2004;79(2):186–94.
16. McFetridge B, Gillespie M, Goode D, Melby V. An exploration of the handover process of critically ill patients between nursing staff from the emergency department and the intensive care unit. Nurs Crit Care. 2007;12(6):261–9.
17. Smith AF, Pope C, Goodwin D, Mort M. Interprofessional handover and patient safety in anaesthesia: observational study of handovers in the recovery room. Br J Anaesth. 2008;101(3):332–7.
18. Collins SA, Bakken S, Vawdrey DK, Coiera E, Currie LM. Agreement between common goals discussed and documented in the ICU. J Am Med Inform Assoc. 2011;18(1):45–50.
19. Horwitz LI, Meredith T, Schuur JD, Shah NR, Kulkarni RG, Jenq GY. Dropping the baton: a qualitative analysis of failures during the transition from emergency department to inpatient care. Ann Emerg Med. 2009;53(6):701–10.e4.
20. Fox RA, Ingham Clark CL, Scotland AD, Dacre JE. A study of pre-registration house officers' clinical skills. Med Educ. 2000;34(12):1007–12.
21. Wu AW, Folkman S, McPhee SJ, Lo B. Do house officers learn from their mistakes? JAMA. 1991;265(16):2089–94.
22. Yao DC, Wright SM. National survey of internal medicine residency program directors regarding problem residents. JAMA. 2000;284(9):1099–104.

23. Brown JP. Closing the communication loop: using readback/hearback to support patient safety. Jt Comm J Qual Saf. 2004;30(8):460–4.
24. Greenberg CC, Regenbogen SE, Studdert DM, Lipsitz SR, Rogers SO, Zinner MJ, et al. Patterns of communication breakdowns resulting in injury to surgical patients. J Am Coll Surg. 2007;204(4):533–40.
25. Barenfanger J, Sautter RL, Lang DL, Collins SM, Hacek DM, Peterson LR. Improving patient safety by repeating (read-back) telephone reports of critical information. Am J Clin Pathol. 2004;121(6):801–3.
26. Clark CJ, Sindell SL, Koehler RP. Template for success: using a resident-designed sign-out template in the handover of patient care. J Surg Educ. 2011;68(1):52–7.
27. Wayne JD, Tyagi R, Reinhardt G, Rooney D, Makoul G, Chopra S, et al. Simple standardized patient handoff system that increases accuracy and completeness. J Surg Educ. 2008;65(6):476–85.
28. Pronovost P, Berenholtz S, Dorman T, Lipsett PA, Simmonds T, Haraden C. Improving communication in the ICU using daily goals. J Crit Care. 2003;18(2):71–5.
29. SBAR technique for communication: a situational briefing model. Cambridge, MA: Institute for Healthcare Improvement. (http://www.ihi.org/IHI/Topics/PatientSafety/SafetyGeneral/Tools/SBARTechniqueforCommunicationASituationalBriefingModel.htm)
30. Haig KM, Sutton S, Whittington J. SBAR: a shared mental model for improving communication between clinicians. Jt Comm J Qual Patient Saf. 2006;32(3):167–75.
31. Leonard M, Graham S, Bonacum D. The human factor: the critical importance of effective teamwork and communication in providing safe care. Qual Saf Health Care. 2004;13 Suppl 1:i85–90.
32. Van Eaton EG, Horvath KD, Lober WB, Rossini AJ, Pellegrini CA. A randomized, controlled trial evaluating the impact of a computerized rounding and sign-out system on continuity of care and resident work hours. J Am Coll Surg. 2005;200(4):538–45.
33. Bates DW. Using information technology to reduce rates of medication errors in hospitals. BMJ. 2000;320(7237):788–91.
34. Bates DW, Gawande AA. Improving safety with information technology. N Engl J Med. 2003;348(25):2526–34.
35. McHugh GA, Thoms GMM. Supervision and responsibility: The Royal College of Anaesthetists national audit. Br J Anaesth. 2005;95(2):124–9.
36. Vaughan D. The challenger launch decision. Chicago: University of Chicago Press; 1996.

# Chapter 4
# Graduate Medical Education and Patient Safety

Jeanne M. Farnan and Vineet M. Arora

> *"The study of error is not only in the highest degree prophylactic, but it serves as a stimulating introduction to the study of truth."*
>
> Walter Lippmann

## Introduction

Graduate medical education (GME) and the training of resident physicians is an integral part of healthcare systems around the world. In the USA alone, there are more than 100,000 residents in approximately 8,500 training programs providing care to over 17 million patients [1]. Teaching hospitals must fulfill and balance two, at times competing, objectives: producing competent independent physicians after the period of training is over and delivering safe care to patients. The former requires that trainees be provided greater opportunities for independent and autonomous decision making, while the latter requires greater supervision and oversight by faculty. Occasional public concerns about being cared for by "student doctors" notwithstanding, literature shows that teaching hospitals overall fulfill these objectives well and have better patient care outcomes [2].

The issue of potential hazardous impact of resident education on patient safety gained national attention in 1984 with the death of Libby Zion, an 18-year-old woman who died in a New York hospital of what was determined to be an adverse drug reaction. The grand jury investigation highlighted risks to patient safety caused by resident fatigue and inadequate clinical supervision [3]. As a result, in 1989, New York State established a limit on resident duty hours to 80 hour per week to

J.M. Farnan, M.D., M.H.P.E. (✉) • V.M. Arora, M.D., M.A.P.P.
Department of Medicine, University of Chicago, 5841 South Maryland, AMB W216, Chicago, IL, 60637, USA
e-mail: jfarnan@medicine.bsd.uchicago.edu; varora@medicine.bsd.uchicago.edu

A. Agrawal (ed.), *Patient Safety: A Case-Based Comprehensive Guide*,
DOI 10.1007/978-1-4614-7419-7_4, © Springer Science+Business Media New York 2014

address the issue of resident fatigue. This became the basis of national duty hour restriction to 80 hour per week by the Accreditation Council on Graduate Medical Education (ACGME) in 2003 [4]. The primary goal of the ACGME policy was to reduce fatigue and improve the safety of care while improving resident well-being and education [5]. In 2004, in the European Union the Working Time Directive was applied to the training of junior doctors, limiting trainees to 56 work hours per week, with other stipulations for consecutive hours worked [6]. Several years later, the US Congress chartered the Institute of Medicine (IOM) to further investigate the issues around the interface of resident training and patient safety. In 2008, the IOM published a follow up report titled "Resident duty hours: enhancing sleep, supervision, and safety" that proposed further reduction in resident work hours [7].

The rationale for this significant policy change has been that the reduced number of duty hours should lead to less fatigue, improved performance, and therefore safer care and published reports do demonstrate improved clinical outcomes [8] and improved resident satisfaction [9] with a reduction in duty hours. However, the most striking and concerning unintended consequence of duty hour restrictions is the discontinuity of care and increase number of handoffs during shift change—both of which have serious implications for patient safety [10]. Since the IOM report also proposed further reduction in *consecutive* work hours, the resulting changes in team structure to accommodate these new proposals may further exacerbate the issues related to handoff and communication. Fortunately, there has been much discussion lately around the impact of handoff communication on patient safety, with the Joint commission incorporating handoff as a National Patient Safety Goal [11] and numerous societies convening to create a "Transitions of Care" consensus policy statement [12, 13]. The topic of handoff and communication and related improvement strategies are also discussed in detail in Chap. 3.

It is concerning that the primary focus of attention of regulation and policy change has been the reduction of resident fatigue through duty hour restriction and relatively little attention has been paid to the quality and quantity of clinical supervision of trainees. Since the traditional approach to resident education remains based on an "apprenticeship" model, i.e., learning while delivering care under the guidance of experienced faculty physicians, clinical supervision plays a critical role in both ensuring the education of the trainees as well as the quality and safety of care. The IOM committee in its deliberations argued that "supervision is the single most important element upon which this education model depends" [7]. The original grand jury indictment in the Libby Zion case had concluded that "the most serious deficiencies can be traced to the practice of permitting…interns and junior residents to practice medicine without supervision" [3]. Residents themselves also identify inadequate supervision as one of the most common causes of medical errors [14].

Clinical supervision has been defined as "the provision of guidance and feedback on matters of personal, professional, and educational development in the context of a trainee's experience of providing safe and appropriate patient care" [15]. The issue of clinical supervision has yet to be examined with respect to the nature of the attending–resident supervision relationship and the identification of factors which encourage or discourage residents from seeking attending physician input into clinical decisions, impact on resident education, and patient outcomes. Most attending physicians have received no training in being a clinical supervisor and increasing workload on attending physicians (partly as a result of the duty hour restrictions) may also inhibit their ability to function as an effective clinical supervisor.

This chapter discusses various patient safety issues pertinent to resident supervision through the lens of two case studies. The chapter also presents practical solutions to improve supervision and communication that can be used by any teaching hospital with a clinical training program. We believe that the key lessons will also be helpful to healthcare organizations in designing strategies for safe supervision in other types of teaching programs such as supervision of mid-level providers and nursing and pharmacy student trainees.

## Case Studies

### Case 1: Poor Outcome Due to Suboptimal Supervision and Failure to Call for Help

#### Clinical Summary

*With the monthly service change, Dr. A is assuming care for a new panel of patients on a housestaff-covered Internal Medicine service. She reaches out to her colleague Dr. R to learn about the patients she will be covering and the trainees that she will be supervising. After discussing the specific clinical scenarios for each patient, Dr. R informs Dr. A that her resident Judy is an outstanding trainee, early in her second year, and on her first inpatient rotation as the senior resident. Judy is currently being considered for a chief residency position, one of high honors in the residency program, and has two intern physicians working with her who are competent and effective. Dr. A is reassured by this information and arranges a time to meet Judy on the team's first on-call day together. During their meeting, Dr. A informs Judy to "Call me if you need me" and then also states that she will be out at a personal function that evening and closes the conversation with "I am sure you are going to do great!"*

*As Judy begins her evening, she is called by the Emergency Department (ED) for an admission of a patient who is hypoxic and tachypneic. Flustered by the many pages and calls she is receiving, Judy informs the ED she will be sending her intern down shortly. Uncertain about the best management for this patient, Judy quickly performs an Internet search to try to come up with a management plan. Her pager continues to alarm, and the ED becomes more insistent as the patient continues to further decompensate. Judy turns to her resident colleagues who are on-call with her, polling them for their advice. Time continues to pass and Judy frantically searches for a pulmonary fellow as the ED informs her that the patient is rapidly declining. She sends her intern to the ED again to obtain laboratory and radiographic studies.*

*Dr. A arrives early next morning to round on the new panel of patients admitted overnight. She congratulates Judy on a good night stating, "I didn't hear from you, so things must have gone well!" Judy informs her that they will need to see only nine*

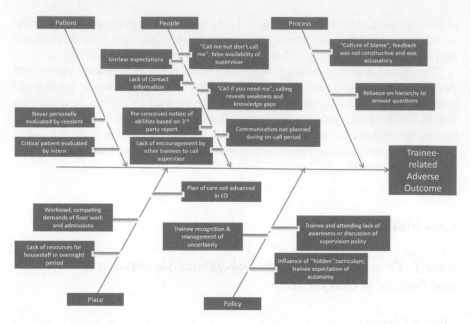

**Fig. 4.1** Case 1: Fishbone diagram depicting contributory factors in the trainee-related adverse outcome

*new patients, one less than the full panel of ten. When questioned, Judy informs that they had admitted a tenth patient; however that patient went into respiratory failure requiring intubation and admission to the medical ICU. Surprised, Dr. A demands to know why she wasn't notified about this development and Judy sheepishly explains her behaviors of the past evening. Visibly disappointed, Dr. A informs Judy that her behavior is negligent and reflects poor judgment. Judy collects herself as rounds begin, with Dr. A informing her "I will certainly expect better next time."*

### Analysis and Discussion

This clinical case scenario is drawn from interviews of resident physicians describing their struggles during training, specifically in the context of describing effective and ineffective supervisory experience on a teaching rotation on an Internal Medicine service. Contributing factors to the trainee-related adverse outcome and associated strategies for improvement are discussed below and in Fig. 4.1.

## *Clinical Supervision*

The case above underscores that suboptimal supervision and failure to call for help combined with heavy individual workload can lead to adverse patient outcomes.

Adequate clinical supervision is fundamental to both ensuring safe care to patients and providing appropriate training to residents. In addition, in the event of a trainee-related adverse outcome, the attending physicians in supervisory capacity may be held accountable for patient outcomes as an on-call duty may be sufficient to establish a patient–physician relationship and duty to supervise [16]. Also, since the sponsoring hospitals employ the physicians-in-training for clinical care, they may be held vicariously liable for adverse outcomes caused by residents acting in accordance with their job description [16].

Therefore, teaching hospitals are required to have appropriate policies and procedures in place to provide adequate clinical supervision. Often these institutional policies are informed by the general program requirements of the Residency Review Committee (RRC) of ACGME which address issues such as certification, training, and availability of clinical supervisors. The 2008 IOM report recommended that trainees have immediate access to an on-site residency-approved supervisor at all times, including nights and weekends [7]. The most recent ACGME guidelines also recommend tailoring the amount of supervision based on the needs of trainees as well as encourage evaluation and development of a trainee's ability to supervise junior colleagues such as interns and medical students [17]. Voluntary oversight organizations of residency training, such as the Association of American Medical Colleges (AAMC) have recommended that programs must balance appropriate faculty supervision with graded resident responsibility.

Clinical supervision, or lack thereof, has been tied to adverse patient outcomes and near misses, with a recent case review of five malpractice firms revealing nearly 54 % of suits filed secondary to inadequate supervision [16]. Problems arise when residents are faced with situations of decision-making uncertainty requiring escalation in care, transitions such as discharge or transfer, and ethical dilemmas such as end of life issues [18]. As seen in the case above, residents tend to utilize a hierarchy of assistance, deferring to peers and more senior trainees before contacting their supervising attending physician because of perceived barriers which may result in delays in the delivery of indicated care and patient harm [18]. This deference to the existing hierarchy, while potentially a source of peer-learning, can also act as a barrier to discussion of errors and a true team-based approach to care [19]. Table 4.1 describes various barriers and facilitators to seeking supervision by trainee physicians.

## Measuring Clinical Supervision

So, if appropriate clinical supervision is vital to patient safety as well as trainee education, how does one measure the adequacy of supervision? It is somewhat easier to measure supervision in procedural care such as surgical training by assessing attending physician's physical presence and direct involvement in procedures. For nonprocedural care, typically, supervision is measured by chart review indicating attending physician involvement which is subjective and non-reliable. Factors which have shown promise in quantifying the supervision include the physical presence of

**Table 4.1** Barriers and facilitators to seeking supervision

| Domain | Major categories | Representative resident comments |
|---|---|---|
| Barriers to seeking attending advice | Conflict with decision-making autonomy | *"it was a pain to kind of run by things with [the attending]because it would influence things too much and then you wouldn't get a chance to make up your own mind and figure it out"* |
| | Fund of knowledge expectations | *"I wouldn't turn to [the attending] for advice unless it's…. just something that I didn't know the answer to..something I should know"* |
| | Existence of defined hierarchy | *"…between the ICU resident or the other residents, I usually talk to them before I would make a decision to go up the chain"* |
| | Fear of repercussion | *"I mean [the attending] said I could call him in the middle of the night if I needed anything but I am not going to do that. I am not going to wake him up…"* |
| Facilitators to seeking attending advice | Need for escalation of care | *"it wasn't anything that critical that needed to be addressed that night, if it had been I would have been totally comfortable calling my attending because she made it a point to know that it was fine in calling"* |
| | Options in decision-making | *"I feel I can call the attendings if I have questions above my head or especially if there are a couple of options of what to do"* |
| | Clinical experience | *"…but if it were more a clinical judgment thing and I hadn't had that situation I would ask [the attending]"* |

the supervisor, overall contribution of the supervisor to the patient's care and to the resident's understanding of the case, and the amount of time spent in supervision [20]. These factors have been compiled into an instrument, the Resident Supervision Index, in a study published by the Department of Veterans Affairs, and initial testing has shown promise with respect to feasibility and reliability of the instrument as a valid measure of resident supervision [20].

Various studies have demonstrated that increased supervision can change clinical assessments, diagnoses, and treatment decisions and possibly improve patient outcomes. Increasing the intensity of supervision in already supervised activities has been found to have an equivocal or a positive impact on the trainee's educational experience and patient outcomes [21, 22]. Further research is needed to examine how augmenting supervision during previously unsupervised rotations, for example, during the overnight period, impacts trainee satisfaction and the delivery of patient care. In addition, given the recent ACGME requirements of ensuring adequate supervisory abilities of peer supervisors, ongoing work continues to create validated instruments to measure the quality of a trainee's ability to supervise more junior colleagues [17].

## Best Practices in Clinical Supervision

There is increasing interest in learning the best practices for clinical supervision that balance the dual role of trainee autonomy and good clinical outcomes. Depending upon the situation, clinical oversight may range from monitoring routine activities to intervening to provide direct patient care [23]. Research suggests that trainees prefer a collaborative approach to supervision so that they are treated as adult learner and are provided specific and focused constructive feedback [24].

One study based on qualitative analysis of the resident interview transcripts revealed that often two extreme models of supervision are practised. In the first model, residents described the attending physician as "micro-manager" dictating the plan of care and allowing few autonomous decisions. In the opposite model, residents described the "absentee" attending physician who is distanced from patient care and allows the residents almost exclusive decision-making power [25]. The micromanaging attendings prevent residents from fully developing their own clinical skills and may generate a sense of resident apathy. On the other hand, the absentee attendings can generate a sense of abandonment and exacerbate decision-making uncertainty and may have detrimental effects on patient care.

Therefore, it is of paramount importance that effective strategies for providing clinical supervision are established. The basic principles for effective supervision are based on a relationship between the supervisor and the trainee in which uncertainty is recognized and addressed early, autonomy is preserved, and communication is planned and easily available. The communication practices should highlight the importance of supervision at times that are critical to patient safety such as transitions between levels of care or clinical deterioration in the condition of the patient.

We recommend the following as a general approach to best practices in supervision. First, encourage the role of the supervisor as an active participant. Instead of passively waiting to be contacted by their trainee, the supervisor should actively reach out to housestaff to assess their level of need. Second, since trainees often initiate the contact, it is critical that they are able to recognize their own clinical uncertainty and decision-making limitations. Third, recognize that there may be cultural and institutional barriers which prevent trainees from seeking the involvement of the attending-level supervisor, especially at an earlier juncture in the patient's care (Table 4.1). This concept is referred to as the "hidden curriculum" and is defined as the set of influences that function at the level of organizational structure and culture, including implicit rules to survive, customs, and rituals [26, 27]. For example, a third-year resident who is about to graduate from the residency programs may be perceived as "weak" by herself and by her peers if there is a recurrent need to communicate with attending physicians regarding patient management issues. The leadership of the training program as well as the sponsoring hospital must provide a cultural environment where trainees and attending physicians can engage in optimal supervision without the fear of retribution. Fourth, a blanket approach to the supervisory process should be discouraged as adequate supervision depends upon the trainee's knowledge and skills, clinical specialty as well as specific context of the clinical situation [28]. Whereas some subspecialties have more explicit supervisory guidelines, for example, anesthesiology, obstetrics and gynecology, and emergency medicine, others, such as internal medicine, pediatrics, and others, do not as explicitly outline the requirements for attending presence or even define who is a qualified supervisor. Finally, resident trainees should also be learning skills in supervising their junior residents and medical student.

## SUPERB/SAFETY Model

The SUPERB/SAFETY model, developed on the basis of a qualitative analysis of the interviews of Internal Medicine residents, is a good bidirectional frame work for clinical supervision (Table 4.2). It allows both supervisors and trainees to identify explicit ways to engage in the supervisory discussion [29]. Effective strategies for attending physician provision of supervision are summarized with the acronym SUPERB: Set expectations for when to be notified, Uncertainty is a time to contact, Planned communication, Easily available, Reassure fears, and Balance supervision and autonomy. Effective strategies for residents to solicit faculty supervision are summarized with the acronym SAFETY: Seek attending physician input early, Active clinical decisions, Feeling uncertain about clinical decisions, End-of-life care or family/legal issues, Transitions of care, and You need help with the system/hierarchy.

We also strongly recommend that institutions establish explicit parameters for residents to contact attending physicians, specifically the "must-contact" clinical scenarios. These scenarios should recognize that clinical uncertainty should be a stimulus for seeking attending input.

## Case 2: Adverse Outcome Related to Duty Hour Restrictions and Poor Handoff

### Clinical Summary

*Jill, a second-year Internal Medicine resident, is frantically trying to sign-out all of her patients at the end of a post-call day. During a rough on-call night, Jill spent a significant amount of her time in a meeting with the family of Mrs. H. After an extensive discussion, Mrs. H's family decided to make her DNR/DNI given her chronic, debilitating respiratory condition. Jill made sure that her interns had completed their work and rushed them out the door as their ACGME-mandated shift was quickly coming to an end. With an eye on the clock, Jill rushes to print out her team's written sign-out in order handoff to Megan, the resident on-call for the coming evening. Jill realizes that there isn't any computer paper to print the new updates she made to the electronic sign-out form. Watching the clock to be sure to sign-out on time, Jill scribbles quick updates on the most recent copy she had in her*

**Table 4.2** SUPERB/SAFETY model

| **SUPERB: Guide for Attending Supervision** | |
|---|---|
| Set expectations for when to be notified | *I want you to contact me if a patient is being discharged, transferred, dies, or leaves AMA* |
| Uncertainty is a time to contact | *It is normal to feel uncertain about clinical decisions. Please contact me if you feel uncertain about a specific decision* |
| Planned communication | *Let's plan on talking ~10 p.m. on your call night and before you leave the each day. If you get busy or forget, I will contact you* |
| Easily available | *I am easy to reach by page, or you can use my cell phone or my home phone* |
| Reassure resident not to be afraid to call | *Don't worry about waking me up, or that I will think your question is silly. I would rather know what is going on* |
| Balance supervision & autonomy for resident | *I want you to be able to make decisions about our patients, but I also know this is your first month as a resident so I will follow closely (Tailor to experience level)* |
| **SAFETY: Resident Guide for Attending Input** | |
| Seek attending input early | *Involving your attending early can often prevent delays in appropriate care. They are also legally responsible for the patients you care for* |
| Active clinical decisions | *Contact your attending if an active clinical decision is being made (surgery, invasive procedure, etc.)* |
| Feel uncertain about clinical decisions | *It is normal to feel uncertain about clinical decisions. You should contact your attending if you feel uncertain about a specific decision* |
| End-of-life care or family/ legal discussions | *These complex discussions can change the course of care. Families and patients should know that the attending is aware* |
| Transitions of care | *Transitions are risky for patients. Seek attending input for discharge or transfer* |
| You need help with the system/hierarchy | *System difficulties and hierarchy may hinder care. Attendings can help expedite care* |

*pocket and heads to find Megan after Megan fails to respond to her pages. After a few minutes, Jill finds Megan, gowned and gloved and prepared to place a central line in one of her newly admitted patients. Jill rushes into the room and says, "Hey, can I sign out? I really need to run. It's already after 1 p.m. and I am post-call. Plus, I have dinner reservations at 6 p.m. and I need a quick nap beforehand!" Megan, preparing for her line, asks Jill to tie her gown as she begins to inject Lidocaine and doesn't appear to acknowledge Jill's haste. "You look really busy; there is really nothing to do on our patients. Mrs. H, she's the sickest one, but there's nothing to do. I am going to leave a copy of the sign-out over here. My cell number is on there if you have any questions!" Jill shouts as she hurries out the door.*

*Later that evening, Megan and her team are both admitting and cross-covering when multiple nursing pages punctuate the team's work. Megan calls back and talks with the nurse covering Mrs. H. "She doesn't look well" the nurse informs Megan. "She's breathing really heavy and fast." Megan sends her intern to quickly evaluate*

*Mrs. H as she works up the next admission. The night progresses, and Mrs. H's respiratory status continues to decline, with the nurses directly paging Megan numerous times. "She's not my patient, so I don't know what she looked like earlier" Megan states. "I'll be up shortly to evaluate her." Finally, as she leaves the room after evaluating Mrs. H. Megan requests that the nurse call anesthesia as the patient will require intubation and transfer to the ICU. After the patient is stabilized and transferred, Megan and her team retreat to the call room for some much needed rest.*

*Early the next morning, Jill arrives to receive sign-out from Megan and finds her resting in the call room. "So, how was your night?" Jill asks. Megan rolls over and grabs a crumpled copy of the sign-out and hands it to Jill. "It wasn't awful. Mrs. H. was intubated and went to the ICU, but your other patients did well." Jill gasps, "What? Mrs. H! We made her DNR/DNI! It is right here on the sign-out!" Jill looks down at the crumpled paper and quickly realizes that she gave Megan the older version of the sign-out. "Oh no!", Jill cries, "I am going to get in so much trouble!"*

### Analysis and Discussion

This case is also drawn from prior qualitative interviews of resident physicians, specifically in the context of critical incidents occurring secondary to ineffective handoff communication. This scenario demonstrates the conflict generated by the duty hour regulations and tension to complete tasks while the clock is ticking. Contributing factors and associated strategies for improvement are discussed below and in Fig. 4.2.

## *Impact of Duty Hours on Resident Education and Well-Being*

The initial implementation of the resident duty hour regulations in 2003, which limited consecutive hours worked and shift duration, were met with skepticism and an anticipation of negative clinical care consequences. However, data obtained post-2003 have revealed that patient outcomes did not worsen and in some circumstances improved after the limitations were put in place [8, 30, 31]. Literature also shows positive changes in resident's perception of well-being and stress [32]. However, concerns remain that shorter shifts may change the intensity of work and potentially adversely impact resident's educational experience. Further, since the most recent regulation specifically limits PGY-1 shift duration, this may result in increased night work amongst senior residents affecting their well-being and subsequent care delivery [33].

Decreasing work hours without also a reduction in workload [34, 35] may improve errors attributed to fatigue but may increase those secondary to overwork. Several recent studies have evaluated the impact of workload during training and found that, for each additional patient that residents admit during a call cycle, subsequent sleep time decreases and there is decreased ability to participate in required educational activities [34]. In light of the new limitations, without subsequent

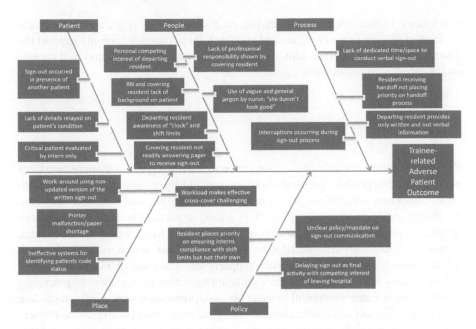

**Fig. 4.2** Case 2: Fishbone diagram depicting contributory factors in the trainee-related adverse outcome

decrease in workload anticipated, these problems may persist, compounding concerns regarding educational quality and opportunity during residency training. Lessons from manufacturing industry and other shift-based specialties warn of the dangers of shift-based work, including resulting errors secondary to attention and impact on personal health and well-being [36, 37]. Aside from workload, other factors to consider include the timing of the performance of complex tasks, the interval between night and day work, and ensuring effective education on sleep hygiene and fitness for duty.

There is the potential that resident education and the subsequent impact on ability to deliver safe and effective clinical care are actually hampered by further duty hour reductions. Inherent in the apprenticeship model of residency training is learning by doing and if in fact residents are doing less, are they learning less? Limiting the training hours may decrease a trainees' exposure to clinical cases, thereby decreasing the overall quality of their clinical education [38]. Prior work done after the implementation of the 80-h work week showed weaker performance of neurosurgical trainees on validated measures of performance [39] and similar findings in other surgical literature notes decrease in operative time and experience after duty hours implementation [40]. Findings in the nonsurgical literature are equivocal, although as discussed above the likely increase in workload or work intensity with shorter shift duration may result in negative educational outcomes for trainees [22, 41]. While the new regulations do include clauses for trainees to violate the restrictions in the setting of a unique case opportunity (e.g., an infrequently performed surgery or evolving/unstable patient), these findings certainly support the assertion that the duty hour solution may not be a one-size-fits-all and will require modification across specialties.

## The "July Effect"

Regardless of the work hour restrictions, concerns remain regarding the transition from undergraduate to graduate medical education, specifically the ability of new interns to rapidly learn new systems, adopt their new professional roles, and simultaneously care for critical and complex patients. The "July effect" or perception that care in teaching hospitals is more dangerous for patients in July secondary to the arrival of a fresh batch of trainees is generally considered to be a one of the most storied medical education urban "legends" [42]. Little literature supports the existence of the "July effect," although many acknowledge that the significant transition from student to practicing intern requires more thoughtful orientation and preparation specifically regarding tasks such as handoff communication and managing uncertainty [42]. Ensuring learner-centered experience-focused orientations coupled with ample availability of more senior and seasoned housestaff are the two strategies suggested to offset any potential impact of the summer season [43].

## Duty Hours and Handoffs

Handoff communication failures clearly contributed to the adverse event in the second case. We can anticipate another increase in the number of care transitions after the implementation of the new regulations and, as such, the ACGME has included explicit language in training and assessment of trainee handoffs. Patients can suffer a multitude of untoward effects secondary to a poor handoff, including readmission, medication errors, or missed tests, and follow-up appointments [44, 45]. Poor transitions occurring even within the hospital, such as transfer to or from a more intensive level of care, may result in medication errors, delay in the delivery of therapies or diagnostic tests, or prolonged length of stay [46]. Handoff education occurs infrequently in the undergraduate medical education environment [47] and, therefore residency-training program must be prepared to provide trainees with content on the importance of effective verbal and written handoff communication. Given that new duty hour limitations will impact service structures and care delivery in residency training, with an increase in the amount of night work and shift-based coverage, programs must ensure the transfer of effective clinical content *and* professional responsibility for patients [12]. Implementing a standardized handoff process, establishing metrics by which to evaluate handoff quality, and involving supervising physicians in the handoff exchange are the best next steps to ensure adequate transfers of care.

# Conclusion and Key Lessons

Residency training is an extremely important and sensitive area in the context of patient safety. First, patients, public-at-large, as well as regulatory and accreditation bodies need to be reassured that the safety and quality of care in a teaching hospital will match or exceed that in the nonteaching hospitals. Second, teaching hospitals are training physicians of the future and the quality of their education will impact their practice for a lifetime and therefore all patient safety efforts of the future. Finally, for attending physicians as well as trainees, hands-on residency training remains the most important conduit providing continuity across generation of physicians—not only of clinical knowledge but also of values of humane and compassionate care.

The following is a summary of the key take home points to be considered by GME training programs and teaching hospitals to ensure both the safety and quality of patient care and education of residents.

- Factors determined to impact adequacy of supervision include the physical presence of the supervisor, the contribution of the supervisor to the patient case, the resident understanding of the clinical scenario, and the overall time spent with the trainee.
- Trainees wish to approach clinical care in a collaborative fashion, and to be treated as adult learners, with constructive and specific focused feedback.
- Paramount to the discussion of supervision is the identification of explicit parameters for contact, specifically the "must-contact" clinical scenarios, and also the easy availability of the supervisor.
- Encourage the role of the supervisor as an active participant; instead of passively waiting to be contacted by their trainee, the attending physician should actively reach out to their housestaff to assess their level of supervisory need.
- Decrements in shift duration, without coincident decrease in workload, may further serve to negatively impact resident well-being and educational quality of residency experience. Resident education, and ability to participate in educational activities, must be considered when implementing strategies to comply with policy.
- Factors to consider in designing effective systems include the timing of complex tasks performed, the interval between night and day work, and ensuring effective education on sleep hygiene and fitness for duty.
- Ensuring learner-centered and experience-focused orientations coupled with ample availability of more senior and seasoned housestaff are two strategies suggested to offset any potential impact of the summer season.
- A standardized handoff process should be utilized which stresses transfer of clinical content and of professional responsibility. Systems should be designed to include protected or overlap time ensure that priority is placed on effective handoff communication.
- Team-based approach to patient ownership should be encouraged to avoid the "not my patient" problem.

# References

1. HCUPNet. Agency for Healthcare Research and Quality (AHRQ). [cited 13 May 2012]. Available from http://hcupnet.ahrq.gov/
2. Kupersmith J. Quality of care in teaching hospitals: a literature review. Acad Med. 2005;80(5):458–66.
3. Bell BM. Resident duty hour reform and mortality in hospitalized patients. JAMA. 2007;298(24):2865–6. author reply 6-7.
4. Duty Hours Language. ACGME. [cited 13 May2012]. Available from http://www.acgme.org/acWebsite/dutyHours/dh_Lang703.pdf
5. Philibert I, Friedmann P, Williams WT. New requirements for resident duty hours. JAMA. 2002;288(9):1112–4.
6. Hope J. EU ban limiting junior doctors to 48-hour working week lifted over public health concerns. Daily Mail. 16 Oct 2009. Available at http://www.dailymail.co.uk/health/article-1220698/EU-ban-limiting-junior-doctors-48-hour-working-week-lifted-public-health-concerns-hospital-EU-Working-Time-Directive-Trusts.html#ixzz1xt1iOQRm. Accessed 15 Jun 2012.
7. Ulmer C, Wolman DM, Johns MME. Resident duty hours: enhancing sleep, supervision, and safety (Committee on Optimizing Graduate Medical Trainee (Resident) Hours and Work Schedules to Improve Patient Safety). Washington, DC: Institute of Medicine (U.S.), National Academies Press; 2009.
8. Volpp KG, Rosen AK, Rosenbaum PR, Romano PS, Even-Shoshan O, Canamucio A, et al. Mortality among patients in VA hospitals in the first 2 years following ACGME resident duty hour reform. JAMA. 2007;298(9):984–92.
9. Barden CB, Specht MC, McCarter MD, Daly JM, Fahey III TJ. Effects of limited work hours on surgical training. J Am Coll Surg. 2002;195(4):531–8.
10. Vidyarthi AR, Arora V, Schnipper JL, Wall SD, Wachter RM. Managing discontinuity in academic medical centers: strategies for a safe and effective resident sign-out. J Hosp Med. 2006;1(4):257–66.
11. The Joint Commission. 2012 Hospital National Patient Safety Goals. [cited 13 May 2012]. Available from http://www.jointcommission.org/assets/1/6/2012_NPSG_HAP.pdf
12. Arora V, Johnson J. A model for building a standardized hand-off protocol. Jt Comm J Qual Patient Saf. 2006;32(11):646–55.
13. Snow V, Beck D, Budnitz T, Miller DC, Potter J, Wears RL, et al. Transitions of Care Consensus Policy Statement American College of Physicians-Society of General Internal Medicine-Society of Hospital Medicine-American Geriatrics Society-American College of Emergency Physicians-Society of Academic Emergency Medicine. J Gen Intern Med. 2009;24(8):971–6.
14. Singh H, Thomas EJ, Petersen LA, Studdert DM. Medical errors involving trainees: a study of closed malpractice claims from 5 insurers. Arch Intern Med. 2007;167(19):2030–6.
15. Kilminster S, Cottrell D, Grant J, Jolly B. AMEE Guide No. 27: effective educational and clinical supervision. Med Teach. 2007;29(1):2–19.
16. Kachalia A, Studdert DM. Professional liability issues in graduate medical education. JAMA. 2004;292(9):1051–6.
17. Nasca T, Day S, Amis S. The new recommendations on duty hours from the ACGME task force. N Engl J Med. 2010;363:e3.
18. Farnan JM, Johnson JK, Meltzer DO, Humphrey HJ, Arora VM. Resident uncertainty in clinical-decision making: a qualitative analysis. Qual Saf Health Care. 2008;17:122–6.
19. Sexton JB, Thomas EJ, Helmreich RL. Error, stress and teamwork in medicine and aviation: cross-sectional surveys. BMJ. 2000;320:745–9.
20. Byrne JM, Kashner M, Gilman S, Aron D, Canno G, Chang B, et al. Measuring the intensity of resident supervision in the department of veterans affairs: the resident supervision index. Acad Med. 2010;85:1171–81.

21. Claridge JA, Carter JW, McCoy AM, Malangoni MA. In-house direct supervision by an attending is associated with differences in the care of patients with a blunt splenic injury. Surgery. 2011;150(4):718–26.
22. Farnan JM, Petty LA, Georgitis E, Martin S, Chiu E, Prochaska M, et al. A systematic review: the effect of clinical supervision on patient and residency education outcomes. Acad Med. 2012;87(4):428–42.
23. Kennedy TJ, Regehr G, Baker GR, Lingard LA. Progressive independence in clinical training: a tradition worth defending? Acad Med. 2005;80(10 Suppl):S106–11.
24. Busari JO, Weggelaar NM, Knottnerus AC, Greidanus PM, Scherpbier AJ. How medical residents perceive the quality of supervision provided by attending doctors in the clinical setting. Med Educ. 2005;39:696–703.
25. Farnan JM, Johnson J, Humphrey H, Meltzer D, Arora VM. The tension between on-call supervision and resident autonomy: from the micromanager to the absentee attending. Am J Med. 2009;122(8):784–8.
26. Hafferty FW, Franks R. The hidden curriculum, ethics teaching, and the structure of medical education. Acad Med. 1994;69(11):861–71.
27. Hundert EM, Hafferty F, Christakis D. Characteristics of the informal curriculum and trainees' ethical choices. Acad Med. 1996;71(6):624–42.
28. Sterkenburg A, Barach P, Kalkman C, Gielen M, ten Cate O. When do supervising physicians decide to entrust residents with unsupervised tasks? Acad Med. 2010;85:1408–17.
29. Farnan JM, Johnson JK, Meltzer DO, Harris I, Humphrey HJ, Schwartz A, et al. Strategies for effective on-call supervision for internal medicine residents: the SUPERB/SAFETY model. J Grad Med Educ. 2010;2(1):46–52.
30. Rosen AK, Loveland SA, Romano PS, Itani KM, Silber JH, Even-Shoshan OO, et al. Effects of resident duty hour reform on surgical and procedural patient safety indicators among hospitalized veterans health administration and medicare patients. Med Care. 2009;47(7):723–31.
31. Fletcher KE, Davis SQ, Underwood W, Mangrulkar RS, McMahon Jr LF, Saint Jr S. Systematic review: effects of resident work hours on patient safety. Ann Intern Med. 2004;141:851–85.
32. Myers JS, Bellini LM, Morris JB, Graham D, Katz J, Potts JR, et al. Internal medicine and general surgery residents' attitudes about the ACGME duty hours regulations: a multicenter study. Acad Med. 2006;81(12):1052–8.
33. Arora VM, Volpp KGM. Duty hours: time to study? J Grad Med Educ. 2011;3(3):281–4.
34. Arora VM, Georgitis E, Siddique J, et al. Association of workload of on-call medical interns with on-call sleep duration, shift duration, and participation in educational activities. JAMA. 2008;300(10):1146–53.
35. Reed DA, Fletcher KE, Arora VM. Systematic review: association of shift length, protected sleep time, and night float with patient care, residents' health, and education. Ann Intern Med. 2010;153(12):829–42.
36. Berger AM, Hobbs BB. Impact of shift work on the health and safety of nurses and patients. Clin J Oncol Nurs. 2006;10(4):465–71.
37. Costa G. The impact of shift and night work on health. Appl Ergon. 1996;27(1):9–16.
38. Volpp K, Friedman W, Romano P, Rosen A, Silber J. Residency training at a crossroads: duty-hour standards. Ann Intern Med. 2010;153:826–8.
39. Jagannathan J, Vates GE, Pouratian N, Sheehan JP, Patrie J, Grady S, et al. Impact of the accreditation council for graduate medical education work-hour regulations on neurosurgical resident education and productivity. J Neurosurg. 2008;110:820–7.
40. Damadi A, Davis AT, Saxe A, Apelgren K. ACGME duty-hour restrictions decrease resident operative volume: a 5-year comparison at an ACGME-accredited university general surgery residency. J Surg Educ. 2007;64(5):256–9.
41. Arora VM, Georgitis E, Siddique J, Vekhter B, Woodruff J, Humphrey H, et al. Association of on-call workload of medical interns with sleep duration, shift duration, and participation in educational activities. J Am Med Assoc. 2008;300(10):1146–53.
42. Clarence H. Braddock III. A Mid-Summer Fog. Nov 2008. AHRQ WebM&M: case & commentary.

43. Young JQ, Ranji SR, Wachter RM, Lee CM, Niehaus B, Auerbach AD. "July Effect": impact of the academic year-end changeover on patient outcomes: a systematic review. Ann Intern Med. 2011;155(5):309–15.
44. Arora VM, Prochaska M, Farnan J, et al. Problems after discharge and understanding of communication with their primary care physicians (PCPs) among hospitalized seniors: a mixed methods study. J Hosp Med. 2010;5(7):385–91.
45. Arora V, Johnson J, Lovinger D, et al. Communication failures in patient signout and suggestions for improvement: a critical incident analysis. Qual Saf Health Care. 2005;14(6):401–7.
46. Hinami K, Farnan JM, Meltzer DO, et al. Understanding communication during hospitalist service changes: a mixed methods study. J Hosp Med. 2009;4(9):535–40.
47. Horwitz LI, Krumholz HM, Green ML, Huot SJ. Transfers of patient care between house staff on internal medicine wards: a national survey. Arch Intern Med. 2006;166:1173–7.

# Chapter 5
# Electronic Health Record and Patient Safety

Ken Ong and Regina Cregin

> *"... irrationally held truths may be more harmful than reasoned errors."*
>
> Thomas Huxley

## Introduction

At the turn of this century, the Institute of Medicine (IOM) released two reports that spurred the nation's interest in the electronic health record (EHR). In 1999, "To Err Is Human" reported that medical errors were responsible for 44,000–98,000 lives lost in hospitals every year [1]. In 2001, in "Crossing the Quality Chasm," the IOM recommended the adoption of information technology (IT) such as EHRs to reduce errors and to improve safety, efficiency, and patient engagement [2]. During the twenty-first century's infancy, EHR adoption by office-based physicians and hospitals was nascent. In 2002, no more than 18 % of office-based physicians had any form of EHR [3]. The earliest published estimate of hospital EHR adoption was 8 % in 2008 [4].

What a difference a decade can make. The American Recovery and Reinvestment Act of 2009 (ARRA), popularly known as "the stimulus plan" was a trigger to push the nation from the flat plateau to the steep slope of the S-shaped curve of

K. Ong, M.D., M.P.H.
Clinical Informatics, New York Hospital Queens, 56-45 Main Street,
Flushing, NY 11355, USA
e-mail: keo9016@nyp.org

R. Cregin, M.S., B.C.P.S., Pharm.D. (✉)
Pharmacy Department, New York Hospital Queens, 56-45 Main Street,
Flushing, NY 11355, USA
e-mail: rgc9002@nyp.org

A. Agrawal (ed.), *Patient Safety: A Case-Based Comprehensive Guide*,
DOI 10.1007/978-1-4614-7419-7_5, © Springer Science+Business Media New York 2014

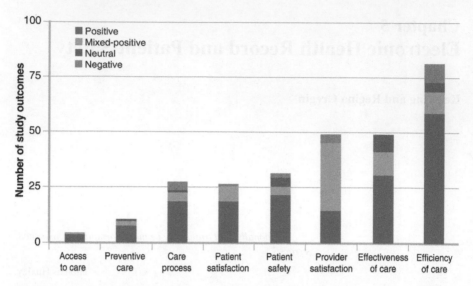

**Fig. 5.1** Evaluations of outcome measures of health information technology, adapted with permission from Buntin MB et al. [6]

EHR adoption. Within the ARRA legislation was the Health Information Technology for Economic and Clinical Health (HITECH) Act which allocates Medicare and Medicaid incentives for EHR adoption to physicians, other healthcare professionals, and hospitals. In order to receive these incentives, office-based physicians and hospitals must deploy CMS-certified EHRs and meet required measures of EHR use, i.e., "Meaningful Use." As of June 2012, 172,186 office-based physicians had registered for Medicare and 84,086 for Medicaid EHR incentives. Some 3,779 hospitals had also registered for their Meaningful Use incentives [5].

The EHR can play a transformative role in health care by improving medication safety, making patient health information available at the point of care, facilitating care coordination, optimizing efficiency, and engaging patients and caregivers. A review of the recent literature concluded that 92 % of the articles on health information technology (HIT) demonstrated net benefit [6]. Outcome measures were positive for efficiency of care, effectiveness of care, patient and provider satisfaction, care process, preventive care, and access to care (Fig. 5.1).

Nonetheless, as Everett Rogers might have predicted, this transforming technology also has unintended consequences:

> "No innovation comes without strings attached. The more technologically advanced an innovation, the more likely its introduction will produce many consequences, both anticipated and latent." [7]

In this chapter, we present three case studies that illustrate some unintended consequences of EHRs and what can be done to prevent them. To protect privacy of patients and institutions, these case studies are fictional; still, they amply demonstrate the potential safety hazards that are relevant to all hospitals using EHR and related technologies.

## Case Studies

### Case Study 1: Indwelling Neuraxial Catheters and Anti-Thrombotic Medication

#### Clinical Summary

*A postoperative patient receiving analgesia medication via the epidural route is recovering in the surgical unit. A member of the surgical team reinitiates the patient's home medications, including the patient's antiplatelet medication that was brought by the patient's family from home because it was not on the hospital's formulary. Administering antiplatelet and anticoagulant medications to patients with indwelling neuraxial catheters can place a patient at risk of spinal hematomas that could result in paralysis or death. The surgical prescriber, unaware that the antiplatelet drug may adversely raise the risk of spinal hematoma, enters the order for the antiplatelet drug in the hospital's computerized physician order entry (CPOE) system with no alerts triggering. The patient receives the antiplatelet medication for 2 days until an anesthesiologist notices the order and immediately discontinues it. However, the epidural catheter must be kept in place for a few extra days to reduce the risk of spinal hematoma while the effects of the antiplatelet agent wear off.*

#### Analysis

Clinical decision support (CDS) is a category of health IT that brings clinically relevant and specific information to patients and caregivers when they need it. The CDS armamentarium grows daily and includes drug–drug and drug–allergy interaction alerts, reminders, order sets, documentation templates, clinical guidelines, and pertinent knowledge references. Bright et al. performed a systematic review of 148 randomized, controlled trials of CDS and concluded that CDS can improve quality of care, patient-safety, and efficiency [8].

However, off-the-shelf CDS systems do not anticipate all the clinical scenarios encountered by prescribers. For example, the concomitant use of anticoagulant/antiplatelet medications with an indwelling neuraxial catheter presents a significant medication safety concern and increases the risk of spinal hematoma and paralysis. This represents a drug (anticoagulant/antiplatelet) and route of administration (epidural) interaction and is not incorporated into most currently available CDS systems that usually contain only drug–drug, drug–allergy, and drug–food interactions. Consequently, concomitant administration of an anticoagulant/antiplatelet medication with indwelling neuraxial catheters is a very real unaddressed risk even for hospitals that have implemented a fully integrated CPOE with CDS.

## Corrective Actions

One solution is the creation of an institution-specific CDS intervention, e.g., a drug–route interaction alert. The development of a CDS intervention is extremely labor intensive and specialized process. CDS interventions can require months of planning and testing, and in some cases policy development. Prior to the creation of a CDS intervention, a multidisciplinary team consisting of clinicians must determine what information is delivered, who receives that information, what format the intervention should take, what the best channel will be [e.g., EMR, personal health record (PHR), or other], and when it should be presented in the workflow. These "5 Rights" are the same decisions that need to be made for any CDS intervention [9, 10]. The actual creation of the CDS intervention is based on the skill of the IT department and their ability to work within the clinical realm. Institution-specific CDS interventions must be constantly maintained and updated to ensure all future workflow and formulary changes are incorporated. Maintenance testing must be performed to ensure the CDS intervention fires and behaves as designed. This requires the development of initial and maintenance test scripts and availability of staff that are capable of detecting errors within the behavior of the CDS intervention, as well as finding solutions to identified errors. An integral part of CDS intervention maintenance is the respect for change control processes within the IT department. What is perceived as a "simple change" in the system can result in the deactivation of a custom CDS intervention, thus putting patients at risk.

For example, to address the system vulnerability in this case study, an institution-specific CDS intervention would be developed to screen all patients with active medications that utilize the epidural route who are also prescribed anticoagulant/antiplatelet agents that are prohibited with their use and vice versa (Figs. 5.2, 5.3, 5.4, and 5.5). The rule would be triggered whether the drug being prescribed is on the hospital formulary or not. This screen is not a simple yes/no attestation, as it is complicated further by the dose and/or route of administration of the offending agent. For example, anticoagulants in prophylactic dosages may not be prohibited with neuraxial analgesia, while the same anticoagulant prescribed as therapeutic dosages could be. Prior to the creation of the CDS intervention, published guidelines are consulted with multiple scenarios vetted within different hospital committees to determine how the clinician would trigger the CDS intervention, what alerts and information would be presented, and whether or not a clinician would be allowed to proceed (i.e., hard versus soft stop). The creation of this CDS intervention requires policy change and intense education of all stakeholders. Based on our experience, from start to finish, the deployment of such decision support rules/CDS intervention may take up to 2 years to develop, test, and implement.

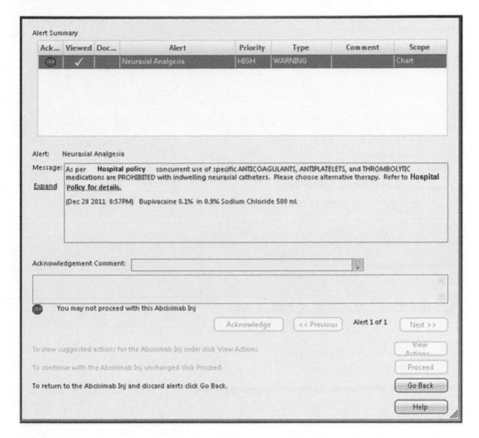

**Fig. 5.2** Scenario 1—Hard stop for medications that are prohibited by hospital policy

## Case Study 2: Incorrect Heparin Dose

### Clinical Summary

*A patient presents to the hospital with a pulmonary embolism. The patient is pre-scribed a therapeutic dose of low-molecular weight heparin (LMWH) at 1 mg/kg/dose. A safety feature in the hospital's CPOE system requires that body weight be documented in the patient's chart before an order for an anticoagulant is placed. The hospital's CPOE system permits documentation of body weight in either kilo-grams or pounds. The patient weighs 88 kg but the weight is incorrectly entered as 88 pounds (the equivalent of 40 kg). Therefore, as the LMWH order is entered as 1 mg/kg/dose, the system calculates the dose to 40 mg. The pharmacy also verifies the order based on the documented, though incorrect, weight. The subtherapeutic dose is administered to the patient for several days until a pharmacist notices the error during clinical rounds. The order is immediately corrected.*

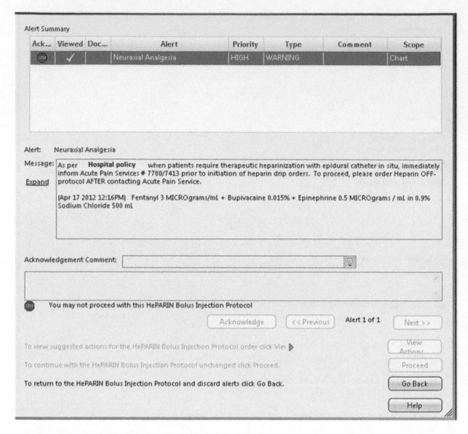

**Fig. 5.3** Scenario 2—Hard stop for "HePARIN drip protocol" with directive of how to treat

## Analysis

The hospital's CPOE system allowed staff to document weight in either pounds or kilograms. The system would calculate the kilogram equivalent whenever a weight was documented as pounds and vice versa. The low dose of the LMWH would not be routinely picked up by the pharmacy during the verification process because this medication is also used in lower dosages for other indications such as deep vein thrombosis prophylaxis. This event exemplifies the numerous other reports of "near misses" and errors when CPOE documentation permits recording weight in both pounds and kilograms [11].

## Corrective Actions

To prevent this kind of error, the best practice would be to restrict documentation of weight in the CPOE to metric units only. In October 2011, in a statement to prevent errors with oral liquids, the Institute for Safe Medication Practices (ISMP)

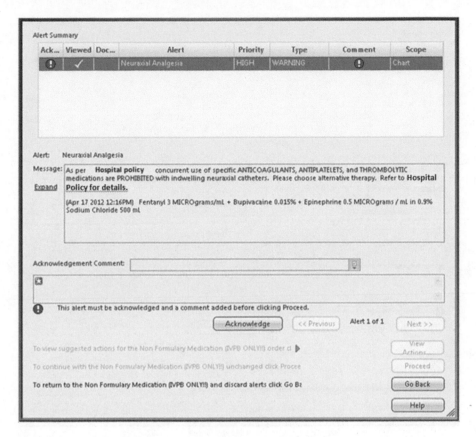

**Fig. 5.4** Scenario 3—Soft stop for medications that are non-formulary

recommendations included measuring weights only in kilograms and not pounds [11]. However, this seemingly simple solution has substantial implementation challenges. It might require approval from various committees as this change affects almost every clinical department within an institution. Before enforcing this measure, the hospital would have to ensure that all scales within the institution have the ability to weigh in kilograms, with the kilogram option locked as the default unit of measure for all scales that also had the ability to weigh in pounds. All orders, flow sheets, and clinical documentation within the CPOE system would have to be reviewed to ensure that the ability to document in pounds was no longer available. Nonetheless, once the root cause is fixed, future error can be prevented.

## Case Study 3: Conflicting Chemotherapy Orders

CPOE systems can simplify the medication ordering process as many data entry fields can limit the ordering of medication dose, route, and frequency only as specific and discrete values. However, similar to many other EHRs, the electronic

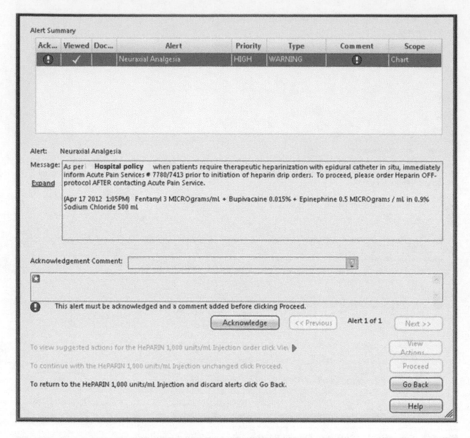

**Fig. 5.5** Scenario 4—Soft stop for HePARIN DRIP (for off protocol use)

medication orders within our hospital's CPOE system also contain a free-text comment field where the prescriber can note any additional information about conditional administration parameters, e.g., administer antibiotic after blood cultures are drawn. While providing flexibility in physician ordering, the free-text comment field also has the potential to be misused.

## Clinical Summary

*Assume an order for chemotherapy arrives in the pharmacy on a Friday afternoon for an inpatient with a cancer diagnosis. All required fields of the order are completed, i.e., dose, route, frequency, and administration time or rate. In the free-text comment field of the order the oncologist enters "administer over 48 h for two doses." In contrast, the administration times selected with the drop down menu reads "administer each dose over 24 h for two doses." The only recourse is for pharmacy to hold the order until they can clarify the order with the oncologist.*

**Analysis**

Some hospitals have pharmacy and CPOE systems from different vendors. When this is the case, little if any data from CPOE system passes into the pharmacy system. A pharmacist looking at the pharmacy system would not see a free-text field entered in CPOE if that field is not interfaced to the pharmacy system.

**Corrective Actions**

To prevent cases like the example given, the free-text comment field would be blocked from any medication order form that did not routinely require the documentation of under what circumstances a medication should be withheld. The oncology staff would receive refresher training focused on CPOE order entry for chemotherapy.

The ultimate resolution is switching the pharmacy system to that of the same vendor of the inpatient CPOE system. This allows all information to flow over to the pharmacist's view including all comments. The pharmacist could then validate each comment on every medication order.

## Discussion

### *Unintended Consequences of Health IT*

While the advantages of EHRs are clear, there is a growing appreciation that there can be safety concerns with the increasing use of EHRs.

In a review of five CPOE implementations, Campbell et al. cataloged nine categories of unintended consequences (Table 5.1) [12]. The following provide examples of such unintended consequences:

- More/new work for clinicians: Especially for physicians new to CPOE, order entry may take more time compared with the paper-based ordering systems.
- Workflow issues: With more efficient workflows, fewer providers may be involved in order entry resulting in fewer checks and balances for reviewing orders.
- Never ending system demands: Clinicians and others constantly want to modify or improve the system. While modifications to the EHR to improve patient safety and quality are welcome, the care and maintenance of an EHR does require resources and funding.
- Paper persistence: Today, more often than not the EHR resides side by side with the paper chart in a hybrid environment. Clinicians have to query two different charts, which may or may not reconcile with one another.
- Changes in communication patterns and practices: Care team members may assume there is less need to communicate directly with each other when orders are entered electronically.
- Emotions: Physicians may become flustered by EHRs with complex screens and difficult-to-understand user interface.

**Table 5.1** Examples of unintended consequences of CPOE, adapted from Campbell EM et al. [12]

| Unintended consequence | Frequency (%) $n=34$ |
| --- | --- |
| More/new work for clinicians | 19.8 |
| Workflow issues | 17.6 |
| Never ending system demands | 14.8 |
| Paper persistence | 10.8 |
| Changes in communication patterns and practices | 10.1 |
| Emotions | 7.7 |
| New kinds of errors | 7.1 |
| Changes in the power structure | 6.8 |
| Overdependence on technology | 5.2 |
| Total | 100 |

- New kinds of errors: CPOE may eliminate illegible paper-based orders but may enable very legible errors when incorrect options are selected from drop-down lists.
- Changes in power structure: Hospital's standardized order sets and document templates can engender feelings of compromised physician autonomy.
- Overdependence on technology: Ironically, while a number of clinicians have negative or frustrating feelings when a new EHR or CPOE is being installed, after successful adoption most of them express serious concerns for medication safety during downtime when automated clinical decision support is not available.

In 2012, the Institute of Medicine released a report on health IT and patient safety [13]. It captured both the potential benefits and safety concerns for various health IT applications including CPOE, CDS, bar code medication administration, and patient engagement tools (Fig. 5.6).

## The Sociotechnical Model

Health IT-related adverse events arise in a milieu of interacting components. To better understand and manage health IT-related safety concerns, an emerging concept is the model of the "sociotechnical system" [13]. This model is also an instrument for root cause analysis (RCA) that accommodates and addresses particularities of health IT (Fig. 5.7). The sociotechnical system perspective recognizes that many variables can cause adverse events and a systems approach is necessary to prevent future adverse events. The five elements in the sociotechnical system are technology, people, process, organization, and external environment.

*Technology* includes hardware, software, and connectivity. Technology incorporates the equipment, machine, and systems categories used in various versions of Cause and Effect diagrams (Ishikawa/Fishbone diagram) [14, 15]. For example, in our case study of the incorrect weight-based heparin dose, the relevant technology

a

### Computerized Provider Order Entry (CPOE)

An electronic system that allows providers to record, store, retrieve, and modify orders (e.g., prescriptions, diagnostic testing, treatment, and/or radiology/imaging orders).

#### Potential Benefits

Large increases in legible orders

Shorter order turnaround times

Lower relative risk of medication errors

Higher percentage of patients who attain their treatment goals

#### Safety Concerns

Increases relative risk of medication errors

Increased ordering time

New opportunities for erros, such as:

- Fragmented displays preventing a coherent view of patients' medications
- Inflexible ordering formats generating wrong orders
- Separations in functions that facilitate double dosing

Disruptions in workflow

### Clinical Descision Support (CDS)

Monitors and alerts clinicians of patient conditions, prescriptions, and treatment to provide evidence-based clinical suggestions to health professionals at the point of care

#### Potential Benefits

Reduction in:

- Relative risk of medication errors
- Risk of toxic drug levels
- Time to therapeutic stabilization
- Management errors of resuscitating patients in adult trauma centers
- Prescriptions of nonpreferred medications

Can effectively monitor and alert clinicians of adverse conditions

Improve long-term treatment and increase the likeihood of achieving treatment goals

#### Safety Concerns

Rate of detecting drug-drug interactions varies widely among different vendors

Increases in mortality rate

High override rate of computer generated alerts (alert fatigue)

**Fig. 5.6** (**a, b**) Potential benefits and safety concerns for health IT

components would include stand-alone and bed-embedded weight scales, electronic order sets, electronic flow sheets, and other CDS interventions.

The *people* component includes not only specific roles and individuals but their knowledge and training as well. Patients, pharmacists, nurses, physicians, physician assistants, front desk staff, and others can all be actors in any given analysis. The case study of conflicting chemotherapy orders underscored the value of targeted refresher training for users as needed for new EHR functionality.

**b**

| Bar-Coding |
| --- |
| Bar-coding can be used to track medications, orders, and other health care products. It can also be used to verify patient identification and dosage. |

| Potential Benefits |
| --- |
| Significant reductions in relative risk of medication errors associated with:<br>• Transcription<br>• Dispensing<br>• Administration errors |

| Safety Concerns |
| --- |
| Introduction of workarounds for example, clinicians can:<br>• Scan medications and patient identification without visually checking to see if the medication dosing and patient identification are correct<br>• Attach patient identification bar-codes to another object instead of the patient<br>• Scan orders and medications of multiple patients at once instead of doing it each time the medication is dispensed |

| Patient Engagement Tools |
| --- |
| Tools such as patient portals, smartphone applications, email, and interactive kiosks, which enable patient to participate in their health care treatment |

| Potential Benefits |
| --- |
| Reduction in hospitalization rates in children<br>Increases in patients' knowledge in treatment and illnesses |

| Safety Concerns |
| --- |
| Reliability of data entered by:<br>• Patients<br>• Families<br>• Friends or |

**Fig. 5.6** (continued)

*Process* or workflow is a sequence of steps required to complete a task or function. In health care, there are a suite of defined processes for each service, level, and setting of care. For example, in the indwelling neuraxial catheter and anti-thrombotic medication case study, the team member who inserts and removes the indwelling catheter is different from the one who writes the medication orders. Catheter removal and anti-thrombotic medication orders are also done at different times.

*Organization* refers to the internal policies, procedures, and culture [16]. Organization in sociotechnical system analysis goes beyond the extant policies and procedures of a physician practice, hospital, or healthcare system. It includes software purchase, design, and interface decisions. For example, in the indwelling neuraxial catheter case study, the organization category includes the hospital's current Policy and Procedure "Patient Controlled Analgesia and Neuraxial Analgesia." Similarly, in the incorrect heparin dose case study, the design decision to permit entering weight in pounds or kilograms was an organizational configuration

**Fig. 5.7** Sociotechnical
system that underlies health
IT adverse events

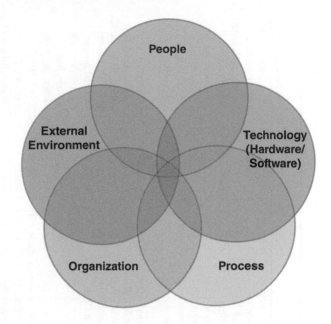

decision. In the conflicting chemotherapy orders case study, one organizational characteristic at the hospital is that the basic EHR training is promoted during new employee orientation; however, advanced training for new EHR functionality is not routinely given.

Providers and healthcare organizations are constantly interacting with the *external environment*, e.g., public and private payers, Joint Commission, National Committee for Quality Assurance (e.g., Patient-Centered Medical Home Recognition), state and local health department, and last but not least CMS's Meaningful Use programs. All these external environmental factors influence the design, implementation, and usage of EHR systems.

An analysis of the three case studies described above in the context of the socio-technical system model reveals a multitude of root causes and actions to prevent future recurrence (Table 5.2).

Although, the sociotechnical system is a valuable tool for RCA *after* an error has occurred, there are two additional tools that can be used *prospectively:* Failure Modes and Effects Analysis (FMEA) and EHR usage metrics. A comprehensive reference guide on FMEA is available online at the Web site of the Veterans Administration's National Center for Patient Safety (http://www.patientsafety.gov/ SafetyTopics/HFMEA/HFMEA_JQI.html) [17]. EHR usage metrics can be monitored using "run charts" to find problems and track their resolution [18]. These metrics can include percent system uptime, mean response time (measured in tenths of a second), percentage of orders entered electronically, percentage order sets used, percent alerts that fire, percent alerts overridden, system interface efficiency, and miscellaneous or free-text orders (which bypass clinical decision support) [19].

**Table 5.2** The sociotechnical system and the three medication errors

| Medication error | People | Technology (hardware/software) | Process | Organization | External environment |
|---|---|---|---|---|---|
| Case study 1: Indwelling neuraxial catheters and anti-thrombotic medication | Nurses, physicians, and physician assistants on the care team rotate each shift in the hospital | Clinical decision support for commercial EHRs typically provides drug–drug, drug–allergy, and drug–food interaction checking but not for drug–device/route of administration interactions | The prescriber may not know the status of the indwelling catheter at time of writing the anti-thrombotic or anticoagulant | Hospital Policy and Procedure "Patient Controlled Analgesia and Neuraxial Analgesia" does not address the drug interaction | National guideline on from the American Society of Regional Anesthesia and Pain Medicine[a] influences the policy/clinical decisions |
| Case study 2: Incorrect heparin dose | Clinicians have different preferences for recording weight | The EHR was configured to accept either kilograms or pounds for weight. Weight is captured in electronic flow sheets and used to auto-calculate drug dosing in order sets | Nurse enters patient weight in EHR. Physician enters the medication order in CPOE | During the implementation of CPOE, a configuration decision was made to permit weight entry in either kilograms or pounds | The USA is one of the few industrialized nations in the world who still uses the imperial measurement system, e.g., pounds, feet, inches, and pints |
| Case study 3: Conflicting chemotherapy orders | A few physicians were not adequately trained on using drop-down menus for prescribing dose, route, and frequency for chemotherapy in CPOE | The pharmacy system was from a different vendor than that of the CPOE system. The comments field was not passed from the CPOE system to pharmacy | Only oncologists are permitted to enter orders for chemotherapy | Basic EHR training is promoted during new employee orientation. Advanced training for new EHR functionality was not routinely given | ARRA/HITECH promotes the Meaningful Use of the EHR by all professional and hospital[b] |

[a]American Society of Regional Analgesia. Regional Anesthesia in the Patient Receiving Antithrombotic or Thrombolytic Therapy: American Society of Regional Anesthesia and Pain Medicine Evidence-Based Guidelines (Third Edition), Reg Anesth Pain Med. 2010;35:64–101
[b]EHR Programs, Centers for Medicare and Medicaid Services. (http://www.cms.gov/Regulations-and-Guidance/Legislation/EHRIncentivePrograms/index. html?redirect=/ehrincentiveprograms; last accessed August 3, 2012)

The "Issues Log" is another tool to collect and manage unintended consequences of health IT [20]. A good sample issues log can be downloaded from the HealthIT. gov Web site [21]. Our hospital employs a hospital intranet-based application that is utilized by all hospital staff to enter IT-related problems. The IT department tracks all issues entered, e.g., who reported the problem, when it occurred, what system it occurred in, and what happened. Each problem is assigned an identifying number and is tracked until resolved.

## *Usability*

Usability is a critically important consideration from the technology category that deserves elaboration. Simply put, usability is how easy a technology is to learn and use. Other related terms include human factors and user-centered design. Shneiderman promotes eight rules for interface design (Fig. 5.8) [22]. Ultimately, we believe a more usable EHR is a safer EHR. While providers can change processes, training, and organization, rarely can they improve the usability of their EHRs. Complaints abound from clinicians about the poor usability of many EHRs. The concerns expressed include the excessive number of clicks to find information, nonintuitive graphic user interface, and lack of integration or interoperability between clinical systems. With the sheer volume and complexity of information in patient care today, poor usability can compromise decision-making and patient safety.

In order to minimize potential adverse impact of EHR on patient safety, the IOM report on patient safety and health IT made a number of significant recommendations [13] including:

- Specify the quality and risk management process requirements that health IT vendors must adopt, with a particular focus on human factors, safety culture, and usability.
- Establish a mechanism for both vendors and users to report health IT-related deaths, serious injuries, or unsafe conditions.

Additionally, the Office of the National Coordinator has proposed new EHR certification rules that would promote safety-enhanced design that mandate that developers adopt user-centered design and document software quality management [23]. If finalized, these rules are important first steps in building more usable and safer EHRs.

## Conclusion: Lessons Learned

- There is mounting evidence of the role of EHRs in improving safety and quality of care.
- Like any innovation, use of EHRs in clinical practice can lead to unanticipated and potentially adverse consequences on patient safety.

| Principles | Characteristics |
|---|---|
| Strive for Consistency | Similar tasks ought to have similar sequences of action to perform, for example:<br>• Identical terminology in prompts and menus<br>• Consistent screen appearance<br>Any exceptions should be understandable and few |
| Cater to universal usability | Users span a wide range of expertise and have different desires, for example:<br>• Expert users may want shortcuts<br>• Novices may want explanations |
| Offer informative feedback | Systems should provide feedback for every user action to:<br>• Reassure the user that the appropriate action has been or is being done<br>• Instruct the user about the nature of an error if one has been made<br>Infrequent or major actions call for substantial responses, while frequent or minor actions require less feedback. |
| Design dialogs to yield closure | Have a beginning, middle, and end to action sequences:<br>Provide informative feedback when a group of actions has been completed |
| Prevent errors | Systems should be designed so that users cannot make serious errors, for example:<br>• Do not display menu items that are not appropriate in a given context<br>• Do not allow alphabetic characters in numeric entry fields<br>User errors should be detected and instructions for recovery offered<br>Errors should not change the system state |
| Permit easy reversal of actions | When possible, actions (and sequences of actions) should be reversible |
| Support internal locus of control | Surprises or changes should be avoided in familiar behaviors and complex data-entry sequences |
| Reduce short-term memory load | Interfaces should be avoided if they require users to remember information from one screen for use in connection with another screen |

**Fig. 5.8** Eight golden rules for interface design. Adapted from Shneiderman B, Plaisant C, Cohen M, Jacobs S. Designing the user interface: Strategies for effective human-computer interaction. Boston, MA: Addison-Wesley; 2009 (reprinted with permission)

- Applying the sociotechnical system model to performing RCA on EHR-related adverse events can identify problems and point to their resolution. Analysis often reveals more than one cause that may involve people, technology, process, organization, and/or the external environment.
- Monitoring usage metrics and maintaining an issues log are vital tools for performance improvement in health IT. After implementation, all health IT requires care and sustenance.
- Ultimately, if implemented judiciously and maintained assiduously, health IT can indeed transform the delivery, quality, and safety of health care.

# References

1. Kohn LT, Corrigan JM, Donaldson MS, editors. To err is human: building a safer health system. Washington, DC: Institute of Medicine; 1999.
2. Institute of Medicine. Crossing the quality chasm: a new health system for the 21st century. Washington, DC: The National Academies Press; 2001.
3. Jamoom E, Beatty P, Bercovitz A, et al. Physician adoption of electronic health record systems: United States, 2011. NCHS data brief, no 98. Hyattsville, MD: National Center for Health Statistics. 2012.
4. DesRoches CM, Worzala C, Joshi MS, Kralovec PD, Jha AK. Small, nonteaching, and rural hospitals continue to be slow in adopting electronic health record systems. Health Aff. 2012;31(5):1092–9.
5. Medicare and medicaid electronic health records incentive program payment and registration report, June 2012. https://www.cms.gov/Regulations-and-Guidance/Legislation/EHRIncentivePrograms/Downloads/June2012_MonthlyReport.pdf. Last accessed 31 Jul 2012.
6. Buntin MB, Burke MF, Hoaglin MC, Blumenthal D. The benefits of health information technology: a review of the recent literature shows predominantly positive results. Health Aff. 2011;30(3):464–71.
7. Rogers EM. Diffusion of innovations. 5th ed. New York: Free Press; 2003.
8. Bright TJ, Wong A, Dhurjati R, et al. Effect of clinical decision-support systems a systematic review. Ann Intern Med. 2012 Jul;157(1):29–43.
9. Osheroff JA, Teich JM. Improving outcomes with clinical decision support: an implementer's guide. 2nd ed. Chicago, IL: Scottsdale Institute, AMIA, AMDIS and SHM; 2012. ISBN 13: 978-1-938904-20-2. http://ebooks.himss.org/product/improvingoutcomes-clinical-decision-support. Last accessed 29 Sep 2012.
10. Approaching clinical decision support in medication management. Agency for Healthcare Research and Quality. http://healthit.ahrq.gov/images/mar09_cds_book_chapter/CDS_MedMgmnt_ch_1_sec_2_five_rights.htm. Last accessed 29 Sep 2012.
11. ISMP statement on use of metric measurements to prevent errors with oral liquids. Institute for Safe Medication Practices, 2011. http://www.ismp.org/pressroom/PR20110808.pdf. Last accessed 29 Sep 2012.
12. Campbell EM, Sittig DF, Ash JS, et al. Types of unintended consequences related to computerized provider order entry. J Am Med Inform Assoc. 2006;13(5):547–56.
13. Institute of Medicine. Health IT and patient safety: building safer systems for better care. Committee on Patient Safety and Health Information Technology. Washington, DC: The National Academies Press; 2012.
14. Ishikawa Diagram. Wikipedia. http://en.wikipedia.org/wiki/Ishikawa_diagram#The_5_Ss_.28used_in_service_industry.29. Last accessed 3 Aug 2012.

15. Institute for Healthcare Improvement. Cause and effect diagram. http://www.ihi.org/knowledge/Pages/Tools/CauseandEffectDiagram.aspx. Last accessed 3 Aug 2012.
16. Sittig DF, Singh H. A new socio-technical model for studying health information technology in complex adaptive healthcare systems. Qual Saf Health Care. 2010;19 Suppl 3:i68–74. doi:10.1136/qshc.2010.042085.
17. DeRosier J, Stalhandske E, Bagian JP, Nudell T. Using health care failure mode and effect analysis: the VA National Center for patient safety's prospective risk analysis system. Jt Comm J Qual Improv. 2002;27(5):248–67.
18. Institute For Healthcare Improvement. The run chart: a simple analytical tool for learning from variation in healthcare processes. http://www.ihi.org/knowledge/Pages/Publications/TheRunChartASimpleAnalyticalToolforLearningfromVariationHealthcareProcesses.aspx. Last accessed 20 Aug 2012.
19. Sittig DF, Campbell E, Guappone K, et al. Recommendations for monitoring and evaluation of in-patient computer-based provider order entry systems: results of a Delphi survey. Annu Symp Proc. 2007;671–5.
20. Identify unintended consequences. HealthIT.gov. http://www.healthit.gov/unintended-consequences/content/identify-unintended-consequences.html. Last accessed 20 Aug 2012.
21. Sample issues log. HealthIT.gov. http://www.healthit.gov/unintended-consequences/sites/default/files/issue-log.xls. Last accessed 20 Aug 2012.
22. Shneiderman B, Plaisant C, Cohen M, Jacobs S. Designing the user interface: strategies for effective human-computer interaction. Boston, MA: Addison-Wesley; 2009.
23. Office of the National Coordinator for Health Information Technology (ONC), Department of Health and Human Services. Health information technology: standards, implementation specifications, and certification criteria for electronic health record technology, 2014 edition; Revisions to the permanent certification program for health information technology. Proposed rule. Fed Regist. 2012;77(171):54163–292.

# Chapter 6
# Clinical Ethics and Patient Safety

Erin A. Egan

> *"Do not be ashamed of mistakes and thus make them crimes."*
>
> Confucius

## Introduction

Patient safety and ethics are both fields that seek to operationalize fundamental values in health care[1]. In both areas, broad values drive practical responses in clinical settings. There are two common sites of overlap. First is to ensure safety practices in areas where clinical ethical concerns arise frequently. Clinical ethics is an everyday practice in all settings but ethical conflicts are most common in hospitals. Areas like end-of-life-care have a strong component of clinical ethics and are high-risk areas for errors and compromise of patient safety. The other area of overlap is professional ethics and the commitment to patient safety. Given that patient safety is grounded in ethical principles and the resultant ethical responsibility of health care professionals to serve and protect patients, commitment to patient safety is a professional ethics responsibility.

---

[1] The term "clinical ethics" describes an area of practice that a provider may have special training in, or the skill set any provider uses in addressing ethical issues in say to day practice. Fully defining and describing the term is outside the scope of this chapter but further information can be found at sites like http://www.asbh.org/publications/content/core.html. Last accessed 22 Apr 2012.

E.A. Egan, M.D., J.D. (✉)
Neiswanger Institute for Bioethics and Health Policy, Loyola University, 5600 S. Quebec,
Suite 113A, Greenwood Village, CO 80111, USA
e-mail: drerinegan@aol.com

A. Agrawal (ed.), *Patient Safety: A Case-Based Comprehensive Guide*,
DOI 10.1007/978-1-4614-7419-7_6, © Springer Science+Business Media New York 2014

**Table 6.1** Principles of ethics and applications in patient safety

| Autonomy—respect for a person's right to control their own body | Beneficence—the duty to provide benefit |
|---|---|
| A central principle of quality care, essential to ensuring patient centered care | Establishing standard practices that promote benefit |
| Examples: adequate informed consent to prevent errors in procedures (wrong-site procedures), preventing unwanted care (proper DNR orders) | Examples: standardizing pre-op antibiotic procedures to maximize efficacy, pharmacy-assisted medication dosing to ensure maximum benefit |
| **Non-maleficence** | **Justice** |
| Establishing practices to prevent harm. Failure to use these practices compromises patient safety | Standardization of practices and procedures ensures equitable treatment across social and societal strata |
| Examples: infection control/hand washing, procedure "time-outs" | |

## Principles of Medical Ethics

The fundamental principles of clinical ethics in the context of care provided in the USA are beneficence, non-maleficence, autonomy, and justice (Table 6.1) [1]. Beneficence is the principle of providing benefit. Non-maleficence is the principle of doing no harm. In ethics these principles are applied to sort out the implications of different courses of action by weighing the values at stake. Both are applicable to patient safety efforts.

This conceptual structure is valuable as a means to categorize and target patient safety efforts along ethical principles. Ultimately, the basis of formally addressing patient safety deficits is to provide benefit and prevent harm. Using ethical precepts facilitates the process of understanding the goals of patient safety improvement in a tangible and concrete sense.

Clinical ethics is a practice or a skill set, meaning that it is a clinical process utilized in the context of patient care. As with any clinical practice there are standards for best practices, and variable level of adherence to those best practice standards. Instead of viewing ethics as a nebulous intellectual endeavor, clinical ethics should assist in solving problems and effectuating desirable outcomes.

## Case Study 1

### Clinical Summary: Decision About the End-of-Life Care

*Mary is a 93-year-old woman presenting to the emergency department (ED). A family member went to check on her at home, where she has lived independently since*

*the death of her husband 5 years prior, and found her agitated and confused. Mary has had decreasing mental status in the ED and appears to be septic. She is intubated and transferred to the ICU. She has several complications including a heart attack, and after 5 days, she shows no signs of being ready to come off the ventilator. The medical team consults family members regarding her "code status." One son thinks she has a living will asking to be "Do not resuscitate (DNR)," but isn't sure where it might be. A daughter says she had spoken with her mother after her husband's death and her mother said she "wouldn't want to be kept alive on machines." The third son says that his conversations with mom about religion have focused on the inherent value of life and he believes she would "want everything done." Two of the children want to withdraw ventilation, while the third wants to proceed to placement of a tracheostomy and PEG tube.*

This scenario is unfortunately all too commonly encountered by almost all physicians with an increasing frequency in hospital across the country. The case demonstrates that the nexus between ethics and safety occurs at two points: the role of clinical ethics practice to promote safety and quality and ensuring that safety mechanisms need to be in place to prepare for and prevent injury related to foreseeable ethical conflicts. Many commonly encountered clinical ethics conflicts follow a similar pattern, and it is important to recognize that patient safety concerns with an ethical component are common, can be predicted, and should be addressed by the same strategies as any other clinical patient safety issue.

## Case Analysis

### Ethics and Law

The law often plays a role in the analysis of ethical issues in health care. An essential step in clinical ethics, particularly at the end of life, is to determine what is permissible. The law itself is not the fundamental basis of either quality or ethics. The law sets rough boundaries within which many practices may be "legal," but says little about what is ethical. Safety failures may result in legal consequences, but a guiding principle of safety promotion is identifying problems and correcting them before a patient is actually injured. Therefore, effective patient safety practices should prevent legal involvement. That being said, the law does have a practical impact in setting standards and influencing change; therefore, knowing the guidelines of the law is essential.

In many areas of healthcare legislation where ethical issues are addressed the law tries to put into place a process that will ideally yield an ethical outcome. Legal solutions tend to be rigid, while clinical solutions need to be flexible. Knowledge of the law is a necessary ingredient for effective ethical decision making, but it is not sufficient in and of itself.

**Table 6.2** Practical comparison of principles in patient safety, ethics, and law

| Principle | Ethics | Patient safety | Law |
| --- | --- | --- | --- |
| Autonomy | Respect for a person physically as well as emotionally<br>Respect for a person's values | Injury caused by providing care that a patient didn't want<br>Injury caused by failing to provide desired care | Assault and battery for unwanted physical interference<br>Negligence[a] claim based on failure to obtain proper consent<br>Negligence for failing to provide necessary care |
| Beneficence | The intent to provide benefit | Failing to ensure practices that promote benefit<br>A safety promotion plan in fact causes harm<br>Causing injury by improperly implementing a beneficial plan | Negligence in providing the standard of care<br>Negligence in creating a hazard despite benevolent intent<br>Negligence for failing to implement a hospital policy or practice that would have prevented harm |
| Non-maleficence | The duty to prevent harm- "first do no harm" | Inadequate safeguards to prevent foreseeable harm<br>Failure of safeguards to prevent harm- existing safeguards are inadequate or are improperly implemented<br>Harm from the intended safeguard itself (i.e., delay in provision of a medication because the medication is not available immediately on the floor—no override or not stocked on floor) | Negligence based on failing to protect a patient (the claim is more severe as the foreseeability of the harms increases)<br>Negligence in failing to uphold hospital policies, negligence/incompetence in execution<br>Negligence in failing to provide competent care, negligence in creating a hazard |

[a]Negligence is a general term for failing to meet the standard of care. The basic elements of any negligence claim is the presence of a duty, a breach of the duty (the failure to meet the standard of care in meeting the duty), harm caused by the breach, the determination of the type, and value of the injury caused

Autonomy, beneficence, and non-maleficence all have legal relevance. Autonomy translates into informed consent. In malpractice cases, beneficence and non-maleficence are relevant to establishing the presence of a duty, the standard of care for the elements of the duty, and whether the duty was met. Table 6.2 demonstrates the parallels between ethics, safety practices, and examples of potential legal causes of action.

Specific to this case, the law clearly recognizes the authority of a person to refuse all unwanted health care, even when it would be life-prolonging or life-saving.[2] The United States Supreme Court has established that life-saving or life-sustaining treatment can be withheld or withdrawn from incompetent (including unconscious) patients, and that States may define the necessary authority required for this decision to be made for an incompetent/unconscious patient [2]. The law has several ways of approaching decision making for an incompetent patient. Often these laws are state specific and healthcare providers need to be familiar with the laws in their own state.[3]

The first step in this case, before invoking the relevant law for the incompetent patient, would be to understand the nuances of determining a patient's decision making capacity. Medical decision making capacity is a fundamental requirement for informed consent to be valid and in most US jurisdictions and is based on four abilities: (a) ability to understand the relevant information about the proposed test or treatment, (b) ability to appreciate the nature of one's situation and the consequences of one's choices, (c) ability to reason about the risks and benefits of potential options, and (d) ability to communicate a choice [3]. Only when these abilities are absent can be patient be deemed incompetent to make a clinical decision.

**Ethics and Patient Safety**

Within the framework defined by the legal parameters, basic quality and safety principles can be applied to the clinical scenario. As defined by the Institute of Medicine, care needs to be safe, effective, patient-centered, timely, efficient, and equitable [4]. Defining what these mean in the clinical context operationalizes the principles.

Safety in this scenario is not making the wrong decision: premature termination of support would be unsafe, but continuing unwanted care is also unsafe. The injury from withdrawing support prematurely is obvious. The injury from continuing unwanted support is also substantial. Freedom from unwanted invasion of one's body is a fundamental societal value as well as a fundamental healthcare value [4].

Patient-centered decision making at the end of life or at any time is critical. Ethics exists only within a clinical context and that context is unique to the patient. It is easy to get distracted by what is safe for the providers or the institution. Withdrawal of support over the objection of a family member has potential consequences for the providers and institution. The perception may develop that the "safest" course is to maintain the status quo (continue the current level of support) or to err on the side of continued medical care in the face of a dispute. However,

---

[2]Several cases have addressed this issue. For an example of the legal reasoning see Bouvia v Superior Court, 179 Cal. App 3d 1127 (1986).

[3]An example of the implementation of New York state's 2010 Family Healthcare Decisions Act in an academic medical center can be found at http://www.amc.edu/academic/bioethics/documents/AMC_FHCDA_Article.pdf. Accessed 13 Jul 2013.

patient-centered care emphasizes that safety in this situation is compromised as much by providing unwanted medical intervention as it is by withdrawing support prematurely.

Improved end-of-life care is often also discussed as an issue of wasted money and wasted resources [5]. These are substantial societal as well as individual concerns. Failure to resuscitate has obvious consequences. However, unwanted resuscitation has immense consequences as well and is an increasingly common issue [6]. The idea that unwanted resuscitation may have legal consequences is developing.[4]

## Solutions

Healthcare providers and institutions may use several strategies to prevent conflict such as presented in this case. First, discussing a patient's wishes regarding their treatment preferences is a standard part of medical care that should be addressed with every patient before an end-of-life situation arises or patient loses decision-making capacity. This patient was unable to express her wishes on admission, but there is ample opportunity in most patients' care to determine a patient's wishes. Second, adequate documentation of a discussion and patient's decision is critical.

In this case a discussion may have been had at some point, but none of the family members are clear what the content of that discussion might have been. Many of the prominent, high-profile media cases have revolved around what a patient said and to whom. Nancy Cruzan and Teresa Schiavo both made statements about how they saw medically dependent life support, but the statements were sporadic, varied in different conversations, and had ambiguous meaning when applied to their actual conditions at the end of their lives [2, 7]. Open conversation within families and among loved ones makes a person's wishes clear and hopefully prevents conflict. A clearly documented statement prevents misunderstanding of a patient's wishes and helps ensure safe end-of-life care. Palliative care is a fundamental issue in end-of-life care and mismanaged palliative care has numerous safety implications. Clear, adequately documented end-of-life care wishes make palliative care safer and in being safer it can be administered more effectively.

Traditionally, patients' preferences for life-sustaining treatments are documented and communicated using patient-generated advance directives or medical orders such as "DNR (Do Not Resuscitate)" regarding cardiopulmonary resuscitation. Unfortunately, these practices have been found to be largely ineffective at altering end-of-life treatments [8]. Advanced directives, such as living wills, are generally unhelpful in clinical settings because of vague instructions and the lack of certainty

---

[4] An early case, Wright v. Johns Hopkins Health Systems Corp., 353 Md. 568, 585–86 (1999), found no liability for unwanted resuscitation but there have been more legal challenges in other states and the claim is gaining favor in ethics and legal discussions of the issue.

as to when to act on them. While medical orders, such as DNR, may appear more helpful due to their specificity, they address a narrow decision regarding resuscitation and do not provide guidance regarding other issues around end-of-life care such as the use of intravenous nutrition and antibiotics. Another barrier to effective DNR orders has been that they need to be rewritten in each setting and at each transition of care. Only a credentialed physician can write an order at a given hospital, so the same physician may not be able to write a valid order at another facility. A new set of orders has to be written with each transition: outpatient to inpatient, nursing home to hospital, hospital to hospice etc. Each set of orders should be complete and should replace all prior orders. Often, if the end-of-life care discussion isn't well documented and/or a conversation about end-of-life care preferences can't be discussed immediately, the patient may be made "full code" until such a discussion can be had. Further, because of the need to renew the DNR order at each visit, often the conflict between loved ones including the patient may need to be revisited and re-inflamed with each transition.

To address the limitations of the traditional practices for communicating patient treatment preferences, there have been attempts to create a set of orders that travels with the patient, is valid in every setting, and can be relied upon by providers in every setting. These are commonly called Physician Orders for Life Sustaining Treatment (POLST).[5] A fundamental benefit of the POLST approach is that the POLST form translates patient's treatment preferences into medical orders. It is designed for patients of any age with advanced illnesses or frailty. The POLST form expands upon CPR status to include orders based on preferences about a range of life-sustaining treatments, e.g., antibiotics, artificially administered nutrition like tube feeding (Fig. 6.1). Some states have variant names, for example, Colorado uses the term Medical Orders for Scope of Treatment or MOST. The mechanisms by which these are valid is dependent on the State, but typically the State legislature enacts the use of the form, often as part of the medical decision-making act that describes medical durable powers of attorney and living wills. The dominant advantage of POLST, namely a single discussion and resulting document can result in an order that clearly indicates the patient's preferences, is obvious. Further, POLST can be relied on safely by anyone, including EMS personnel. Providers' concerns for their own legal safety in failing to resuscitate someone are negated by a POLST document in a State that recognizes it.

Ultimately, for a patient without decision-making capacity with end-of-life issues, a plan of care needs to be decided upon. If there is an advance directive, it needs to be found. A patient's own wishes, expressed by them in writing, are invaluable. Under Federal law a patient must be asked on admission whether they have an advance directive and must be given information about it [9]. If there is no advance directive and no durable power of attorney for health care, a decision maker must be

---

[5]POLST.org describes the orders and summarizes which states have enacted formal POLST type laws. Available at http://www.polst.org. Accessed 13 Jul 2013.

HIPAA PERMITS DISCLOSURE OF POLST TO OTHER HEALTH CARE PROVIDERS AS NECESSARY

EMERGENCY MEDICAL SERVICES AUTHORITY · EMSA · CALIFORNIA

## Physician Orders for Life-Sustaining Treatment (POLST)

EMSA #111 B
(Effective 4/1/2011)

**First follow these orders, then contact physician.** This is a Physician Order Sheet based on the person's current medical condition and wishes. Any section not completed implies full treatment for that section. A copy of the signed POLST form is legal and valid. POLST complements an Advance Directive and is not intended to replace that document. Everyone shall be treated with dignity and respect.

| | |
|---|---|
| Patient Last Name: | Date Form Prepared: |
| Patient First Name: | Patient Date of Birth: |
| Patient Middle Name: | Medical Record #: *(optional)* |

**A** Check One — **CARDIOPULMONARY RESUSCITATION (CPR):** ***If person has no pulse and is not breathing.*** ***When NOT in cardiopulmonary arrest, follow orders in Sections B and C.***

☐ **Attempt Resuscitation/CPR** (Selecting CPR in Section A requires selecting Full Treatment in Section B)
☐ **Do Not Attempt Resuscitation/DNR** (Allow Natural Death)

**B** Check One — **MEDICAL INTERVENTIONS:** ***If person has pulse and/or is breathing.***

☐ **Comfort Measures Only** Relieve pain and suffering through the use of medication by any route, positioning, wound care and other measures. Use oxygen, suction and manual treatment of airway obstruction as needed for comfort. ***Transfer to hospital only*** *if comfort needs cannot be met in current location.*

☐ **Limited Additional Interventions** In addition to care described in Comfort Measures Only, use medical treatment, antibiotics, and IV fluids as indicated. Do not intubate. May use non-invasive positive airway pressure. Generally avoid intensive care.
☐ ***Transfer to hospital only*** *if comfort needs cannot be met in current location.*

☐ **Full Treatment** In addition to care described in Comfort Measures Only and Limited Additional Interventions, use intubation, advanced airway interventions, mechanical ventilation, and defibrillation/cardioversion as indicated. ***Transfer to hospital*** *if indicated. Includes intensive care.*

**Additional Orders:** ____________________

**C** Check One — **ARTIFICIALLY ADMINISTERED NUTRITION:** ***Offer food by mouth if feasible and desired.***

☐ No artificial means of nutrition, including feeding tubes.
☐ Trial period of artificial nutrition, including feeding tubes.
☐ Long-term artificial nutrition, including feeding tubes.

Additional Orders: ____________________

**D** — **INFORMATION AND SIGNATURES:**

**Discussed with:** ☐ Patient (Patient Has Capacity) ☐ Legally Recognized Decisionmaker

☐ Advance Directive dated ______ available and reviewed → Health Care Agent if named in Advance Directive:
☐ Advance Directive not available Name: ____________
☐ No Advance Directive Phone: ____________

**Signature of Physician**
My signature below indicates to the best of my knowledge that these orders are consistent with the person's medical condition and preferences.

| | | |
|---|---|---|
| Print Physician Name: | Physician Phone Number: | Physician License Number: |
| Physician Signature: *(required)* | | Date: |

**Signature of Patient or Legally Recognized Decisionmaker**
By signing this form, the legally recognized decisionmaker acknowledges that this request regarding resuscitative measures is consistent with the known desires of, and with the best interest of, the individual who is the subject of the form.

| | | |
|---|---|---|
| Print Name: | | Relationship: *(write self if patient)* |
| Signature: *(required)* | | Date: |
| Address: | Daytime Phone Number: | Evening Phone Number: |

SEND FORM WITH PERSON WHENEVER TRANSFERRED OR DISCHARGED

**Fig. 6.1** Example POLST form

chosen. In this scenario, in most states the three children would have equal authority. Some states would treat the situation differently if the patient had a living spouse. If the children have equal authority then a facilitated family meeting is the main tool for resolution. Most often these are effective, especially if all the interested parties are available, in person, and appropriate support is provided. In this case the presence of a religious advisor may be beneficial since one child's concerns are based on religion. Usually there is an informal "majority rules" approach, but if there is one outspoken member of the minority position the hospital counsel and administration should be involved. However, the guiding principles of the discussion and the plan of care should be the basic quality improvement principles with a focus on safety, efficacy, and patient-centeredness.

## Discussion

There are core competencies in end-of-life care as well as in most ethical aspects of medical practice. The American Council for Graduate Medical Education (ACGME)'s Residency Review Committee prescribes areas of expected competence in ethical practice in several contexts [10]. A minimal level of ethical knowledge and clinical skill is part of professional practice. Analogous expectations exist for most clinical practitioners in their respective codes of ethics and clinical competencies.

After evaluating the role of patient safety and the relationship to ethical practice, the case scenario may be simplified. Like most situations where clinical ethics is involved, this will be an intense and painful discussion for this family no matter how well it is handled, and it may be that consensus is not possible. The value of a clinical ethics approach, especially if a clinical ethicist is involved, is to make the discussion productive and effective. Ultimately, whenever there is an ethical conflict there is a potential injury resulting from a "wrong" decision. There is no single ethically right decision for every situation, but in any situation there is a need to reach a resolution.

## Case Study 2

### Clinical Summary: Physician's Persistent Non-compliance with Hand-washing

*A large tertiary care medical center has been able to recruit a well-known cardiologist who has several large grants. The presence of this physician at this institution is important to the mission of the institution, and the grants are important to the department and the institution. The physician is well liked by patients, colleagues, and other healthcare team members. It has been brought to the physician's attention several times that she forgets to wash her hands or use sanitizer, but she indicates that "washing her hands isn't her priority, taking care of patients is." There is an outbreak of Clostridium difficile in the hospital, affecting cardiology patients disproportionately. A number of clinicians have raised concerns that the hand-washing practice of this physician is contributing to the outbreak, and a data review indicates a much higher rate of infection among her patients.*

Disruptive physician behavior is a problem across all provider levels and care settings. The term "disruptive physician" usually applies to physicians who are impaired at work, abusive, or sexually and personally inappropriate. However, the American Medical Association Code of Medical Ethics defines disruptive behavior as "personal conduct, whether verbal or physical, that negatively effects or may potentially negatively affect patient care [11]." Failure to adhere to clearly beneficial patient safety practices, such as hand washing, negatively effects patient care and should be addressed as a disruptive behavior. It is noteworthy that while often disruptive behavior points toward physicians, this behavior is found in all levels of clinicians across all care settings.

## Case Analysis

Approaching quality improvement and patient safety issues in this case from an ethical perspective, the principles of beneficence and non-maleficence describe the underlying philosophy. The healthcare system as a whole, and all of the members within it, have a duty to promote welfare and avoid harm. In a very real sense, quality improvement is inherently an ethical issue. Failing at any opportunity to confer benefit or prevent harm affects quality adversely, but it also compromises the ethical obligations inherent to providing health care.

There are many professional codes of ethics, specific to various professions within health care. Despite the variation of skills and practices, most professional codes of ethics are similar in regards to basic ethical principles. The physician code of ethics will be used for the purposes of discussion, but most professional codes of ethics could be used with similar conclusions.

Competence is a universal ethical obligation [11, 12]. While this seems too self-evident to warrant discussion, clearly established safety practices take a notoriously long time to implement uniformly. For example, despite the clear benefits of prescribing aspirin after a myocardial infarction, removing Foley catheters as early as possible to prevent UTI's, or ensuring that advance directives are known and available, these practices have been adopted slowly by providers and have only taken uniform hold with targeted hospital initiatives along with the regulatory pressure of Medicare Core Measures and The Joint Commission [13, 14]. The need for strong incentives and disincentives to ensure uniform practice indicates that knowledge alone doesn't ensure competent practice.

Similarly, the ascendance of the best interest of the patient and protecting the patient from harm is ubiquitous. It would hold that a practice that has been shown to have overwhelming benefit and has essentially no risks would be adhered to without reservation. Failure to do so would seem to be a breach of the central ethical foundations of health care. Despite this, physicians (and other healthcare providers) routinely deviate from known safety practices, but many of these physicians would be indignant to be labeled "unprofessional" or "disruptive."

Hand washing is one such practice with immense positive clinical impact. Improved hand washing consistently lowers morbidity and mortality from infectious disease in the hospital [15]. Still, failure to adhere to this unequivocal best practice is common and hand washing/hand-sanitizing rates are embarrassingly as low as 50 % [16].

In this scenario there is a prominent provider clearly ignoring safety practices. In clinical situations where quality and safety practices are of more ambiguous benefit, it is easy to see why they would be even more difficult to enforce. However, in the case of hand washing, there is no potential argument that the practice in question has adverse effects or the evidence of benefit is equivocal. Otherwise, she is a good doctor in terms of patient care, patient satisfaction, peer interactions, and contribution to the field.

This scenario may appear implausible to lay people as no provider should refuse to wash her hands despite prompting, and no institutional culture should permit it. However, we know that providers do refuse to wash their hands, and institutions tolerate it. The literature on how and why is addressed elsewhere, while this discussion focuses on the responsibility issue [17].

If a physician is presented with the evidence that she is doing harm and refuses to change her practice, she should be removed from patient interaction. This requires integrity from an institution but is a position that institutions are increasingly willing to take. The response to disruptive physician behavior is discussed extensively in medical codes of ethics [11] and among executive and healthcare administrators [18]. The most hopeful outcome is that, when presented with direct evidence of harm and a conversation in the context of ethical and professional responsibilities, the physician will change. Changing her habit may take time, but most providers would respond to evidence that they are injuring patients. When behavior is tied to principles of conferring benefit and preventing harm, professionals tend to be very committed to trying to improve. However, if the provider remains defiant and uncooperative the institution is obligated to prevent contact between this physician and patients, and the provider's clinical privileges should be suspended or terminated.

## Discussion

A central implication of the duty to protect patients is to advocate for safety and to inculcate a culture of safety. "Culture change" is a catchphrase in quality improvement because it is a critical element of change. A culture that insists on safety practices creates that reality, and a system that refuses to create that reality will not develop a true culture of safety. Failure to participate in such a culture fails ethical duties, and the participation in this culture needs to empower people in any setting to expect safe practices from people regardless of the setting.

Creating a culture of safety, empowering everyone in a system to ensure quality and safety, enforcing best practices, and seeking systematic improvement are not only patient safety goals but also ethical professional practice goals. What may not be recognized by the providers is the relationship between ethics, duty, quality and safety.

The seminal IOM report, To Err is Human [19], had widespread impact with the assertion that thousands of people die from healthcare errors, more than from breast cancer, or motor vehicle accidents or AIDS. However, these conclusions were not based on new data, but instead reframed existing data. The data had been available for some time but had never been concretely translated by providers into the idea that preventable mistakes kill patients. The reframing of the existing data had overwhelming impact. The information wasn't new, but the paradigm for understanding it was redefined. The IOM report made it clear that healthcare institutions *cause* death. People who are in the hospital die for no other reason than that they are in the hospital.

Addressing the issue of death caused by preventable errors directly and explicitly made the issue of patient safety central to ethical and professional behavior. The connection between beneficence, non-maleficence, quality, and safety was made very clear. Once providers and institutions saw errors in tangible terms as preventable harm, the perception shifted and quality became an ethical issue. Healthcare providers want to be altruistic. They want to help people. They want to prevent harm. As soon as quality and safety were understood in these terms, providers and institutions became more committed to quality and safety. Enlightenment wasn't the trigger, responsibility was.

## Conclusion

Ethics, quality, and safety are interrelated concepts. There are issues of quality improvement in clinical ethics and developing strategies in clinical ethics practice that parallel quality improvement initiatives in other areas. There is also the inherent ethical obligation to be committed to quality improvement and improving patient safety. The strongest motivators in quality and safety have recognized the fundamental ethical responsibility and underlying motivation of providers to take care of patients.

### Key Lessons Learned

- The role of quality measures to improve patient safety is as relevant in ethics practice as it is to any clinical practice. Standards for best practices in ethics are available, and adherence to them promotes the safe and effective care of patients.
- Providers need to be knowledgeable and committed to safety in areas of clinical ethics practice.

- Promoting and participating in patient safety measures is an ethical obligation. Failure to adhere to known safety practices is a failure to meet the professional standards of ethics.
- Repeated failure to follow patient safety guidelines is inherently disruptive behavior and should be treated as such when considering consequences for failure to protect the safety of patients.

# References

1. Beachamp TL, Childress JF. Principles of biomedical ethics. 4th ed. New York, NY: Oxford University Press; 1994.
2. Cruzan v. Director, Missouri Department of Health. 1990; 497 U.S. 261.
3. Appelbaum PS, Grisso T. Assessing patients' capacities to consent to treatment. N Engl J Med. 1988;319(25):1635–8.
4. Institute of Medicine. Crossing the quality chasm: a New Health System for the 21st century. Washington, DC: National Academy Press; 2001.
5. Morrison RS, Penrod JD, Cassel JB, et al. Cost savings associated with United States hospital palliative care consultation programs. Arch Intern Med. 2008;168(16):1783–90.
6. Dull SM, Graves JR, Larsen MP, Cummins RO. Expected death and unwanted resuscitation in the prehospital setting. Ann Emerg Med. 1994;23(5):997–1002.
7. Greer, GW. Circuit Judge (2000-02-11). "In re: the guardianship of Theresa Marie Schiavo, Incapacitated, File No. 90-2908GD-003". Florida Sixth Judicial Circuit. Retrieved 2006-01-08. p. 9–10.
8. Hickman SE, Nelson CA, Perrin NA, et al. A comparison of methods to communicate treatment preferences in nursing facilities: traditional practices versus the physician orders for life-sustaining treatment program. J Am Geriatr Soc. 2010;58(7):1241–8.
9. Patient Self Determination Act, Omnibus Budget Reconciliation Act of 1990 Pub.L. 101-508, 104 Stat. 1388.
10. ACGME Core Competencies Definitions. Available at http://www.gahec.org/cme/Liasions/0) ACGME%20Core%20Competencies%20Definitions.htm. Accessed 13 Jul 2013.
11. American Medical Association, Code of Medical Ethics, Opinion 9.045 Physicians with Disruptive Behavior. Available at http://www.ama-assn.org/ama/pub/physician-resources/medical-ethics/code-medical-ethics/opinion9045.page. Accessed 22 Apr 2012.
12. American medical Association, Code of Medical Ethics, Principles of Medical Ethics. Available at http://www.ama-assn.org/ama/pub/physician-resources/medical-ethics/code-medical-ethics/principles-medical-ethics.page? Accessed 1 July 2012.
13. The Joint Commission. Core Measure Sets. Available at http://www.jointcommission.org/core_measure_sets.aspx. Accessed 13 Jul 2013.
14. Medicare process of Care Measures see http://www.cms.gov/Medicare/Quality-Initiatives-Patient-Assessment-Instruments/HospitalQualityInits/HospitalProcessOfCareMeasures.html (last accessed 22 Apr 2012) and for Medicare Outcome Measures see http://www.cms.gov/Medicare/Quality-Initiatives-Patient-Assessment-Instruments/HospitalQualityInits/OutcomeMeasures.html. Last accessed 22 Apr 2012.
15. Guideline for hand hygiene in health-care settings. Available at http://www.cdc.gov/mmwr/preview/mmwrhtml/rr5116a1.htm. Accessed 13 Jul 2013.
16. Center for Innovation in Quality Patient Care, Johns Hopkins Medicine. Available at http://www.hopkinsmedicine.org/innovation_quality_patient_care/areas_expertise/infections_complications/hand_hygiene/. Accessed 13 Jul 2013.

17. Wachter RM, Pronovost PJ. Balancing "no blame" with accountability in patient safety. N Engl J Med. 2009;361(14):1401–6.
18. McDonald O. Disruptive physician behavior. QuantiaMD and American College of Physician Executives; 2011. Available at https://www.quantiamd.com/home/qrc_disruptive. Accessed 22 Apr 2012.
19. Institute of Medicine. To err is human. Washington, DC: National Academy Press; 2000.

# Part II
# Examples

# Chapter 7
# Medication Error

Mei Kong and Abdul Mondul

> *"All things are poison, and nothing is without poison. Only the dose makes a thing not poison."*
>
> Paracelsus

## Introduction

Medications are the most common source of medical errors both in hospitals and in the ambulatory care setting, causing harm to at least 1.5 million people per year [1]. Medications account for approximately "1 out of every 131 outpatient deaths and 1 out of 854 inpatient deaths" [2], a total of 7,000 estimated potentially preventable deaths per year.

Adverse drug events (ADEs), in particular, related to ineffective patient education regarding medications and monitoring of drug therapies are accountable for up to 66 % of the adverse events after patients are discharged from the hospital [3]. Further, the use of high risk medications such as warfarin, insulin, and digoxin especially in elderly patients account for 33 % of the ADEs treated in emergency departments (EDs) and 41 % of ADEs leading to hospitalizations [4].

A medication error is defined as "any preventable event that may cause or lead to inappropriate medication use or patient harm while the medication is in the control

M. Kong, R.N., B.S., M.S.N. (✉)
Assistant Vice President, Patient Safety and Employee Safety, New York City Health
and Hospitals Corporation, 125 Worth Street, Room 402, New York, NY 10013, USA
e-mail: meikong2010@hotmail.com

A. Mondul, M.D.
Associate Medical Director and Patient Safety officer, Lincoln Medical Center,
Weill Medical College at Cornell University, 234 E 149th Street, 8th Floor Office 8-30,
Bronx, NY 10451, USA
e-mail: moondul@hotmail.com

A. Agrawal (ed.), *Patient Safety: A Case-Based Comprehensive Guide*, 103
DOI 10.1007/978-1-4614-7419-7_7, © Springer Science+Business Media New York 2014

of the health care professional, patient, or consumer. Such event may be related to professional practice, health care products, procedures, and systems [5]." An adverse drug event (ADE) is defined as an injury or harm to the patient that is caused by medication usage [6]. It is important to note that not all medication errors lead to ADEs and not all ADEs are medication errors.

According to the National Coordinating Council for Medication Error Reporting Program (NCC-MERP), medication errors are categorized into the following nine categories depending on the level of patient harm [5]:

Category A: Circumstances or events occur that have the capacity to cause error (no error).

Category B: An error occurred, but the error did not reach the patient (error, no harm).

Category C: An error occurred that reached the patient but did not cause patient harm (error, no harm).

Category D: An error occurred that reached the patient and required monitoring to confirm that it resulted in no harm to the patient and/or required intervention to preclude harm (error, no harm).

Category E: An error occurred that may have contributed to or resulted in temporary harm to the patient and required intervention (error, harm).

Category F: An error occurred that may have contributed to or resulted in temporary harm to the patient and required an initial or prolonged hospital stay (error, harm).

Category G: An error occurred that may have contributed to or resulted in permanent patient harm (error, harm).

Category H: An error occurred that required intervention necessary to sustain life (error, harm).

Category I: An error occurred that may have contributed to or resulted in patient death (error, death).

Medication errors may occur at any of the five stages of the medication management process namely (1) ordering/prescribing, (2) transcribing and verifying, (3) dispensing and delivering, (4) administering, and (5) monitoring and reporting. It is estimated that 39 % of the errors occur during prescribing, 12 % during transcribing, 11 % during dispensing at the pharmacy, and 38 % during administering [7]. As illustrated in Fig. 7.1, most medication errors occur as a result of multiple vulnerabilities and failures in the continuum of the medication management process (the Swiss cheese concept) [8].

Similar to other adverse events, medication errors can arise from human errors and/or systems failures. Human errors include problems in practice (e.g., taking short cuts), training deficiencies, undue time pressure, distractions, and poor perception of risk. Systems failures can be related to products, procedures, or processes [2]. The most common medications associated with severe ADEs and mortality include central nervous system agents, anti-neoplastics, and cardiovascular drugs. The types of errors that contribute to patient death involve the wrong dose (40.9 %), the wrong drug (16 %), and the wrong route of administration (9.5 %) [9].

The American Hospital Association lists the following as the common types of medication errors [10]:

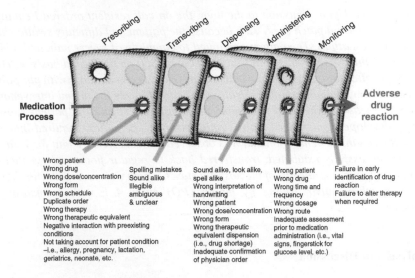

Fig. 7.1 Medication points of failures

- Incomplete patient information such as allergies, other medicines they are taking, previous diagnoses, and lab result
- Unavailable drug information such as a lack of up-to-date warnings from the Food and Drug Administration (FDA)
- Miscommunication of drug orders which can involve poor handwriting, confusion between drugs with similar names, misuse of zeroes or decimal points, confusion of metric and other dosing units, and inappropriate abbreviations
- Incorrect labeling as a drug is prepared and repackaged into smaller units
- Environmental factors such as heat, lighting, noise, and interruptions
- Failure to follow established facility policies and procedures

## Case Studies

### Case Study 1: Respiratory Depression Caused by Opioid Overdose

#### Clinical Summary

*A 56-year-old patient with a history of metastatic esophageal cancer was admitted for progressive enlargement of a left neck mass leading to dysphagia and severe pain related to bone metastasis. He had been taking non-steroidal anti-inflammatory drugs (NSAIDs) at home with partial pain relief. In the ED, he was treated with intramuscular (IM) Ketorolac and was admitted for pain management and hypercalcemia. Upon admission to the floor, the on-call resident ordered Fentanyl 50 mcg transdermal patch every 72 h because the patient had difficulty swallowing oral pain medications. The patient continued to complain of severe pain and additional morphine was administered subcutaneously (SQ) on as needed basis. Forty-eight hours after the admission, the patient was found to be comatose with pin-point pupils and slow, shallow breathing. Naloxone hydrochloride 0.4 mg/ml intravenous push (IVP) was administered to reverse the opioid effect and the patient subsequently developed generalized tonic–clonic seizures. The patient required intubation and mechanical ventilation and was observed in intensive care unit for 7 days. He was successfully extubated, transferred back to regular floor, and eventually discharged home. For the rest of the hospital stay, his pain was managed with short-acting morphine elixir 10 mg by mouth (PO) every 4 h with breakthrough coverage.*

## Analysis and Discussion

This case study illustrates a number of errors related to opioid prescribing for pain management. First, Fentanyl patch is a long-acting agent (the onset of action is up to 48 h); therefore it should not be used to treat acute pain especially in opioid-naïve patients. Second, the patient received a combination of SQ morphine along with Fentanyl leading to opioid overdose. Fifty micrograms of Fentanyl is equivalent to 135–224 mg of daily oral morphine equivalency. The prescribing physician should have been more aware and careful about the potential risks of prescribing opioids in high doses. At the same time, neither the pharmacist nor the nurses raised an alarm about the dose of pain medications being received by this patient. Finally, the rapid reversal of opioids may lead to seizures and other withdrawal symptoms. Hence, Naloxone should have been diluted and given in 0.04 mg/ml boluses, one-tenth of the IVP dose given to the patient.

Literature shows that opioid analgesics rank among the drugs most frequently associated with ADEs [11]. The most serious and sometimes fatal side effect is respiratory depression which is generally preceded by sedation. The reported incidence of respiratory depression in postoperative patients is about 0.5 % [11]. All patients receiving opioids must be adequately assessed and reassessed for pain and for previous history of opioid use/abuse to identify potential opioid tolerance or intolerance. There is commonly a lack of knowledge about potency differences among opioids, especially equivalence between short-acting and long-acting/sustained release opioid; therefore, sufficient time should be allowed to assess the patient's response to an initial dose before increasing the dosage or prescribing long-acting opioids. When converting from one opioid to another, or changing the route of administration from oral, IV, or transdermal, a pharmacist or pain management expert should be consulted if available or a conversion support system should be used to calculate correct doses [12]. When opioids are administered, the potential for opioid-induced respiratory depression should always be considered, especially in opioid-naive patients.

**Fig. 7.2** Tall Man Lettering

> **TaxOL and TaxOTERE**
>
> **oxyCONTIN and oxyCODONE**
>
> **DOBUTamine and DOPamine**
>
> **levoTIRACetam and levoFLOXacin**

## Case Study 2: Wrong Drug Dispensing and Administration Due to Similar Sounding Names

### Clinical Summary

*On the oncology unit, Dr. Sure ordered Taxol (paclitaxel) 260 mg IV (175 mg/m² × 1.5 m² body surface area = 262.50 mg) for Ms. Jones for her advanced stage breast cancer. After a review of the order, the pharmacist mistakenly dispensed 260 mg of Taxotere (docetaxel). The nurse reviewed the order and thought what was sent up by pharmacy was the correct medication and administered Taxotere 260 mg. The usual adult dose for Taxotere is 60–100 mg/m² IV [13]. Due to this error the patient received the wrong medication at three times the usual dose and died 4 weeks later from neutropenic sepsis and hepatic failure.*

### Analysis and Discussion

Both Taxol and Taxotere are used for breast cancer, are from the same family of medications, the taxanes, but have different pharmacokinetics and side-effect profiles. There is an increased risk of serious (possibly fatal) reactions when receiving higher doses of Taxotere, such as severe neutropenia, neurosensory symptoms, asthenia, fluid retention, trouble breathing, chest pain or tightness, fast or irregular heartbeat, or abdominal swelling [13].

Since Taxol and Taxotere are look-alike, sound-alike, and spell-alike drugs, they need additional safeguards for differentiation. A simple and frequently used solution to improve safe use of such medications is to use Tall Man Lettering (Fig. 7.2) to highlight the dissimilar letters in two names [14]. The Institute for Safe Medication Practices (ISMP), FDA, the Joint Commission, and other safety organizations have promoted the use of tall man letters as a means of reducing confusion between similar drug names [15]. This methodology can be used throughout the medication process including on CPOE ordering screens, computer-generated pharmacy labels, pharmacy computer drug selection screens, shelf labels, automated dispensing cabinet screens, computer-generated medication administration records, and even preprinted order sheets.

On the pharmacy dispensing side, the use of separate storage areas and different color containers could have helped to distinguish these two otherwise similar sounding medications. The nurse unfortunately also missed the opportunity to avert the error from reaching the patient. Had a bar-coded medication administration (BCMA) system to ensure the "five rights" of medication administration (right drug, dose, route, patient, and time) been in place at the bedside, the system would have detected that the medication being administered does not match the medication ordered thus averting this high-risk error [16]. Additionally, most institutions mandate verification by two nurses prior to administration of a high-alert medication such as a chemotherapeutic agent which had not been implemented at this hospital due to staffing constraints.

## Discussion

There are five essential strategies in improving medication safety. These include:

1. Use of Information Technology

   IT applications such as electronic health records (EHRs) and computerized physician order entry (CPOE), especially when augmented by a point-of-care clinical decision support (CDS) system have been demonstrated to reduce medication errors and improve patient safety [17–19]. Advantages of CPOE include legibility, prompt pharmacy review, links to drug–drug interactions, decision algorithms, easier ADE identification, less risk for look-alike/sound-alike medication errors, and improved medication reconciliation. Another advantage of CPOE is the capacity to embed CDS tools in the form of order sets, guidelines, or protocols [17, 18]. In addition to the safety of medication ordering, IT tools also improve efficiency of the process through the automation of medication preparation and packaging via the use of robotic dispensers.

   Another technology that has been demonstrated to improve medication safety is the bed-side bar-coded medication administration (BCMA) system. In this system, the nurse scans the bar-coded bracelet on the patient's wrist band to ensure that the medication(s) will be administered to the right patient. The nurse also scans the unit dose of the medications. The system compares each medication with the physician's orders and alerts the nurse to any mismatch of patient identity or of the name, dose, or route of administration of the medication. BCMA systems have been shown to lead to a 54–87 % reduction in medication errors during the administration step [20, 21].

   It is important to note that technology is not a panacea and can have unintended and potentially adverse consequences on safety. A widely quoted 2005 study found that CPOE implementation in an academic tertiary care children's hospital during an 18-month period actually resulted in an unexpected *increase* in mortality rate [22]. A commercial CPOE program that was designed for adult general medical–surgical usage was quickly implemented across this pediatric facility without appropriate customization, workflow configuration, and testing/user training. The study found an unexpected increase in mortality coincident with CPOE implementation and concluded that technology must not replace the critical thinking process necessary to make appropriate treatment choices. Other reports have also demonstrated that safe implementation of CPOE requires ongoing assessment of the system integration process with the human interface, as well constant monitoring and evaluation of medication error rates and mortality [23].

Other risks of CPOE include "alert fatigue" and an overreliance on the automated decision process sometimes substituting critical clinical thinking. "Alert fatigue" occurs when physicians receive too many alerts of questionable perceived value leading them to override the alerts. Therefore, the CPOE implementation should carefully consider the number and types of alerts that are turned on in a system [24, 25].

2. Addressing Health Literacy and Engaging Patients and Families
   Medication error prevention requires collaboration among different members of the healthcare team as well as with the patients and their families. The team should recognize the higher risk of medication error in patients with lower literacy levels as they may not have the skills necessary to effectively navigate the medication use process and are more likely to misinterpret the prescription label information and auxiliary labels [26, 27].

3. High Alert Medications
   High alert medications, such as anticoagulants, opioids, sedatives, insulin, chemotherapy, and electrolytes, can cause an immediate and life-threatening condition for the patient even when administered in usual doses. Due to high risk for patient harm, institutions should take additional steps to identify and mitigate risks to patient safety from such medications. Some steps include (1) standardize protocols and dosing; (2) establish order sets for the physicians with automated alerts; (3) dispense medications from pharmacy only and utilize auxiliary color-coded labels indicating high-alert medications; (4) establish monitoring parameters on assessing, reassessing, and documentation of patient responses to the medications; (5) train staff on early recognition of potential adverse events and how to rescue patients, including antidotes that are available; (6) employ redundancies such as automated or independent double checks, a procedure in which two clinicians separately check each component of prescribing, dispensing, and verifying the high-alert medication before administering it to the patient [28, 29].

4. Medication Reconciliation
   Medication reconciliation is the process of comparing a patient's medication orders to all of the medications that the patient has been taking. It should be done at the points of transition in care where new medications are ordered or existing orders are rewritten. Transitions in care include changes in setting, service, practitioner, or level of care. The medication reconciliation process comprises of the five steps below:

   (a) Develop a list of current medications, e.g., home medications
   (b) Develop a list of medications to be prescribed

(c) Compare the medications on the two lists
(d) Make clinical decisions based on the comparison
(e) Communicate the new list to appropriate caregivers and to the patient

Studies have shown that more than half the patients experience one or more unintended medication discrepancy at the time of a hospital admission and nearly 40 % of these have the potential for moderate to severe harm [30]. It is easy to overlook medications that may cause an adverse event, especially when combined with new medications or different dosages, so an effective medication reconciliation process across care setting can help prevent errors of omission, drug to drug interactions, drug–disease interactions, and other discrepancies [31].

5. Foster Pharmacy Collaboration
   Pharmacists can advise physicians in prescribing medications and enhance both physicians' and patients' understanding of medications [32]. Pharmacist participation during rounds with the medical teams on a general medicine unit contributed to a 78 % reduction in preventable ADEs (from 26.5 to 5.6 per 1,000 hospital days) by providing support at the time when decisions about therapy are made [33]. In addition, increased collaboration with the team resulted in increase in interventions during rounding, such as dosing-related changes and recommendations to add or modify a drug therapy [33].

## An Interdisciplinary Approach to Medication Error Prevention

In this section, we describe the role of various health team members in preventing medication errors and improving safety.

### The Prescribers' Role

Prescribing is an early point at which medication errors can arise. For safer prescribing, ordering physicians should stay knowledgeable with current literature review, consult with pharmacists and other physicians, as well as participate in continuing professional education. It is critical that prescribers evaluate the patient's overall status and review all existing therapies before prescribing new or additional medications to ascertain possible antagonistic or complementary drug reaction(s). Medication orders should be complete, clear, and unambiguous and should include patient name, generic name, brand name (if a specific product is required), route and site of administration, dosage form, dose, strength, quantity, frequency of administration, prescriber's name, and indication. In some cases, a dilution rate and time of administration should be specified. The desired therapeutic outcome for each medication should be expressed when prescribed. It is important not to use inappropriate abbreviations such as "QD," "BID," etc. The prescriber should educate the patient/caregivers about the medication prescribed, special precautions or observations, and potential anticipated side effects such as dry mouth or the first-dose hypotension. Finally, the prescriber should follow up and evaluate the need for continued therapy for individual patients on a regular basis [34].

## The Pharmacists' Role

The pharmacist, in collaboration with other team members, should be involved in assessing therapeutic appropriateness, medication interactions, and pertinent clinical and laboratory data for all orders. Pharmacists need to be familiar with drug distribution policies and procedures to ensure safe distribution of all medications and related supplies. They also maintain orderliness and cleanliness in the work area where medications are prepared and should perform one procedure at a time with as few interruptions as possible. They should observe how medications are actually being used in patient care areas to ensure that dispensing and storage procedures are followed as recommended. A review of medications that are returned to pharmacy is important as such review processes may reveal system breakdowns or problems that resulted in medication errors (e.g., omitted doses and unauthorized drugs). Pharmacists also play a key role in counseling patients/caregivers and verifying that they understand why a medication was prescribed, its intended use, any special precautions that might be observed, and other needed information [34].

## The Nurses' Role

Nurses play a key role in medication safety because they are the final check point in the medication process before the medication is actually administered to the patient [34]. Also, by virtue of their direct involvement in patient care activities, nurses are in the best position to detect and report medication errors. Nurses need to review medications with respect to desired outcomes, therapeutic duplications, and possible drug interactions. They must review and verify all orders before medication administration and ensure that the drug dispensed matches the order in all respects. It is the standard practice for a nurse to verify the "five rights"—the right patient, drug, time, dose, and route—at the bedside prior to administration. It is essential for a nurse to observe patients for medication responses and reactions, especially after the first dose. Nurses also play a key role in the education of patients and family to ascertain that they understand the use of their medications and any special precautions or observations that might be indicated [34].

## The Patients' and Caregivers' Role

The most important role for the patient and the family is to keep an up-to-date list of all medications. They should learn to recognize their pills—what they look like in size, shape, and color, and the indication and potential side effects.

Teaching patients is <u>not simply</u> preparing a list of pills with days and times attached; it should also include information about their diseases and the indication for medications. Patients should be asked to repeat-back or demonstrate-back to make sure they understood that which was taught.

# Conclusion

Medication errors are frequent, often harmful but with good systems and processes largely preventable. Equally important in medication safety is the role of organizational culture that promotes transparency in reporting and a non-punitive response to human errors. Only through an open and honest discussion of errors and systems failures, changes can be made to improve performance and prevent harm [35].

## Lessons Learned

- Medication errors occur at all phases of the medication process.
- Even seemingly simple medication errors are multifactorial, frequently involving more than one process and more than one line of responsibility.
- Many medication errors occur due to poor communication. A collaborative approach and better communication and interaction among members of the healthcare team and the patient are essential.
- Information technology (IT) solutions such as CPOE and BCMA are critical elements of an overall organizational strategy to prevent errors.
- Developing an organizational culture of safety, so that leaders and staff are committed to safety and are preoccupied with potential errors, is a vitally important piece in improving medication safety. A safety culture embraces open communication and empowers staff to report concerns.
- Finally, we must always respect the power of medication and never underestimate its potential to cure but also to harm patients.

# References

1. Medication errors injure 1.5 million people and cost billions of dollars annually. The National Academy of Science. 2006. Available at http://www8.nationalacademies.org/onpinews/newsitem.aspx?RecordID=11623. Last accessed 12 Nov 2012.
2. Institute of Medicine. To err is human: building a safer health system. Washington, DC: National Academy; 1999.
3. Forster AJ, Murff HJ, Peterson JF, et al. The incidence and severity of adverse events affecting patients after discharge from the hospital. Ann Intern Med. 2003;138:161–7.

4. Budnitz DS, Pollock DA, Weidenbach KN, et al. National surveillance of emergency department visits for outpatient adverse drug events. JAMA. 2006;296:1858–66.
5. National Coordinating Council for Medication Error Reporting and Prevention. What is a medication error? Available at http://www.nccmerp.org/aboutmederrors.htm. Last accessed 12 Nov 2012.
6. Gandhi TK, Weingart SN, Borus J, et al. Adverse drug events in ambulatory care. N Engl J Med. 2003;348:1556–64.
7. Leape LL, Bates DW, Cullen DJ, et al. Systems analysis of adverse drug events. ADE Prevention Study Group. JAMA. 1995;274:35–43.
8. Reason J, Carthey J, de Leval M. Diagnosing 'vulnerable system syndrome': an essential prerequisite to effective risk management. Qual Health Care. 2001;10(Suppl II):ii21–5.
9. Phillips J, Beam S, Brinker A, et al. Retrospective analysis of mortalities associated with medication errors. Am J Health Syst Pharm. 2001;58:1835–41.
10. American Hospital Association. Improving medication safety. Available at http://www.aha.org/advocacy-issues/tools-resources/advisory/96-06/991207-quality-adv.shtml. Last accessed 12 Nov 2012.
11. Davies EC, Green CF, Taylor S, et al. Adverse drug reactions in hospital in-patients: a prospective analysis of 3695 patient-episodes. PLoS One. 2009;4:e4439.
12. The Joint Commission Sentinel Event Alert Issue 49. Safe use of opioids in hospitals. August 8, 2012. Available at http://www.jointcommission.org/assets/1/18/SEA_49_opioids_8_2_12_final.pdf. Last accessed 1 Dec 2012.
13. Taxotere® Prescribing Information. Bridgewater, NJ: Sanofi-Aventis U.S. LLC; September 2011. Available at http://www.taxotere.com/oncology/dosage_administration/administration.aspx. Last accessed 12 Nov 2012.
14. Institute for Safe Medication Practices: ISMP Medication Safety Alert!—ISMP updates its list of drug name pairs with TALL man letters. November 18 2010. Available at http://www.ismp.org/Newsletters/acutecare/articles/20101118.asp. Last accessed 12 Nov 2012.
15. Filik R, Purdy K, Gale A, et al. Drug name confusion: evaluating the effectiveness of capital ("Tall Man") letters using eye movement data. Soc Sci Med. 2004;59:2597–601.
16. Marini SD, Hasman A. Impact of BCMA on medication errors and patient safety: a summary. Stud Health Technol Inform. 2009;146:439–44.
17. Walsh KE, Landrigan CP, Adams WG, et al. Effect of computer order entry on prevention of serious medication errors in hospitalized children. Pediatrics. 2008;121:e421–7.
18. Sittig DF, Stead WW. Computer-based physician order entry: the state of the art. J Am Med Inform Assoc. 1994;1:108–23.
19. Bates DW, Leape LL, Cullen DJ, et al. Effect of computerized physician order entry and a team intervention on prevention of serious medication errors. JAMA. 1998;280:1311–6.
20. Paoletti RD, Suess TM, Lesko MG, et al. Using bar-code technology and medication observation methodology for safer medication administration. Am J Health Syst Pharm. 2007;64:536–43.
21. Agrawal A. Medication errors: prevention using information technology systems. Br J Clin Pharmacol. 2009;67:681–6.
22. Han YY, Carcillo JA, Venkataraman ST, et al. Unexpected increased mortality after implementation of a commercially sold computerized physician order entry system. Pediatrics. 2005;116:1506–12.
23. Koppel R, Metlay JP, Cohen A, et al. Role of computerized physician order entry systems in facilitating medication errors. JAMA. 2005;293:1197–203.
24. Isaac T, Weissman JS, Davis RB, et al. Overrides of medication alerts in ambulatory care. Arch Intern Med. 2009;169:305–11.
25. van der Sijs H, Aarts J, Vulto A, et al. Overriding of drug safety alerts in computerized physician order entry. J Am Med Inform Assoc. 2006;13:138–47.
26. Kirsch I, Jungeblut A, Jenkins L, et al. Adult literacy in America: a first look at the findings of the National Adult Literacy Survey. Washington, DC: US Department of Education; 1993.

27. Agency for Healthcare Research and Quality (AHRQ). Pharmacy Health Literacy Center. Available at http://www.ahrq.gov/pharmhealthlit/. Last accessed 12 Nov 2012.
28. The Joint Commission Sentinel Event Alert Issue 11: high-alert medications and patient safety. November 19, 1999. Available at http://www.jointcommission.org/assets/1/18/SEA_11.pdf. Last accessed 1 Dec 2012.
29. Cohen MR, Kilo CM. High-alert medications: safeguarding against errors. In: Cohen MR, editor. Medication errors. Washington, DC: American Pharmaceutical Association; 1999.
30. Cornish PL, Knowles SR, Marchesano R, et al. Unintended medication discrepancies at the time of hospital admission. Arch Intern Med. 2005;165:424–9.
31. Barnsteiner J. Critical opportunities for patient safety and quality improvement: Medication reconciliation. Patient safety and quality: an evidence-based handbook for nurses. Agency for Healthcare Research and Quality. U.S. Department of Health and human services, AHRQ publication No. 08–0043, Section V-38. April 2008.
32. Leape LL, Cullen DJ, Clapp MD, et al. Pharmacist participation on physician rounds and adverse drug events in the intensive care unit. JAMA. 1999;282:267–70.
33. Kucukarslan SN, Peters M, Mlynarek M, et al. Pharmacists on rounding teams reduce preventable adverse drug events in hospital general medicine units. Arch Intern Med. 2003;163:2014–8.
34. ASHP (American Society of Health-System Pharmacists) Guidelines on Preventing Medication Errors in Hospitals, pp. 208–16. Available at http://www.ashp.org/DocLibrary/BestPractices/MedMisGdlHosp.aspx. Last accessed 12 Nov 2012.
35. Marx D. Patient Safety and the "Just Culture": a primer for health care executives. New York, NY: Columbia University; 2001.

# Chapter 8
# Medication Reconciliation Error

Robert J. Weber and Susan Moffatt-Bruce

*"Error is the discipline through which we advance."*

William Ellery Channing

## Introduction

A substantial body of evidence from international literature points to the potential risks to patient safety posed by medication errors and the resulting preventable adverse drug events. In the USA, medication errors are estimated to harm at least 1.5 million patients per year, with about 400,000 preventable adverse events [1]. In Australian hospitals about 1 % of all patients suffer an adverse event as a result of a medication error [2]. In the UK, of 1,000 consecutive claims reported to the Medical Protection Society from 1 July 1996, 193 were associated with prescribing medications [3]. Medication errors are also costly—to healthcare systems, to patients and their families, and to clinicians [4, 5]. Prevention of medication errors has therefore become a high priority worldwide.

Literature suggests many of the medication errors occur during care transition points such as hospital admission, transfer, and discharge due to multiple changes in medication regimens and inadequate communication among physicians, nurses, and pharmacists [4]. In a systematic review, 54–67 % of all admitted patients were found to have at least one discrepancy between home medications and the

R.J. Weber, Pharm.D., M.S., (✉)
The Ohio State University Wexner Medical Center,
Room 368 Doan Hall, 410 West 10th Avenue, Columbus, OH 43210, USA
e-mail: Robert.Weber@osumc.edu

S. Moffatt-Bruce, M.D., Ph.D.
Quality and Operations, Wexner Medical Center at The Ohio State University,
130 Doan Hall, 410 West 10th Avenue, Columbus, OH 43210, USA
e-mail: Susan.Moffatt-Bruce@osumc.edu

A. Agrawal (ed.), *Patient Safety: A Case-Based Comprehensive Guide*,
DOI 10.1007/978-1-4614-7419-7_8, © Springer Science+Business Media New York 2014

medication history obtained by admitting clinicians, and that in 27–59 % of cases; such discrepancies have the potential to cause harm [6–8].

In response to these mounting safety concerns, the Joint Commission (TJC), in 2006, mandated that all accredited facilities must "accurately and completely reconcile medications across the continuum of care." After careful consideration, TJC has continued to maintain medication reconciliation as a National Patient Safety Goal as of 2012 (http://www.jointcommission.org/assets/1/18/NPSG_Chapter_Jan2013_HAP. pdf). The Institute of Healthcare Improvement (IHI) has also incorporated performing medication reconciliation as a part of its 100,000 Lives Campaign. Another impetus for medication reconciliation is the growing interest in innovative models of care delivery, such as accountable care organization (ACO) and patient-centered medical home (PCMH) where patients have a direct relationship with a provider who coordinates a cooperative team of healthcare professionals, takes collective responsibility for the care provided to the patient, and arranges for appropriate care with other qualified providers as needed. One key element of PCMH accreditation by the National Committee for Quality Assurance is the ability to coordinate care via managing information, such as medication lists, efficiently across providers and settings, preferably using the current health information technology such as EHR. Clearly an effective medication reconciliation process would be vital to achieve a successful implementation of PCMH.

Medication reconciliation is one of the most important safety practices to reduce medication errors during care transitions and can be defined as "comparing a patient's current medication orders to all of the medications that the patient had been taking before the transition," e.g., comparing and reconciling admission medication orders with the home medications. To ensure patient safety, it is important to recognize that the broad definition of "medications" includes prescription drugs as well as "over-the-counter" drugs and herbals, etc., because these may have important interactions with each other. For the purpose of medication reconciliation, medications are defined by the Joint Commission as "any prescription medications, sample medications, herbal remedies, vitamins, nutraceuticals, vaccines, or over-the-counter drugs; diagnostic and contrast agents used on or administered to persons to diagnose, treat, or prevent disease or other abnormal conditions; radioactive medications, respiratory therapy treatments, parenteral nutrition, blood derivatives, and intravenous solutions (plain, with electrolytes and/or drugs); and any product designated by the Food and Drug Administration (FDA) as a drug" [9].

Recent experience suggests that inadequate reconciliation accounts for 46 % of all medication errors and up to 20 % of all adverse drug events (ADEs) among hospitalized patients [10]. Further, medication errors can be reduced by more than 76 % when medication reconciliation is implemented at admission, transfer, and discharge [11].

There are five essential steps to medication reconciliation: determining a current list of medications; developing a listing of medications to be prescribed; comparing the two lists; making clinical decisions based on the two lists; and finalizing and communicating the list of medications to the patient and other clinicians. Table 8.1 lists the steps in the medication reconciliation process in a clinical scenario where a patient is admitted from home for surgery, goes through several steps in care transitions, and is discharged home.

**Table 8.1** Example of care transitions and steps in the medication reconciliation process

| Care transition | Medication reconciliation process step | Example |
|---|---|---|
| Hospital admission | Determine a current list of home meds prior to admission | Interview with family and call to patient's pharmacy show she also takes hydrochlorothiazide 25 mg PO once daily, Senna 2 tab PO at bedtime |
| | Compare and reconcile home meds and admission orders | Admission medication list holds ASA, Alendronate, and Atorvastatin. Other home meds are continued. New meds: antibiotic along with thrombosis prophylaxis |
| Transfer from one level of care to another | Compare and reconcile meds on the surgical floor and the step down unit | Antibiotic discontinued, thrombosis prophylaxis still continued, hydrochlorothiazide increased to 50 mg PO once daily, Lisinopril changed to atenolol 25 mg PO once daily. Other meds as before |
| Hospital discharge | Transition of care to home discharge—reconcile of hospital medication with home medications | Thrombosis prophylaxis discontinued, ASA, Atorvastatin, Alendronate continued on discharge. Home medication list as follows: oral antibiotic, ASA 325 mg PO once daily, Atorvastatin 40 mg PO once daily, hydrochlorothiazide 50 mg PO once daily, atenolol 25 mg PO once daily, Alendronate 70 mg PO weekly, multivitamin PO once daily, Senna 2 tab PO at bedtime |
| Outpatient follow-up with the same or different provider | Communicate of medication list to patient and providers | Medication list is reviewed with the patient and their family, along with finalized listing stored in the patient's medical record, and the patient's home pharmacy. Specific attention is paid to the increased dose of hydrochlorothiazide, discontinuation of Lisinopril, addition of atenolol. Side effects of atenolol reviewed with patient. Community pharmacist called with new medication list |

Case: 74-year-old community-dwelling female admitted for surgery; per patient history, current home medications on admission include ASA 325 mg PO once daily, Atorvastatin 40 mg PO once daily, Lisinopril 10 mg PO once daily, Alendronate 70 mg PO weekly, multivitamin PO once daily

The goal of this chapter is to provide a case-based approach to understanding the root cause of and solutions to preventing medication reconciliation errors. In addition, key "take home" points will be presented that will provide the reader with a mental "toolkit" to prevent medication reconciliation errors.

The two cases presented in this chapter represent hypothetical cases that may occur in any hospital or ambulatory setting. Case 1 occurs in a hospital that utilizes an electronic health record (EHR) with computerized medication reconciliation; Case 2 occurs in a hospital that is partially computerized and does not have computerized physician order entry (CPOE). The summary of the root cause analyses and the solutions to prevent future error are based on "real life" discussion of a typical sentinel event root cause analysis (RCA) group formed as part of a hospital's quality improvement process. Throughout this chapter, suggestions for improving safety in the medication reconciliation process are provided that can be applied to any healthcare setting.

## Case Studies

### Case 1: Digoxin Toxicity Due to Inadequate Discharge Medication Reconciliation

#### Clinical Summary

*M.K. is an 85-year-old female with a past history of congestive heart failure (HF), atrial fibrillation, asthma, and chronic renal failure who is admitted (ADMISSION 1) with acute exacerbation of HF, fatigue, and loss of appetite. M.K.'s medications prior to admission include digoxin 0.25 mg once a day; metoprolol XL 100 mg once a day, ramipril 2.5 mg PO once a day; multivitamin 1 tab once a day, tylenol 325 mg PO four times a day as needed for joint pain, and albuterol inhaler two puffs every 6 h as needed for shortness of breath. A laboratory value of significance on admission is a serum digoxin concentration of 2.4 ng/ml (range 0.9–2.4 ng/ml). M.K's digoxin is held, and a decision is made by the medical team not to continue digoxin in the future due to concern for digitalis toxicity. The patient is successfully treated with diuresis (furosemide, metolazone) and is prepared for discharge home where her daughter will administer her medications. Three days after hospital discharge the patient is readmitted (ADMISSION 2) with the family stating "my mother is seeing things." A STAT digoxin level measures 3.4 ng/ml and the patient is treated with digoxin immune fab. On review of the past admission (ADMISSION 1) by the attending physician and discussion with M.K.'s family, it is found that the digoxin was inadvertently continued with the home medication regimen, causing digitalis toxicity and ADMISSION 2.*

*Figure 8.1 graphically depicts a timeline for this case study. As illustrated, during the patient's hospital stay, there were several occasions where digoxin on the discharge medication list could have been reviewed, verified, and checked for accuracy.*

#### Root Cause Analysis

The leading question for the RCA team was: why was digoxin continued at home in a patient with suspected digoxin toxicity? Fundamentally, this was a failure of the medication reconciliation process, especially at discharge and the RCA revealed the following contributing factors (1) suspected digoxin toxicity was not documented as a problem in the EHR during ADMISSION 1; (2) digoxin was "held," and not

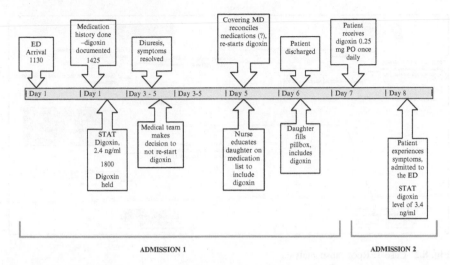

**Fig. 8.1** Timeline for Case 1: Digoxin toxicity due to inadequate discharge medication reconciliation

discontinued during admission medication order entry; (3) decision to discontinue digoxin during ADMISSION 1 was not documented in the daily progress notes; (4) discharge planning discussion on ADMISSION 1 did not include medications, and there was no discussion about discontinuing digoxin; (5) family or patient were not made aware of high normal digoxin level on ADMISSION 1; (6) physician not directly related to the case was covering on a weekend when the decision was made to discharge M.K from ADMISSION 1. Therefore, the discharging physician, who was not completely familiar with the patient's hospital course and medical history, completes the computerized medication reconciliation on ADMISSION 1 and does not notice digoxin was held; (7) nurse caring for MK provided the family with computerized discharge instruction sheet for ADMISSION 1 and did not notice that digoxin is continued.

As a result, the patient's discharge medication list contained digoxin 0.25 mg once a day. M.K.'s family arranges medication at home according to the discharge instructions from ADMISSION 1 and resumes MK.'s digoxin.

Clearly, the fundamental failure in this patient involved inadequate medication reconciliation at various stages of transition and a lack of communication among various caregivers. Multiple healthcare professionals were managing the transitions of care for this patient and no one had the comprehensive "big picture" of the patient's problems on ADMISSION 1. While the patient's main problem was exacerbation of CHF, an important clinical problem was a high-normal digoxin serum concentration. The significance of digoxin level was downplayed, despite the fact that the medical team intended to discontinue the digoxin. The documentation of digoxin discontinuation was also overlooked in the EHR. During the medication

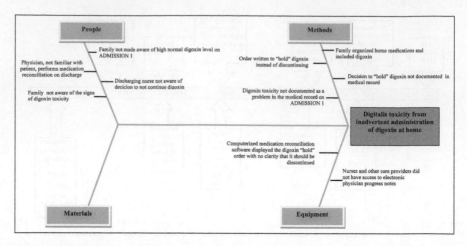

**Fig. 8.2** Case 1: Root cause analysis

reconciliation process, a covering medical resident, simply ordered the admission list of medications and added metolazone. This mistake occurred since the physician may not have properly understood or incorrectly used the functionality of computerized medication reconciliation in the EHR.

This case also represents breakdown in communication between the discharging nurse and the patient's family. There was no discussion with the patient's family on admission regarding concerns with digoxin; as a result the patient's family was not aware of any problems when M.K.'s daughter restarted digoxin. Nurses caring for the patient did not notice the digoxin had not been restarted, indicating a breakdown in communication on the daily care plan. There was no communication with the patient that the digoxin was a concern. The patient was capable of understanding this information and should have been warned of the potential for digoxin toxicity.

Figure 8.2 represents the various process breakdowns that precipitated the medication reconciliation error.

## Steps for Error Prevention

The most significant prevention step involves improving communication among caregivers and with patients and family so that everyone is on the "same page" in terms of the patient's correct medication list. Additionally, improving the design and user interface of the EHR would also help. For example, the digoxin was continued primarily because the order was "held" in the computer system versus being discontinued. The system design improvement may consist of a "forcing function" upon discharge so the discharging physician must make a deliberate decision to discontinue or continue a medication. Additionally, an EHR must have interoperability such that the same medication information is available to all caregivers, and ideally a copy of the medication list is "exported" to the patient's personal health record for access at home.

## Case 2: Anticoagulant Omitted Upon Transfer to a Rehabilitation Facility Leading to PE

### Clinical Summary

*B.A., an 83-year-old woman, has undergone hip fracture surgery and is ordered "fondaparinux 2.5 mg subcutaneous once daily" postoperatively. Preprinted standing orders for postoperative hip fracture treatment are not available on the nursing unit when B.A returns to the floor, and the fondaparinux was written as an individual order along with other postoperative medications. B.A.'s postoperative course is uneventful, and she is transferred to a rehabilitation facility on postoperative day 3. On postoperative day 7 (day 4 at the rehabilitation center) she complains of shortness of breath, chills, sweating, malaise, and rapid heart rate, along with right calf swelling, redness, and pain. She is transferred to the hospital and the emergency room physician discovers that fondaparinux was not continued on the transfer to the rehabilitation facility. B.A. is admitted for a possible deep vein thrombosis (DVT) / pulmonary embolism (PE) from inadequate anticoagulation prophylaxis. B.A. is placed on therapeutic anticoagulation (intravenous heparin 800 units/h), venous Doppler studies prove positive for DVT, and a nuclear lung scan to detect a PE is not conclusive. After a 10-day hospital stay that is complicated by a fall, pain control issues, and difficulty in achieving a therapeutic warfarin dose, B.A. recovers fully and is transferred back to an assisted living facility.*

*Figure 8.3 illustrates the timeline for this event. The absence of anticoagulation for 4 days and immobility placed B.A. at risk for a postoperative DVT.*

### Root Cause Analysis

The primary RCA question in this case is: why was fondaparinux omitted from the transfer medication list? The RCA revealed the following contributory factors for this error of omission from the medication list (1) specific directions for fondaparinux were not included on the original postoperative order (e.g., "continue for 7 days for prophylaxis"); (2) the standard order set for hip fracture repair was not available due to supply problems at the hospital's printer and therefore not used; (3) the admission medication list was used to create the discharge/transfer medication list; as a result fondaparinux was omitted from B.A's discharge medication list; (4) the rehabilitation facility did not conduct a thorough medication "intake" and screening for DVT prophylaxis in B.A.; and (5) DVT prophylaxis was missed by the admitting physician as well as the pharmacist filling prescription orders in the rehabilitation center.

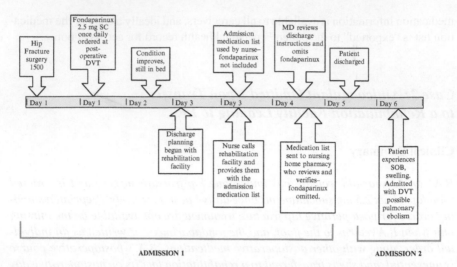

**Fig. 8.3** Timeline for Case 2: Anticoagulant omitted upon transfer to a rehabilitation facility leading to PE

Figure 8.4 represents the variations in practice that caused the error in case 2.

**Steps for Error Prevention**

A major initiative that may possibly prevent this error from occurring in the future is the computerization of order entry. In this case, a computerized standing order for postoperative hip fracture medications would have included the duration of the fondaparinux therapy, and this order would have been included on the computerized medication reconciliation list.

# Discussion

The case studies in the chapter clearly illustrate the importance of performing consistent and accurate medication reconciliation in various settings to ensure patient safety. A key to error-free medication reconciliation is obtaining an accurate history of prescription medications as well as over-the-counter products such as vitamins, nutraceuticals, and herbal products. A detailed medication history produces an accurate home medication list; this accuracy carries through a patient's hospital stay or ambulatory course and results in an accurate medication list on any transition of care. Gathering information for a thorough medication history may be time consuming, involving phone calls to pharmacies, and other providers. Prescription

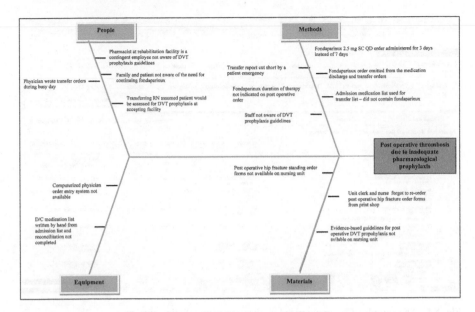

**Fig. 8.4** Case 2: Root cause analysis

claims data, sometimes interfaced with an EHR, can be used to determine home medication but adherence should be interpreted cautiously [12]. An alternative to physicians conducting the medication history includes nurses, pharmacists, medical students, and pharmacy students obtaining medication histories. Froedert Hospital in Milwaukee, Wisconsin used pharmacists to conduct medication histories and perform medication reconciliation with success [13]; an American academic medical center used nurses with the specific function of managing medications at the transition of care with success in preventing reconciliation errors [14].

Common causes of errors in the home medication list include (1) patients failing to bring the prescription bottles to the hospital or doctor's visit; (2) limited access to vital information (e.g., labs test results.) in the care provider's office or other care area (e.g., the emergency room) to adequately interpret the home medication list; (3) untrained or inexperienced personnel documenting the home medication list in a hospital or physician's office; and (4) unclear labeling of home medication bottles [15].

We suggest the following key considerations to clinicians to resolve and reconcile medications on a patient's home or hospital drug list. Does this medication duplicate any medications from the home medication list? Will prescribing this medication confuse the patient? Is this medication prescribed resulting in too many medications for the patient to accurately track and take? Poly-pharmacy, or a high number of medications for a patient, is a well-documented contributing factor to hospital readmissions [16]. The focus of prescribing medications during the hospital stay should be to simplify the discharge medication list to minimize medication

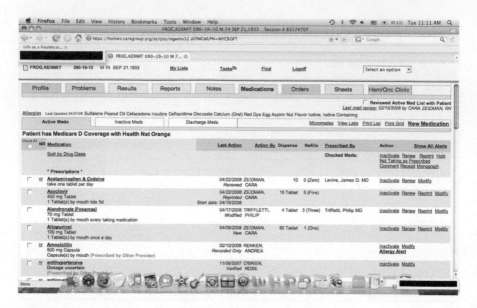

**Fig. 8.5** An example of a computerized medication reconciliation system

errors in compliance, adherence, and self-administration. Similarly in the ambulatory setting, the focus of prescribing medications is to keep the list as simple as possible and maintain adherence and treatment goals. However, simplifying the medication list offers a unique challenge to clinicians, since a patient's condition may be worsening, resulting in various combinations of medications and changing medication dosages and frequencies.

The two cases in this chapter demonstrate discrepancies in the discharge medication list. Proper discharge medication reconciliation requires that the physician, in consultation with other clinical team members, the patient, and their family, makes the decision to modify, continue, or discontinue hospital medications to generate the discharge medication list. Using an EHR's functionality, medication reconciliation can be completed with a lesser risk of error. Figure 8.5 shows an example of an electronic medication reconciliation form [17]. The prescriber can choose the action (inactivate, renew, or modify) for each medication to generate the final medication list. However, the prescriber may mistakenly choose an action or not know what each action means. Using Fig. 8.5 again as an example—does the term "inactivate" mean discontinue the medication, hold the medication, or neither? Also, institutions, clinics, and physicians' offices must have clear guidelines as to which level of provider (e.g., pharmacists, nurse, medical assistant, physician) can access the system to perform reconciliation.

Patients' proper understanding of their medication regimen is one of the most important factors in preventing medication errors [18]. This step may be more difficult when dealing with a vulnerable population (elderly, developmentally delayed, differing levels of literacy) and will require using resources to increase understanding (e.g., pictures, patient-friendly terminology to describe the instructions). The final medication list should be shared with patients, their families, and other clinicians involved in the care. For example, in the digoxin administration error, while the discussion of the patient's medication regimen with the daughter took place, leading questions should have been asked to include: Does this medication list look correct to you? Do you know why each medication is being prescribed? Do you have an adequate supply of each of these medications? Has your mother had any problems with these medications in the past? Discussing any of these questions may have drawn suspicion to the continuation of the digoxin.

With the growing adoption of EHRs by various healthcare organizations, electronic medication reconciliation systems are now commonplace. A study evaluating the impact of an electronic medication reconciliation system in an acute inpatient hospital found a substantial reduction in the unintended discrepancies between home medications and admission order [19]. In another study evaluating a computerized medication reconciliation system, over 60 % of those physicians surveyed felt that medication reconciliation was important, and the computerized approach to reconciliation promoted efficiency [20]. Researchers found that while compliance with medication reconciliation was not necessarily related to the functionality, or its ease of use, or availability, it was closely correlated to the prescriber's historical compliance to medication reconciliation using a paper system. This point brings out the importance of culture and its influence in preventing medication reconciliation errors. Clinical and administrative leaders must strive to build a culture of safety where medication reconciliation is considered a key process to promote patient safety and caregivers are held accountable for failing to adhere to this safety practice.

## Key Lessons Learned

- Develop an interdisciplinary approach to obtaining a patient's medication history by assigning specific responsibilities to gathering and documenting medication information.
- Develop a policy and procedure for systematic review and use of a computerized (or manual) system for medication reconciliation. Special attention should be paid to approving the types of healthcare personnel allowed to conduct medication reconciliation and assign key responsibilities to complete various tasks in the medication reconciliation process.
- Design communication notes that are shared among all caregivers. In an electronic system improve interoperability of data; in a paper system place information in a specific part of the chart.
- In computerized medication reconciliation, design the system to minimize "free text" data entry of medications to reduce errors.
- Involve the patient and their family in the medication reconciliation process by reviewing carefully the home medication list and assessing patient understanding with special attention to language preference and health literacy.

- Other practical points about managing patient's medications from home include (1) verifying medications by a pharmacist; (2) focus on high-risk patients (elderly, patients with 10 or more medications) as a priority; (3) using electronic resources to aid in drug identification. Two examples of pill identification resources can be found at http://www.rxlist.com/pill-identification-tool/article. htm and http://www.drugs.com/imprints.php.
- Implement leadership strategies to force accountability for medication reconciliation in patient care.

# References

1. Aspden P, Institute of Medicine (US). Committee on identifying and preventing medication errors. Preventing medication errors. Washington, DC: National Academies; 2006.
2. Runciman W, Roughhead E, Semple S, Adams R. Adverse drug events and medication errors in Australia. Int J Qual Healthcare. 2003;15(Suppl):i49–59.
3. Chief Pharmaceutical Officer. Building a safer NHS for patients. Improving medication safety. London: Department of Health, 2004. Available at http://webarchive.nationalarchives.gov. uk/20130107105354/http://www.dh.gov.uk/en/Publicationsandstatistics/Publications/ PublicationsPolicyAndGuidance/DH_4071443. Last accessed 13 Jul 2013.
4. Bates DW, Spell N, Cullen DJ, Burdick E, Laird N, Petersen LA, et al. The costs of adverse drug events in hospitalized patients. Adverse Drug Events Prevention Study Group. JAMA. 1997;277:307–11.
5. Vincent C, Neale G, Woloshynowych M. Adverse events in British hospitals: preliminary retrospective record review. BMJ. 2001;322:517–9.
6. Cornish PL, Knowles SR, Marchesano R, Tam V, Shadowitz S, Juurlink DN, et al. Unintended medication discrepancies at the time of hospital admission. Arch Intern Med. 2005;165(4):424–9.
7. Gleason KM, Groszek JM, Sullivan C, Rooney D, Barnard C, Noskin GA. Reconciliation of discrepancies in medication histories and admission orders of newly hospitalized patients. Am J Health Syst Pharm. 2004;61(16):1689–95.
8. Akwagyiram I, Goodyer LI, Harding L, Khakoo S, Millington H. Drug history taking and the identification of drug related problems in an accident and emergency department. J Accid Emerg Med. 1996;13(3):166–8.
9. 2010 Hospital Accreditation Standards. Available at http://www.jointcommission.org/standards_ information/standards.aspx, Accessed 15 May 2013.
10. Barnsteiner JH. Medication reconciliation: transfer of medication information across settings-keeping it free from error. J Infus Nurs. 2005;28(2 Suppl):31–6.
11. Rozich JD, Howard RJ, Justeson JM, Macken PD, Lindsay ME, Resar RK. Standardization as a mechanism to improve safety in health care. Jt Comm J Qual Saf. 2004;30(1):5–14.
12. Cutler DM, Everett W. Thinking outside the pillbox—medication adherence as a priority for healthcare reform. N Engl J Med. 2010;362:1553–5.
13. Murphy EM, Oxencis CJ, Klauck JA, Meyer DA, Zimmerman JM. Medication reconciliation at an academic medical center: a comprehensive program from admission to discharge. Am J Health Syst Pharm. 2009;66:2126–31.
14. Vira T, Colquhoun M, Etchells E. Reconcilable differences: correcting medication errors at hospital admission and discharge. Qual Saf Health Care. 2006;15:122–6.
15. Weber RJ. Medication reconciliation pitfalls. Available at http://webmm.ahrq.gov/case. aspx?caseID=213. Accessed 19 Feb 2012.

16. Yvonne K, Fatimah BMK, Shu CL. Drug-related problems in hospitalized patients on poly-pharmacy: the influence of age and gender. Ther Clin Risk Manag. 2005;1:39–48.
17. Life as a healthcare CIO. Available at http://geekdoctor.blogspot.com/. Accessed 13 Jul 2013.
18. Villanyi D, Fok M, Wong RY. Medication reconciliation: identifying medication discrepanices in acutely ill older hospitalized patients. Am J Geriatr Pharmacother. 2011;9:339–44.
19. Agrawal A, Wu WY. Reducing medication errors and improving systems reliability using an electronic medication reconciliation system. Jt Comm J Qual Patient Saf. 2009;35:106–14.
20. Turchin A, Hamann C, Schnipper JL, Graydon-Baker E, Millar SG, McCarthy PC, et al. Evaluation of an inpatient computerized medication reconciliation system. J Am Med Inform Assoc. 2008;15:449–52.

# Chapter 9
# Retained Surgical Items

Verna C. Gibbs

> *"Any man may err! Only a fool persists in his error."*
>
> Marcus Tullius Cicero

## Introduction

A retained surgical item (RSI) refers to surgical materiel used during a procedure that is *inadvertently* left in any part of a patient. RSI rather than retained foreign object (RFO) or retained foreign body (RFB) is the preferred term in the current surgical safety vernacular because RFO and RFB may be used to refer to swallowed or inserted objects, irretrievable shrapnel, bullets, and broken miscellaneous parts of toys and weapons [1]. The presence of these objects may require operative intervention and they often can't be removed and therefore are retained but these are not the instruments and tools that healthcare providers have used to heal patients. It is important to realize that an RSI is a surgical patient safety problem.

There are four classes of surgical items: cotton soft goods (sponges and towels), small miscellaneous items (SMIs), sharps, and instruments. In most reports the most frequently retained items have been cotton soft goods, particularly sponges [2]. The RSIs are usually discovered after the development of clinical symptoms such as

V.C. Gibbs, M.D. (✉)
A National Surgical Patient Safety Project to Prevent Retained Surgical Items,
NoThing Left Behind, 270 Collingwood Street, San Francisco, CA 94114, USA

San Francisco VA Medical Center, San Francisco, CA, USA. University of California
San Francisco Medical Center, San Francisco, CA, USA
e-mail: drgibbs@nothingleftbehind.org

A. Agrawal (ed.), *Patient Safety: A Case-Based Comprehensive Guide*,
DOI 10.1007/978-1-4614-7419-7_9, © Springer Science+Business Media New York 2014

pain or the presence of a mass. X-ray examination, especially computerized tomography (CT) scans have been most informative in establishing a diagnosis. The true incidence of RSI is unknown but public data systems from states such as California, Pennsylvania, and New York and regulatory agencies such as The Joint Commission continue to report yearly cases [3]. If this problem is to become a "never event" it is not so important to know how many cases there have been, it is only important to know that the number of cases is still greater than zero.

One obstacle in case reporting has been the difficulty in agreeing on a simple and unequivocal definition of when a surgical item is considered to be retained. The National Quality Forum has defined a list of serious reportable events (SRE) which they term "never events" [4]. Unintended retention of a foreign object in a patient *after surgery* or other invasive procedure is an SRE. The area of conflict in this definition is "after surgery." Many have opined that surgery is over with the closure of the surgical wound which is akin to having a "wheels down" interpretation of when an airplane has landed [5]. Many would argue that an operation or procedure isn't over when the incision is closed. The differences in opinion about what "after surgery" means has led to disagreement and probably over-reporting of RSI events. In 2011, the NQF reviewed the definitions of all the SREs and reexamined the question of when it is "after surgery." The new definition states that surgery ends after all incisions have been closed in their entirety, devices have been removed, final surgical counts have concluded, and the patient has been taken from the operating or procedure room [4]. Just changing the definition will undoubtedly lead to a reduction in the number of reported cases. Other obstacles to case reporting include legal liability and medical and hospital staff reputation concerns.

Efforts to discern risk factors for retention based on patient characteristics such as patient size or operative characteristics such as the type of procedure or operative circumstance have been undertaken in the past [6, 7]. It turns out that there isn't any predictable relationship between the likelihood of retention and the type of item, the number of items used, the size of a wound or cavity, or the type of case or medical specialty [8]. Retention of surgical items has occurred in all types of procedures. Retained sponges have occurred when only ten sponges were used in an elective operation; yet hundreds of instruments can be used during a case and whole instrument retention remains very uncommon. Operating room (OR) policies that stipulate that a surgical item count only needs to be performed in cases where there is a risk of retention are difficult to enforce because it leaves open for judgment just when that risk of retention would exist. It is more insightful to look at OR personnel characteristics and the OR environmental conditions under which people work rather than patient characteristics. Changing the focus from the patient to the providers and environment has revealed failed OR practices and poor communication as the key elements that lead to patient injury and harm from RSIs. This is further evidence for the usefulness of thinking of an RSI as a surgical patient safety problem rather than just another perioperative complication.

RSIs occur because of problems in the OR practices and communication strategies of multiple OR stakeholders [9]. The OR practices are not just the counting processes of the nurses and surgical technologists. Surgeon performance of a wound sweep instead of a methodical wound exam, cursory reading of X-ray images by radiologists, and poor quality radiographic images taken by radiology technologists also contribute to errors which lead to RSI. Anesthesiologists may confound surgical item management by mixing their sponges and instruments with those that are counted and risk managers and administrators may be overly concerned with institutional protection to the exclusion of transparency needed for effective learning. An RSI is reflective of system problems and therefore require systemic solutions. Admonishing circulating nurses to just "count" harder will not address the complex nature of this problem.

## Case Studies

To further illustrate the root causes and preventive strategies for RSI, we describe three retained sponge cases presenting three different perspectives. The fact that these are all cardiothoracic cases is of no particular significance because all types of surgical cases have had retained sponges. It is not important what type of case, where the sponges were lost, what kind of sponge was involved or when the sponge was found as much as it is important to try to understand why and how the sponges were retained and where the failures in the OR practices and communication strategies occurred. It has been difficult to see that there is any pattern or obvious corrective action to take when looking at an individual retained sponge case and since most hospitals have very few of these events, root cause analyses and focused reviews have been unable to uncover real systemic improvements for prevention. We present an alternative analysis.

We have characterized all cases of retention as belonging to one of three essential type of case based upon the status of the surgical counts as recorded at the end of the operation. The three types of cases are no count retention case (NCRC), correct count retention case (CCRC), and an incorrect count retention case (ICRC) [10]. We use the nomenclature of surgical counts but it is equally useful to use this term—surgical count—as a surrogate for some form of sponge management without being specific as to the actual action of counting sponges.

## *Clinical Summary*

### Case 1: No Count Retention Case

*The patient had third-degree heart block and was undergoing placement of a pacemaker in the cardiac catheterization suite which was adjacent to the main operating rooms. All monitors were placed correctly, the patient had oxygen on, and a site on his right chest was prepped and sterilely draped. There were ten 4 × 4 raytex sponges (a neologism for a surgical sponge that contain a radiopaque marker woven into the gauze interstices of the sponge to enable X-ray detection) on the procedure table. An incision was made, a subcutaneous pocket was fashioned, and the pacemaker inserted. There was commotion in the room next door and the physician looked up and could see that the patient was in trouble. The pacemaker patient was completely stable so the physician took a raytex and stuffed it in the newly created pocket and went to the suite next door to assist. About 15 min transpired before the physician came back to complete the pacemaker insertion. The patient was still completely stable. The skin incision was closed and a fluoroscopic image was taken to check the pacemaker position. This looked good and the pacemaker was functioning well. A sterile dressing 4×4 sponge was put over the wound and the patient was discharged home.*

*Over the next 2 months the patient developed redness, tenderness, and edema around and over the area where the pacemaker was. There were no electrocardiographic abnormalities and fluoroscopic imaging of the site showed an intact pacemaker. The patient received a course of antibiotics which initially helped the redness but recurred once the antibiotics were stopped. The physicians decided that the patient had developed an allergic reaction to the metal of the pacemaker and decided to remove it and replace it with another brand. Upon opening the incision and removing the pacemaker, an infected raytex sponge was encountered. The unsuspecting physician was amazed to find the retained sponge because it had not been seen on any of the fluoroscopic images but the patient had not had a formal chest X-ray series (AP and lateral). The sponge was removed and the infection in the incision was treated. The pacemaker was placed on the other side and the patient subsequently did well.*

## Case 2: Correct Count Retention Case

*The patient had severe coronary artery disease and aortic stenosis and underwent a coronary artery bypass graft (CABG) and aortic valve replacement. The cardiac surgery team had worked together before and all were experienced clinicians. It was a long case that started in the morning and went through lunch into the early afternoon. The scrub technologist and the circulating nurse had both been given a morning break and lunch relief by two different relief nurses. Lap pads and 4×4 raytex sponges were used during the case. The raytex were only used as the grafts were being sewn in as a surgeon preference because of a belief that the gauze interstices of the raytex absorbed blood better. The CABG went well and there were no untoward problems. During the aortic valve replacement there was some bleeding, but this was eventually controlled. Lap pads were used during this portion of the case. The patient came off bypass well and there were no problems. As the sternum was closed the nurse informed the surgeon that the closing sponge counts were correct. The chest was then completely closed and the operation concluded. The nurses told the surgeon the final counts were correct and the patient was taken to the ICU. On the first postoperative day after the morning ICU chest X-ray had been taken, a radiologist called the surgeon to inform him that there were radiopaque markers consistent with a raytex sponge in the patient's left chest (Fig. 9.1). The surgeon was completely surprised because he had been told that the final sponge counts had been called correct. The patient was taken back to the OR for removal of the sponge and subsequently did well.*

**Fig. 9.1** Chest X-ray
showing a retained raytex
4×4 sponge

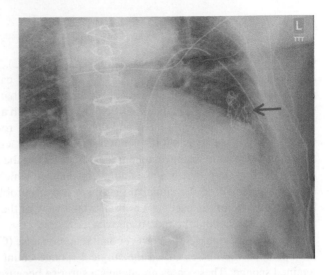

## Case 3: Incorrect Count Retention Case

*The patient had repair of a thoracic aneurysm through a left thoracoabdominal incision. This was a long operation that involved a large volume blood loss and multiple changes of nursing staff giving breaks and lunch relief to each other. At the end of the operation as the abdominal wound was being closed, the nurses informed the surgeon that there was a missing lap pad. The surgeon looked in the abdomen and explored it and said he didn't see anything and continued to close the wound. The nurses explained to the surgeon that it was hospital policy that an intraoperative X-ray had to be obtained when there was an incorrect count so the surgeon agreed and a radiology technologist came to take a film. The surgeon told the technologist they just needed an X-ray of the chest because he had explored the abdomen and there wasn't a lap pad in the abdomen so the technologist just took an AP view of the chest. The image was sent back to the OR and the surgeon looked at it and told the staff he didn't see anything on the film. The nurses continued to look for the lap pad in the trash and receptacles in the room and entered in the OR record "miscount of lap pad, X-ray negative." The missing lap pad was never found. The patient went to the ICU and had a daily chest X-ray as was the usual practice for care after this type of operation. For the first 3 days, the morning chest X-ray was read by the same radiologist. There were no unusual findings reported. On the morning of the fourth day a different radiologist was assigned to read the morning ICU chest X-rays. On that morning the new radiologist called the surgeon to ask him if he knew about the lap pad in the patient's left chest and was wondering why it hadn't been removed. The surgeon shook his head in dismay and said "oh so that's where that missing lap pad was." The patient was taken back to the OR for removal of the lap pad and subsequently did well.*

## Analysis of Case Studies

Case 1 is an example of a no count retention case (NCRC). There was an RSI but no counts or methodology was employed by any surgical care personnel during the procedure to track, manage, or account for the sponges that were being used. These types of cases are common in non-OR environments such as cardiac catheterization labs, procedure rooms, and labor and delivery birthing rooms. Surgical items are being used in these areas but it has not been common practice or a matter of policy to have in place some management process for tracking the items to make sure none are left in the patient. The use of a fluoroscopic image which had only an AP projection falsely reassured the clinician that there was no problem with the pacemaker but the presence of the pacemaker obscured the view of the retained raytex because the sponge was behind the radiopaque device.

Case 2 is an example of a correct count retention case (CCRC), i.e., at the end of the operation the nurses called the surgical sponge count correct yet there was a retained sponge. These cases are always a surprise because everyone thinks things are just fine until the surgical item is discovered hours, days, months, or even years later inside the patient. The item is discovered either because the patient develops symptoms – usually pain related to an infection, or the presence of a mass or an X-ray has been obtained for some other reason which incidentally shows the presence of the surgical item. In CCRC the OR practices that have been used by the nurses and surgeons to track, manage, and account for the surgical item during the case have failed. The practices were employed, the nurses counted, the surgeons did a sweep but neither identified that an item was still in the patient and that the count was in fact wrong. While the counting was underway, no one identified an error and in retrospective analysis frequently no one can determine when the mistake in the counting practice occurred. Often they attribute the cause of the error to distractions or inexperience yet rarely look at the details of the counting practice itself. The surgeon may have performed a "sweep" around the wound but didn't look and feel with intention for surgical items in order to remove them. This is designated as a CCRC based on the count as recorded in the medical record; not that in a post hoc analysis the count was truly incorrect. CCRC demonstrate problems with OR practices.

Case 3 is an example of an incorrect count retention case (ICRC). At the end of the operation, in spite of everyone knowing there was an incorrect count and that an item was missing, no one was able to find it and the patient left the OR with the sponge still inside. All stakeholders acknowledged that the count was incorrect, yet no further actions were taken to find the sponge or prove that it was not still inside the patient. In these cases, the surgical item management practices were working because the team members correctly identified that something was missing, but then other elements failed. The radiology technologist took a poor quality X-ray and the surgeon lacked the knowledge to direct the technologist to obtain additional views and then incorrectly interpreted the film. There was no hospital requirement that radiologists, who are the content experts in radiographic interpretation, review in real-time OR films obtained when looking for missing surgical items (MSI). When the radiologist was looking at the post-procedure films, it wasn't known that there was a missing lap pad since there was no further communication from nursing to surgeons or surgeons to radiologists. Therefore, the radiopaque marker of the lap pad was attributed to other expected surgical material. It was not until a "new set of eyes" looked at the films and identified the persistent radiographic abnormality and questioned what it was. These ICRC usually are problems with communication and involve errors in the exchange of information and knowledge between multiple stakeholders.

It turns out from review of case series from around the country about 80 % of RSI cases are CCRC and 20 % are ICRC which is why nursing personnel so frequently are called to task to review the "counting" practices [10]. But the problems in these CCRC cases aren't that the staff didn't count, they have counted and in many cases they have counted many times, yet somewhere in the process of the counting, an error or errors have been made. Because they don't know with certainty when the error occurred, external cofounders are implicated as causal to the problem with the counting practice that led to the mistaken count. Most common explanations are that there was a distraction or noise or they were hurried or there were breaks and relief changes. Very few to no reports outline exactly what practice is being employed when performing a surgical count. That is, an exact process composed of individual steps that everyone follows that makes up the counting practice in that OR. This is one of the true roots of the problems with counting. There usually isn't one practice of counting but as many practices in place as there are people doing it. It often turns out at the end of the case that the surgical items have indeed been counted but they have not been accounted for. Similarly, surgeons often perform a wound "sweep" which just by the nature of the action may not uncover sponges packed behind pacemakers, stuck between loops of bowel, or lodged in parts of the chest. They do not have specified practices for the performance of a methodical wound examination that is done solely with the intent to find and remove surgical items that are not intended to remain in the patient [9]. It is not the failure of one surgical stakeholder that leads to an RSI but the concatenation of failed practices by multiple stakeholders.

In the 20 % of retention cases that are ICRC, as is illustrated in the case example, the initial practices that were employed by the nurses and surgeons to count and look for the items worked. The nurses told the surgeon they were missing a lap pad and the surgeon looked carefully in the abdomen and didn't find anything. The team then moved to bring in the secondary defender against RSI—the radiology team— and it was here that lack of knowledge and errors in communication set them up to fail. The radiology technologist took only an AP view of the chest rather than an AP and an oblique or a lateral view and the image that was obtained was read only by the surgeon rather than by a radiologist who is the true content expert in radiographic interpretation. The surgeon didn't do a manual exploration of the chest because he assumed the X-ray would provide the necessary information. The nurses never found the sponge and didn't move the missing sponge up the chain of command to notify the nurse manager or risk managers that there was a problem in this case and that the patient still remained at risk. After the patient left the OR, no actions were taken to further confirm that indeed the intraoperative X-rays had been complete (which they weren't) and whether the image was indeed truly "negative." The radiologist reading the postoperative films missed the retained lap primarily because the radiologist wasn't looking for one but also because "everything on the film" wasn't seen which is a radiologists' nightmare [11, 12]. If the radiologist had been told that a lap pad was missing in the case and was never found, it might have been discovered sooner. As it turned out, the lap pad was on all of the postoperative films but it took a "new set of eyes" to see that something was there that shouldn't be there. Usual remedies after a RSI case include policy changes and additional steps to perform in an already overburdened and variable process. Understanding aspects of human fallibility and putting into place stronger communication linkages are different approaches to solve this problem.

We can take this analysis to the development of action plans for systemic remediation. If a hospital has a NCRC or a CCRC, the problem is with the OR practices and all surgical personnel need to change their practices [13]. There are only two real choices here. Either improve the existing practice or get a whole new practice. If it's decided that improvement is the route that is going to be taken, the first step is to look at the practice that is being employed and break down the process steps that make up the practice. There are two primary ways to improve a process—decrease the number of steps in the process or increase the reliability of any individual steps.

## Strategies to Prevent RSIs

### Soft Goods (Sponges and Towels)

Examination of the practices of counting sponges through observational audits and focused reviews led to the development of the Sponge ACCOUNTing system (SAS) by the NoThing Left Behind® project. The SAS is a standardized manual sponge management system that is an improvement practice which simplifies and increases the reliability of the process of accounting for surgical sponges [14].

The SAS requires OR personnel to use a wall-mounted dry erase board to record the sponge counts and requires surgeons to perform a methodical wound exam at the closing count in every case. Nurses and surgical technologists must ensure that all sponges in a case are used only in multiples of ten and at the end of the case, all the sponges are placed in blue-backed hanging plastic sponge holders, each of which has ten pockets. There should be no empty pockets visible at the final count if all the sponges have been accounted for. There are safety practice rules for surgeons and nurses to follow which standardize the practice, reduce individual variation, and are expected to prevent CCRC (Table 9.1). Embodied in the SAS are also communication tools (wall mounted checklist) for nurses, surgeons, and radiology stakeholders to use at point of service so an ICRC can be prevented (Fig. 9.2). Table 9.2 describes the guidelines for planning optimal image quality for suspected RSIs.

**Table 9.1** Sponge ACCOUNTing actions

| Surgeons | Nurses |
|---|---|
| IN COUNT | IN COUNT |
| Only use X-ray detectable sponges | Work only in multiples of 10 |
| Don't cut or alter them | Discover the number of sponges in a pack |
| Avoid use of small sponges in large cavities | See, separate, and say |
| | Document count on dry erase board |
| CLOSING COUNT | CLOSING COUNT |
| Take a "Pause for the Gauze" | Take a "Pause for the Gauze" |
| Perform a Methodical Wound Exam (not just a sweep) before asking for closing suture | Remind surgeon to perform a Methodical Wound Exam |
| | Count sponges in field and in holders |
| Get the sponges out so the nurses can count them | Check back to surgeon the status of the count |
| FINAL COUNT | FINAL COUNT |
| Before leaving the OR look at the sponge holders and see that there are NO EMPTY POCKETS | ALL sponges (used and unused) MUST be in the sponge holders before the patient leaves the OR |
| Verification "Show Me" step | NO EMPTY POCKETS |
| Dictate actions in the operative report | Show the surgeon all the sponges have been ACCOUNTED for |
| | Document count in intra-operative record |

If it's decided that a whole new practice is needed for sponge management rather than an improvement like the SAS, then there are technological adjuncts that use 2D matrix computer-assisted technology which counts sponges or electronic article surveillance technology which can detect the presence of a sponge with a compatible radiofrequency (RF) tag, or radiofrequency identification (RFID) technology which can count and detect sponges that contain an RFID chip.

The computer-assisted technology consists of sponges that have two-dimensional matrix labels annealed to them and a handheld or table-mounted scanning device that can read the labels [15]. Each sponge has a unique identifier that enables the scanner to count different types of sponges. The sponges are counted maintaining "line of sight" for each sponge and the sponges must be removed from the patient and individually passed under the scanner. The scanner has no capacity to "read-through" the patient to detect the presence of a matrix-labeled sponge. In the event of a missing sponge, an X-ray is used to determine if it is in the patient.

The electronic article surveillance system consists of sponges that have a small passive RF tag sewn into a pocket on each sponge and a handheld wand or mat which contains the antennae and detection system [16]. The tag is detected when the handheld wand or mat is activated and a visual and audible signal is registered on a console that a sponge has been detected. The system does not distinguish between sponge types or number of sponges. The signal readout will be the same intensity if there are one or five sponges. In the event of a missing sponge, the mat can be activated to determine if the sponge is in the patient or the wand can be used to wand the patient or scan the trash to find the sponge. This system does not count sponges.

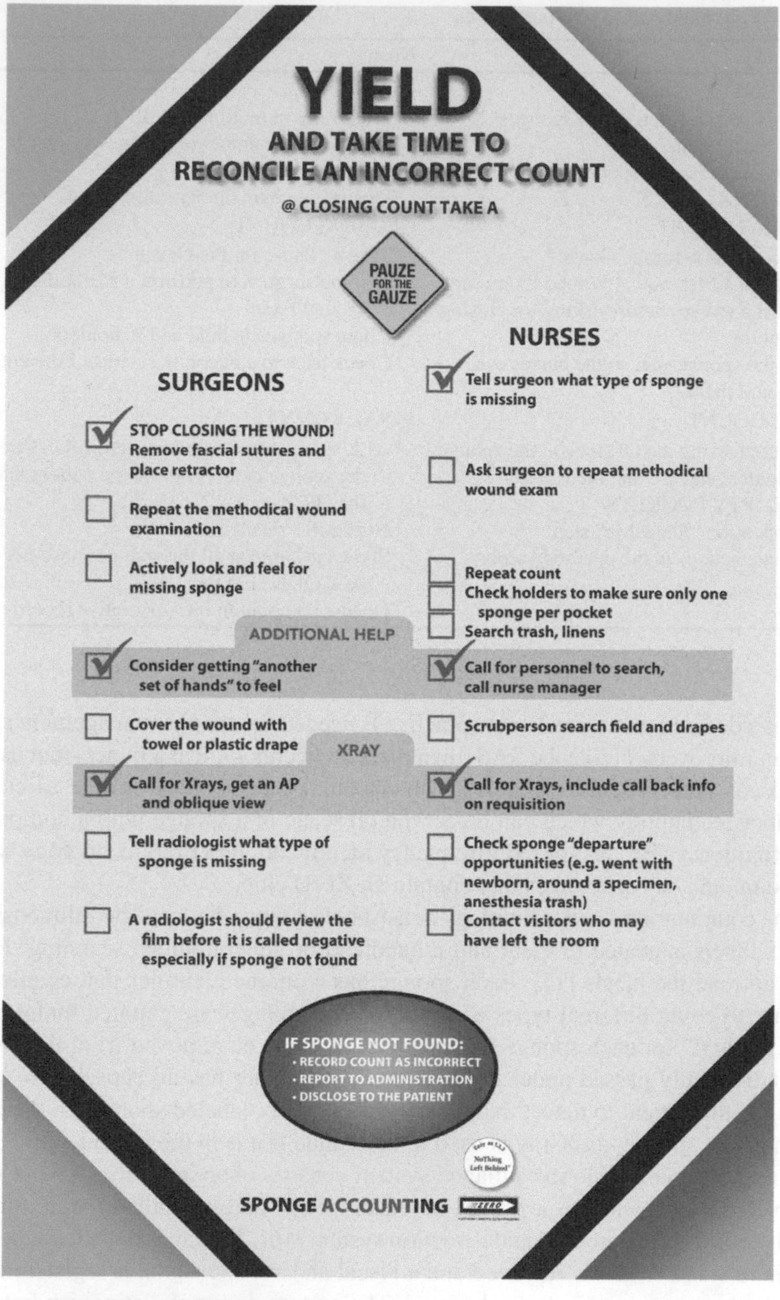


**Fig. 9.2** Multi-stakeholder incorrect count checklist

**Table 9.2** Guideline for obtaining X-rays for suspected retained surgical items (RSI)

| Exam | Views | Region of interest (ROI) | Comments |
|---|---|---|---|
| MSI cranium | AP and lateral | Top of skull to below the mandible and bilateral skin borders | Include face and neck if ENT surgery |
| MSI chest | AP and oblique/ lateral | Apices to costophrenic angles (CPA) and bilateral skin borders | This may require more than one film for the AP projection. The oblique may be a single 14×17 of the ROI |
| MSI abdomen/ pelvis | AP and oblique/ lateral | Diaphragm to pubis and bilateral skin borders | This may require more than one film for the AP projection. The Oblique may be a single 14×17 of the ROI |
| MSI vagina | AP and inlet | Inferior gluteus to above crest and bilateral skin borders. Inlet must show the pelvic ring | Inlet: place 14×17 vertical with 25° caudal angulation. Special attention needed to avoid grid cut-off |
| MSI spine | AP/PA and lateral | C-spine: neck; T-spine: chest; L-spine: abdomen | C-spine: 11×14 T-spine: 14×17 L-spine: 14×17 |
| MSI extremity | AP and lateral | Include above and below ROI and bilateral skin borders | Use large films |

The RFID system has a unique radiofrequency identification chip sewn into each type of sponge and a separate computer console with a scanning bucket or an attached wand into which used sponges are placed [17]. Each sponge has a specific identifying chip and thus sponges of different types pooled together can be distinguished and counted. Used sponges can be put directly into the bucket or into plastic bag-lined kick buckets and the entire plastic bag full of sponges then placed into the scanning bucket. The sponges will all be individually counted. If there is a missing sponge it can be detected with a wand that is attached to the bucket by a long cord. This device offers a complete sponge counting and detection system.

## Small Miscellaneous Items and Unretrieved Device Fragments

SMIs used during procedures includes vessel loops, bovie scratch pads, trocars, parts of instruments or tools like screws, bolts, drill bits and guidewires, sheaths, and tubes. These items have become the second most commonly reported RSI [18]. The metal items are radiopaque while others are non-radiopaque and some are a combination of both in that surgical items composed of multiple parts may have one part that contains a radiopaque marker while another part does not. Many of these non-radiopaque SMIs are made of plastic and are disposable. Rather than try to classify cases by the type of item, we have analyzed cases by the location of the procedural event. This segregates cases into OR cases and non-OR cases.

## OR Cases

If we assume that the devices and SMIs are being used correctly, that is there is not a direct breakage of the device because of the way in which it was used, then there are three essential causes for parts or pieces of surgical items to be retained [19]. The first is because of manufacturer defects present in the tools or instruments when they are made. These defects may not be apparent until the actual device is deployed or used. The more common problem associated with retained SMI and unretrieved device fragment (UDF) is using worn or used equipment that is not recognized at the time of the case or is only recognized when the used equipment breaks or a piece breaks off. The last and probably most frequent problem with retained SMI is related to the plethora of new equipment, devices, and tools that are now used during operations. Many of these devices are unfamiliar and are composed of multiple separable parts. It is difficult for the surgeon at the time of the operation to recognize that there is something missing and the circulating nurse is often too far away from the site to identify a problem which means that the surgical technologist or person in the scrub position must become the content expert in this domain of surgical equipment.

SMI's are usually retained because of failed item management and error detection practices. The scrub person is in the closest position to check the condition of all items passed to and returned from the field [13]. Optimal performance will require knowledge about the tools that are used. The scrub position requires more than just passing instruments back and forth. OR managers will have to adopt standardized practices beyond just counting items, such as having standardized back tables where there is "a place for everything, and everything in its place" so the items and their constituent parts can be properly accounted for. If something is found to be amiss it is most important that if the scrub person "sees something, they will say something" so a concerted search can be undertaken to find the missing parts.

UDF are frequently so small that it is difficult to find them and they will not lead to any apparent harm if left behind. Larger UDFs can cause irritation, infection, obstruction or embolization. It is a matter of clinical judgment on the part of the surgeon to determine whether to try to remove the material or leave it alone. If it is decided to leave the material in the patient, it is important that the patient be informed and a disclosure discussion held between the patient and the surgeon.

## Non-OR Cases

The primary non-OR cases of retained SMI involve procedural areas in the hospital including cardiology suites, radiology areas, and the ICU. Items left in patients from these areas usually include guidewires, sheaths, catheters, introducers, and various tubes. The objects can be either intravascular or in interstitial spaces. These items are usually retained because of problems with provider practices of insertion, usage or removal techniques. If the wires or catheters are left intravascular, interventional radiology has a very good chance of retrieving the items. This should be done as soon as possible after discovery because left in the heart or vessels for too long they become embedded in the intimal surface and can't be removed.

## Sharps

Needles are a frequent source of miscounts and their retention primarily involves practice problems even though these cases are usually ICRC. A small needle is known to be missing but the surgeon makes a clinical decision to intentionally not remove it. Suggestions for practice improvement involve accounting for needles by size and building a needle management policy around the ability to detect and find needles [1]. Needles should be passed back to the scrub person on an instrument and the use of a "safety zone" is highly recommended [13]. Best practices involve safe management of the needles on the back table. If a small needle is lost, it is often not possible to retrieve it. Small needles <15 mm are frequently difficult to see on X-ray, difficult to find in situ, and have not been reported to cause problems in large cavity spaces if lost. If a patient has a retained small needle it is unlikely to cause a problem for future MRIs. They are unlikely to wobble or cause injury and won't heat because they do not form complete loops. We know of patients who have very small needles left in the mediastinum and broken needles left in the pelvis because they have been incidentally noted on CT scans. The most important action in the event of a miscount for a missing needle is to disclose to the patient that there is a possibility that there could be a retained needle and consider obtaining a CT scan which has the necessary resolution to see needles of all sizes. This may or may not change the decision about whether or not it can or should be removed. The best strategy is to focus on strong needle management practices to prevent loss in the first place.

## Instruments

Retention of whole instruments is very rare and is the result of incorrect practices of surgeons and nurses. These cases are uniformly CCRC. Interestingly enough the most commonly retained type of instrument is a retractor and the long, thin malleable retractor is the most common item [14]. This particular instrument is used after performance of the wound exam during fascial closure, so prevention of its retention is highly dependent on instrument accounting practices used by surgical technologists and nurses. If there are mistakes in "the count," there are no further opportunities for identification of the error until the retractor is discovered. X-rays have high specificity and sensitivity to show instruments since most are made of metal. The use of mandatory postoperative X-rays for abdominal and chest cases was an early recommendation [7]; however, this practice has been abandoned in many facilities because most X-rays are negative and the time, X-ray exposure, and cost of obtaining them has not been rewarded with a significant yield. There are special circumstances when mandatory X-rays in lieu of performing instrument counts are useful. These cases include orthopedic and neuro-spine cases where X-rays are performed at the conclusion of the case to check the alignment and positioning of the surgical constructs. In these circumstances the images can be used to also look for the presence of any surgical instruments, but the X-rays must be obtained while the patient is still in the OR and cannot substitute for sponge, sharp or SMI counts. Short of this practice, most hospitals still use various counting protocols to determine that all instruments have been accounted for.

## Conclusion

An RSI is a surgical patient safety problem. These are system problems and can be prevented by multi-stakeholder use of reliable OR practices and effective communication techniques. The operative words here are "reliable" and "effective." These are human undertakings and as such are subject to human error but understanding why people fall into the error traps and learn how to avoid them, makes these events preventable. Much has been written about team-based training programs such as crew resource management as applied to medical units. In the operating theater, nurses and surgeons have a long tradition of working together but not always as a functional team [20]. Enhanced communication strategies and rule-based practice actions can be successful in transforming a rare event into a true never event. In order for the practices to work, they must be employed in every case, every time, and not only in cases where there is a perception of a risk of retention. Enforcing this undertaking alone is the greatest challenge. No matter which route is taken, multiple stakeholders will have to become engaged, work together, and change behavior to develop a safer OR. Engaging surgeons and radiologists, anesthesia personnel, and OR nursing staff in addition to physicians and technological staff throughout the hospital to rethink and change some of their behaviors and practices seem daunting. Not doing otherwise to prevent harm to patients is unacceptable. At the end of every procedure, together we must make sure there is NoThing Left Behind.

## Key Lessons Learned

- Analyze an RSI case to identify practice or communication problems (or both).
- Reduce variation and customization in OR practices and make sure all stakeholders are employing the same standardized practices.
- A policy should be reflective of the actual practice and should be a multi-stakeholder policy since the effort to prevent RSI requires multidisciplinary actions.

- The use of strong communication tools specific to the OR or procedural environment are necessary.
- Leadership rule and policy enforcement has to include medical staff as well as hospital staff.
- Prevention of RSI requires practice change which takes longer than most people expect.
- Consistency yields excellence.

# References

1. Gibbs VC. NoThing Left Behind: a national surgical patient safety project to prevent retained surgical items. Available at http://www.nothingleftbehind.org. Accessed 1 June 2012.
2. Gibbs VC, Coakley FD, Reines HD. Preventable errors in the operating room: retained foreign bodies after surgery—part 1. Curr Probl Surg. 2007;44:281–337.
3. Sentinel event. The Joint Commission. Available at http://www.jointcommission.org/sentinel_event.aspx. Accessed 1 June 2012.
4. National Quality Forum. Available at http://www.qualityforum.org/projects/hacs_and_sres.aspx. Accessed 1 June 2012.
5. Frequently asked questions. The Joint Commission. Available at http://www.jointcommission.org/about/JointCommissionFaqs.aspx?faq#69. Accessed 1 June 2012.
6. Kaiser CW, Friedman S, Spurling KP, et al. The retained surgical sponge. Ann Surg. 1996;224:79–84.
7. Gawande AA, Studdert DM, Orav EJ, et al. Risk factors for retained instruments and sponges after surgery. N Engl J Med. 2003;348:229–35.
8. Cima RR, Kollengode A, Garnatz J, et al. Incidence and characteristics of potential and actual retained foreign object events in surgical patients. J Am Coll Surg. 2008;207:80–7.
9. Gibbs VC. Patient safety practices in the operating room: correct site surgery and nothing left behind. Surg Clin North Am. 2005;85:1307–19.
10. California Hospital Patient Safety News. Multi-disciplinary RSI reduction 2011;3(1):1. Available at http://www.chpso.org. Last accessed 17 June 2012.
11. Whang G, Mogel GT, Tsai J, Palmer SL. Left behind: unintentionally retained surgically placed foreign bodies and how to reduce their incidence. Am J Roentgenol. 2009;193:S79–S89.
12. McIntyre LK, Jurkovich GJ, Gunn MD, Maier RV. Gossypiboma; tales of lost sponges and lessons learned. Arch Surg. 2010;145:770–5.
13. AORN, Inc. Recommended practices for prevention of retained surgical items. In: Perioperative standards and recommended practices. Denver, CO: AORN, Inc., 2010.
14. Gibbs VC. Retained surgical items and minimally invasive surgery. World J Surg. 2011;35:1532.
15. Surgicount Medical website. Available at http://www.surgicountmedical.com. Accessed 13 Jul 2013.
16. RF Surgical Website. Available at http://www.RFsurg.com. Accessed 13 Jul 2013.
17. Clearcount Medical Website. Available at http://www.clearcount.com. Accessed 13 Jul 2013.
18. FDA Public Health Notification. Unretrieved device fragments. 2008. Available at http://www.fda.gov/MedicalDevices/default.htm. Accessed 13 Jul 2013.
19. Daley PM, Brophy T, Steatham J, Srodon PD, Birch MJ. Unretrieved device fragments—the clinical risk of using poor quality surgical instruments. Med Dev Decontam. 2010;14:18–22.
20. Goldberg JL, Feldman DL. Implementing AORN recommended practices for prevention of retained surgical items. AORN J. 2012;95:205–19.

# Chapter 10
# Wrong-Site Surgery

Patricia Ann O'Neill and Eric N. Klein

> *"What are man's truths ultimately? Merely his irrefutable errors."*
>
> Frederich Nietzsche

## Case Studies

### Case 1: Wrong Limb Amputation

#### Clinical Summary

Mr. Jones is 51-year-old diabetic male with a history of chronic ulcerations involving both lower extremities. After 2 days of increasing fatigue, fever, and foul smelling drainage from his right foot he presented to Dr. Michaels' surgery office for evaluation.

Dr. Michaels diagnosed wet gangrene of the right foot extending above the ankle. The left foot had a deep, chronic ulcer on the lateral plantar aspect but was pink with minimal exudate and felt to be viable. Dr. Michaels had an extensive discussion with the patient regarding the need for amputation to control his infection. Mr. Jones reluctantly agreed to the procedure and signed consent for a below knee amputation of the right lower extremity. The surgeon's office assistant booked the operative procedure as an emergency in the local hospital.

P.A. O'Neill, R.N., M.D., F.A.C.S. (✉)
Department of Surgery, SUNY-Downstate/Kings County Hospital Center, Trauma Office 451
Clarkson Avenue, Room C-3211, Brooklyn, NY 11203, USA
e-mail: paoneill05@gmail.com

E.N. Klein, M.D.
Department of Surgery, SUNY Downstate Medical Center, 450 Clarkson Avenue, Brooklyn,
NY 11203, USA
e-mail: eric.klein@downstate.edu

A. Agrawal (ed.), *Patient Safety: A Case-Based Comprehensive Guide*,
DOI 10.1007/978-1-4614-7419-7_10, © Springer Science+Business Media New York 2014

*Six hours later Mr. Jones arrived in the holding area of the operating suite, while the nursing team set up the room and equipment for the amputation. The surgeon arrived shortly after and, while he was changing into his scrubs, the anesthesiologist and circulating nurse brought the patient into the operating room, induced anesthesia, and proceeded to prep and drape the patient.*

*Dr. Michaels entered the operating room, thanked his colleagues for their efficiency, and proceeded with the amputation. After Dr. Michaels cut through all the soft tissues and ligated the major blood vessels, the circulating nurse became anxious and called out to the team. While organizing her paperwork she noted that the surgical consent was for a right below knee amputation but the team was operating on the left leg. There was immediate silence followed by a prolonged period of distress by the members of the operating team. Unfortunately, the procedure had progressed to a point where they were committed to amputation and Dr. Michaels had no choice but to complete the amputation of the left leg. The following morning Mr. Jones underwent a right below knee amputation to treat his gangrenous extremity by another surgeon.*

## Analysis of Errors

Analysis of this case reveals a series of errors and system failures leading to the wrong limb amputation and subsequent bilateral leg amputations, despite the fact that the surgeon had obtained the correct consent (See Table 10.1).

The first error was performed by Dr. Michael's office assistant who inadvertently booked the case as a "left" below knee amputation rather than a "right" amputation. The office assistant routinely booked the surgeon's cases via phone. This error could have been prevented if she had been required to review the consent form at the time of the booking. Similarly, if the individual who received the call and put the case on the OR schedule had had a faxed copy of the signed consent form to review at the time the case was entered, the discrepancy could have been identified and rectified at the time of booking.

Once the patient arrived in the holding area of the operating suite there was no attempt made to confirm the correct procedure by any member of the OR team. At that time, there was no requirement in place for the team to confirm the planned procedure with the patient and the consent form.

In an effort to be efficient, the anesthesiologist and scrub team brought the patient into the operating room while the surgeon was still changing into scrubs. This was a common practice in that OR to minimize turnover time. In addition, there were several other emergency cases still waiting to be done and the team was pressured to move the case along. Once in the room, the team proceeded to prep and drape the wrong extremity according to the OR schedule.

When Dr. Michaels arrived in the OR, he proceeded with the left leg amputation without taking the time to review the consent, confirm the surgical site, or discuss the planned procedure with the other members of the team. The fact that the left leg was already prepped and draped introduced the risk of a perception error and/or confirmation bias, increasing the chances that he would not recognize that the wrong leg was prepped. The fact that Mr. Jones had skin ulcerations involving both lower

**Table 10.1**  Case 1: Timeline of events/risks and solutions

| Risks and failures during the process | Solutions |
| --- | --- |
| The surgeon determines need for right below-knee amputation and obtains appropriate informed consent<br><br>Office assistant books case as "left" below knee amputation instead of "right"<br><br>Wrong procedure placed on OR schedule | Standardize booking process for all operative procedures<br><br>Require that provider/clerk cross-check procedure against a written consent or medical record at time of booking<br><br>Have electronic booking form or, fax the consent or a written booking form to the OR if off-site booking |
| OR team failed to verify the planned procedure with the patient and medical record prior to the patient entering the OR<br><br>The operative site was not marked by the surgeon and confirmed prior to entering the OR<br><br>The opportunity to identify the booking error before entering the OR was missed | Block entry into the OR unless a verification process has been performed with both the patient and consent form by all members of the surgical team<br><br>Assure that the surgeon physically marks the intended operative site and have it confirmed by other members of the team before entering the OR |
| The left leg was already prepped and draped at the time of the surgeon's arrival increasing the chances of a perception error or confirmation bias on the part of the surgeon<br><br>Another opportunity to identify the error in laterality was missed | Assure that the correct operative site is marked and visible before the patient is prepped and draped |
| There was no team discussion performed prior to the start of the operation to reconfirm the planned procedure with the patient and the consent form<br><br>The team proceeded to amputate the wrong leg | Do not allow any incision until a "time-out" process is performed by all member of the operative team<br><br>The process must reconfirm the correct patient, the correct procedure, and the correct side/site and agreed on by all |

extremities was another factor that contributed to the sequence of events. Since both legs were already bandaged upon arrival to the holding area, there was less of an opportunity for a member of the team to identify the discrepancy between the diseased limb and the one booked for amputation.

## Case 2: Death from Wrong-Patient Procedure

### Clinical Summary

*Mrs. Smith was a 68-year-old female with a history of prior left pneumonectomy for lung cancer. She was admitted to the MICU for COPD exacerbation and required endotracheal intubation for respiratory failure. Mr. Wong was the patient in the bed adjacent to Mrs. Smith and was also in respiratory failure requiring mechanical ventilation. During afternoon rounds, the medical team decided to place a central venous catheter in Mr. Wong.*

*The team had difficulty reaching Mr. Wong's wife for consent. Due to the delay, the day resident signed out the procedure to the night-float resident. Shortly thereafter, the night resident gathered the required supplies and began placing a central line via Mrs. Smith's right subclavian vein. During the procedure, the nurse came to the bedside to inquire what the night resident was doing as she was not aware of any planned procedure for her patient. The resident replied that an informed consent for central venous catheter insertion was in the patient's chart and proceeded with the insertion. While the nurse was confirming the consent, the resident called frantically for her to come back because the patient was arresting. A code was called but resuscitation efforts were unsuccessful. The resident realized that she had placed the central line in the wrong patient. A postmortem examination determined that the cause of death was a right-sided tension pneumothorax.*

*The resident was suspended for the remainder of her second year because she failed to adhere to the "Universal Protocol" policy. The nurse was reprimanded for not being more observant and ensuring the safety of her patient. While the resident had excellent medical knowledge and clinical skills, she decided that the stress caused by her mistake was too overwhelming and she decided to pursue a career in the pharmaceutical industry.*

## Analysis of Errors

Similar to Case 1, a series of errors and contributing factors led to the death of Mrs. Smith. These errors could have been interrupted at several points during the process, had appropriate policy and procedure been followed (See Table 10.2). As in Case 1, an informed consent was properly obtained for the correct procedure on the correct patient. Unlike in Case 1, the institution did have a policy in place (the "Universal Protocol") that mandated a "verification" and "time-out" process to identify the correct patient, the correct procedure, and the correct side/site prior to initiating any invasive procedure. However, the policy was not followed.

In her haste to get started, the resident failed to notify the nurse that the procedure was being performed. She failed to verify the patient's identity against the consent obtained earlier by the prior team. Had this been done, the resident would have immediately recognized that the procedure was planned for Mr. Wong.

When Mrs. Smith's nurse was puzzled at seeing a procedure being performed without having prior knowledge, she should have immediately voiced her concern and insisted that the resident stop the procedure until she could verify the correct patient and procedure in concordance with the consent. Once the nurse questioned the procedure the resident should have been cued into recognizing that this was a potential safety issue and subsequently stopped on her own accord until these issues were clarified. Had this been done, the procedure would have been aborted before causing harm to Mrs. Smith.

Other factors that increased the risk for error in this case include the fact that the procedure was planned by the day team but executed by the night team. Shift work and handoffs are occurring with increasing frequency in medicine today. All practitioners need to recognize the increased risk for miscommunication and misinterpretation of

**Table 10.2** Case 2: Timeline of events/risks and solutions

| Risks and failures during the process | Solutions |
| --- | --- |
| Patient Wong was unable to sign own consent leading to delay in procedure<br>Delay required procedure to be "signed-out" to the night float resident<br>Combination of "hand off" and a sedated patient imposed increased risks for patient misidentification | Standardize the process for hand offs<br>Assure accurate transfer of information with special attention to follow up procedures and tasks<br>Need increased provider vigilance when performing high risk procedures in high risk environments |
| Resident initiated the procedure without confirming the correct patient and consent<br>Resident failed to involve the patient's nurse in the process<br>Procedure initiated on the wrong patient | Implement the Universal Protocol for all bedside procedures<br>Protocol requires a verification and time-out process be performed with a second team member prior to the initiation of any invasive procedure in order to assure the correct procedure is performed on the correct patient |
| Patient's nurse raised concern at the initiation of the procedure but failed to insist the procedure be stopped until plan confirmed<br>Opportunity to halt procedure before patient harm missed | Foster an environment where open communication is respected and valued among all members of the healthcare team<br>Empower any member of the team to stop a procedure immediately if there are any patient safety concerns |
| Resident proceeded with procedure on the wrong patient despite nurse's concern causing pneumothorax in a patient with a prior pneumonectomy causing the patient's death | Promote individual accountability for patient safety. Educate providers to stop all procedures immediately if any team member raises a safety concern until the issue is resolved or corrected |

information transmitted during handoff procedures. The transfer of information during handoffs must be structured and complete and all parties must be extra diligent during the process. Time pressures and increased workloads often lead to employees "cutting corners" and by-passing policies to get the work done.

## Discussion

Case 1 has many similarities to the real-life case of Mr. Willie King that occurred at University Community Hospital in Tampa Florida on February 20, 1995. Like the patient in the scenario, Mr. King was left with unnecessary bilateral below knee amputations because the planned surgical procedure was erroneously booked as a left below knee amputation rather than a right below knee amputation. Policies and procedures were not in place to pick up the error before the wrong amputation was performed [1]. The case of Willie King was heavily publicized at the time and although the circumstances of his case are not unique, it is historic in that the notoriety from the King case brought wrong-site surgery (WSS) to the forefront of patient safety initiatives. As a result of its publicity, the Joint Commission initiated its Sentinel Event policy as a method to identify and track the leading causes of medical errors within the USA. This initiative mandated that accredited hospitals analyze and report any unexpected occurrence that resulted in death or serious physical or psychological injury to a patient [2]. In 2002, the National Quality Forum (NQF) followed the Joint Commission's lead and developed its own list of 27 Serious Reportable Events [3].

## Definition

"Wrong-site surgery (WSS)" is most often associated with surgical procedures performed on the wrong side (laterality) of the correct patient. However, the term WSS actually encompasses a broader definition of surgical errors and includes any procedure that is performed on a wrong patient, a wrong procedure performed on the correct patient, and all procedures performed on the correct patient but at the wrong level or the wrong site such as the wrong vertebral level or the wrong finger. The definition of WSS also includes the placement of incorrect implants and prostheses such as when a prosthesis for a left hip is inserted into the right hip or a left corneal implant is placed into the right eye.

## Incidence

The true incidence of WSS is somewhat difficult to determine. It depends on how one defines WSS, how the data is collected, and whether or not mandatory reporting by institutions is required. For instance, Kwann and coauthors evaluated all wrong-site surgeries reported to a single, large, medical malpractice insurer in Massachusetts between 1985 and 2004. Among the 2,826,367 operations performed at the hospitals within that system, there were only 25 wrong-site operations identified from the malpractice claims. This produced an incidence of 1 in 112,994 operations [4]. Based on these results, the authors concluded that WSS is an exceedingly rare event. However, using single payer malpractice claims to determine the rate of wrong-site procedures underestimates its true incidence. For one thing it fails to identify cases in which malpractice claims were never filed. It should be pointed out that Kwann's analysis excluded spine-related procedures. Since spine surgery is one of the specialties at highest risk for WSS, one has to interpret Kwann's results cautiously.

In contrast to Kwann's study, the Physician's Insurance Association of America (PIAA) evaluated claims from 22 malpractice carriers insuring 110,000 physicians from 1985 to 1995. The PIAA study revealed 331 WSS cases and 1,000 closed malpractice claims involving WSS. Their study identified a significantly higher number of cases occurring over a shorter period of time when compared to Kwann's analysis [5].

After the Joint Commission initiated its mandatory reporting in 1995, there were 531 sentinel events involving wrong-site surgeries reported between 1995 and 2006. Similar results were seen in several states that also require mandatory reporting of these events. The State of Minnesota reported 26 wrong-site surgeries during their first year of public reporting and another 31 during their second year [6]. In Virginia, a WSS was reported in 1 of every 30,000 surgeries equating to about 1 case per month and in New York, a WSS was reported in 1 out of every 15,000 surgeries [7]. Thus, wrong-site surgeries are not rare events. Wrong-site surgical procedures ranked the highest among all 4,074 sentinel events reported to the Joint Commission between January 1995 and December 2006 [8].

WSS affects all surgical specialties. Of 126 Joint Commission sentinel cases of WSS reported between 1998 and 2001, 41 % involved orthopedic or podiatric surgery, 20 % general surgery, 14 % neurosurgery, 11 % urologic surgery. The remaining cases included cardiothoracic, ear–nose–throat, and ophthalmologic surgeries [9]. Wrong-site surgical and invasive procedures occur throughout all surgical and nonsurgical settings. Of the 126 cases of WSS reported to the Joint Commission, 50 % of the WSS cases occurred in either a hospital-based ambulatory surgery unit or freestanding ambulatory setting. Twenty-nine percent occurred in the in-patient operating room and 13 % in other in-patient areas such as the Emergency Department or the ICU [8, 10]. Similar results were found by Neily and colleagues in a review of the Veterans Health Administration (VHA) National Center for Patient Safety database. Of 342 reports of surgical events in Neily's study, there were 212 actual adverse events (62 %) and 130 close calls (38 %). One hundred and eight (50.9 %) of the adverse events occurred in the operating room (OR) and 104 (49.1 %) occurred elsewhere [11]. Similar results were reported by the same group in a 2011 follow-up study (See Fig. 10.1) [12]. As with the Joint Commission data, wrong-side surgery procedures in Neily's study were the most common errors performed within the OR while wrong-patient procedures were the most frequent in the non-OR setting. Although intraoperative errors tend to get more publicity, errors performed outside the OR are no less harmful.

## Impact

Cases of WSS that result in significant harm are not only devastating to the patient but also to the families, the caregivers, and the institutions involved. Intense media attention often leads to a loss of public trust in the healthcare system and its providers. Defending these types of errors is nearly impossible and those involved usually pay a significant emotional, professional, and financial price for the event. In Case 2 the young resident had such difficulty dealing with the consequences of her error that she gave up a promising career in medicine (see Chap. 23 on "Second Victim" phenomena). In the case of Willie King, the Florida authorities suspended the surgeon's license for 6 months and fined him $10,000. The Tampa hospital paid Mr. King $900,000 and the surgeon paid an additional $250,000 directly to Mr. King [13].

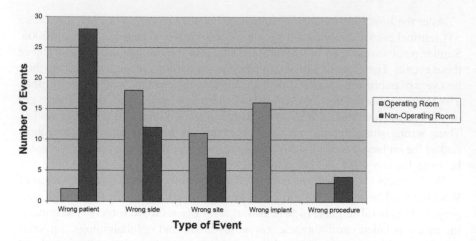

**Fig. 10.1** Comparison of wrong-site procedures performed inside and outside of the operating room based on the Veterans Health Administration patient safety database between July 2006 and December 2009. Of note, wrong-patient procedures outside the operating room outnumbered all other events in either location reprinted with permission from Elsevier

## Preventive Strategies

As previously stated, increased attention has been focused on WSS since 1995 when the Joint Commission initiated its mandatory reporting. Interestingly, however, the problem of WSS was recognized earlier by several medical associations and efforts were actually made to educate practitioners about strategies to reduce these errors. Between 1988 and 2001 several professional and orthopedic societies throughout the UK, Canada, and the USA recognized the seriousness of WSS procedures and initiated several safety campaigns in an effort to reduce their occurrence [14, 15]. Although these efforts were genuine, they had only a moderate impact on reducing the incidence of WSS possibly because they relied on voluntary participation.

The Universal Protocol was implemented on July 1, 2004 and applied to all Joint Commission accredited organizations including ambulatory care facilities and office-based surgery programs [2, 7]. The protocol was also to include special procedure units such as Endoscopy and Interventional Radiology. In 2009, the WHO extended this mandate to require that the "Universal Protocol" be performed for all procedures done outside of the operating room as well [16].

The Universal Protocol consists of three steps: verification, site-marking, and "time out." It requires multiple people to confirm that the correct procedure is being performed on the correct location of the correct patient. Table 10.3 describes the intended process for each of these three steps. If there is a discrepancy in the information provided or a team member has concerns regarding the elements of the case

**Table 10.3** The three steps of the universal protocol for preventing wrong site surgery [17]

*Conduct a preprocedure verification process*

Address missing information or discrepancies before starting the procedure

- Verify the procedure, the patient, and the site
- Involve the patient in the verification process
- Identify the items that must be available for the procedure

*Mark the procedure site*

At a minimum, mark the site when there is more than one possible location for the procedure and when performing the procedure in a different location could harm the patient

- Mark the site before the procedure is performed
- Involve the patient in the site marking process
- The site is marked by a licensed independent practitioner who is ultimately accountable for the procedure and will be present when the procedure is performed

*Perform a time-out*

The procedure is not started until all questions or concerns are resolved

- Conduct a time-out immediately before starting the procedure or making the incision
- All relevant members of the procedure team actively communicate during the time-out
- The team members must agree, at a minimum, on the correct patient, the correct site, and the correct procedure to be done

at any point during these three processes, the procedure should not proceed until the discrepancy is reconciled. It is believed that performing the Universal Protocol will significantly reduce the rates of WSS [16].

## *Root Causes and Potential Solutions*

Unfortunately, even after the initiation of mandatory reporting and implementation of the Universal Protocol, the problem of WSS still exits. At first glance it seems hard to understand why these events occur with such frequency and why they have been so hard to eliminate. It is not a surprise that wrong-site and wrong-side surgeries occur more commonly in the orthopedic, podiatric, neurosurgical, and urological specialties since most of the procedures performed by these specialties involve laterality. However, if laterality was the only risk factor for WSS, then the initiation of "site-marking" would essentially eliminate the problem. Like many other errors in medicine today the causes of WWS are complex and many factors contribute to their occurrence. The most common of these are listed in Table 10.4 [9, 18]. Awareness of these root causes allows institutions and practitioners to become more vigilant during high risk situations and may even prompt the institution or practitioner to create additional preventive measures.

For example, it has been shown that wrong-patient procedures are more prone to occur in fast-moving environments. Eye operations are particularly vulnerable to wrong-patient, wrong-site, and wrong-implant errors because they are short procedures with rapid turnover times. There are usually several patients waiting

**Table 10.4** Common risk factors for wrong-site surgery [9, 18]

Patient-related factors
- Morbid Obesity
- Physical deformity
- Comorbid conditions
- Presence of bilateral disease

Procedure-related factors
- Emergency case or procedure
- Need for unusual equipment or set-up
- Multiple procedures performed
- Multiple surgeons/physicians involved
- Change in personnel
- Room changes

Environmental factors
- Incomplete or inaccurate communication
- Poor booking practices
- Failure to engage patient or family in the processes
- Unusual time pressures

simultaneously at the center for similar procedures involving one or the other eye. The knowledge that such situations increase the risk for error should prompt the team to be more vigilant during their verification and time-out process [7, 19]. Such knowledge may also prompt prevention measures such as scheduling only right- or left-sided procedures on a particular day.

Poor communication and incomplete patient assessment are the two factors that have been shown to contribute most to inadequate patient or site verification. Of 455 wrong-site surgeries reviewed, inadequate communication was deemed to be the root cause in almost 80 % of the cases [7]. Types of communication errors include miscommunication, misinformation, information not shared, and information not understood. These communication errors are often perpetuated by incomplete or inadequate preoperative assessments, such as what occurred in case 1. However, having a process in place by itself will not be effective if the involved individuals do not complete the process appropriately and diligently every time.

Good communication is an active process. It must engage the patient and/or family members in the informed consent and again during the surgical site verification process. A collaborative team approach, with each team member taking individual responsibility to assure the correct patient and site, is the best way to prevent an error due to inaccurate or incomplete information and will serve to catch a "miss" by other members of the team.

There is no doubt that the initiation of the Universal Protocol with a quality "verification" and "time-out" process prevents WSS errors. However, as previously stated, the Universal Protocol, by itself, does not prevent all WSS errors. In a review of 13 cases of WSS from a liability insurance company database, nine of the errors actually originated prior to the patient arriving in the perioperative area. These sources of error included an incorrectly printed MRI (11 %), a referral to a surgeon that specified the incorrect laterality of pathology (11 %), multiple pathologies that were not identified, clarified, or documented during the clinic visit (33 %), and incorrect OR scheduling (44 %). A tenth error originated in the holding area where the surgeon discussed a change in the laterality of a procedure for a patient with bilateral pathology. The patient did not recall consenting to the contralateral procedure because the patient did so after receiving sedation [18].

Another overlooked cause of WSS includes perception errors due to a person's inability to discriminate right from left. A study of Irish medical students in 2008 showed significant variability in the students' ability to distinguish the right hand from the left hand using stick figure illustrations. The errors in discrimination occurred most frequently when the figures were varied between views of the front and back. This emulates the situation in the operating room where patients are often positioned in different orientations. The study also showed that the ability to perform right–left discrimination was significantly worse when figures were viewed from the front than when they were viewed from the back. This is an important finding since most patients are supine on the operating table and thus viewed from the front by the surgeon [20].

There are also risk factors unique to certain subspecialties. Wrong-site procedures have been reported by anesthesiologists in association with increased use of regional anesthesia. Reasons include the fact that nerve blocks are performed prior to the surgical time-out. Since the site for the nerve block is usually away from the operative site, marking of the operative site may not be enough to assure that the anesthesiologist injects the correct site. Edmonds reported two cases of wrong-site peripheral nerve blocks and suggested the creation of a policy that mandates that the anesthetic consent specify the laterality of the surgery and that a separate anesthetic time-out be performed to include participation of the nurse and patient prior to the start of regional anesthesia. Of note, marking of the injection site for regional anesthesia by the anesthesiologist was not advised because a second marking could be a source of later confusion at the time of incision [21].

Dental procedures pose several risks for wrong-site (tooth) surgery. There are currently three major systems that can be used for numbering teeth for identification (1) The Universal/National System, (2) The Federation Dentaire International System, and (3) The Palmer Notation Method. Each of these systems number teeth differently. Thus, a written notation identifying a specific tooth using one system by one practitioner will refer to a different tooth if a different system is used to interpret that notation by another practitioner. Misidentification also occurs in patients in whom teeth are already missing. Correct identification of the remaining teeth is more difficult because the roots or sockets of the missing teeth are often obscured leading to a miscount of the remaining teeth. To avoid these errors, Lee recommends a standardized referral form for oral procedures that includes a diagram of the mouth for marking the desired pathologic tooth. Since there is no practical way to mark teeth at the time of surgery, it is essential that the correct site be marked on a dental diagram or X-ray [22].

Foot surgery is prone to a similar set of errors because patients use a variety of terms to refer to their toes. One study asked 100 patients to label the toes on each

**Table 10.5** The percent of operating room (OR) caregivers who rated their collaboration with other members of the OR team as "high" or "very high"

| Caregiver Position Performing Rating | Caregiver position being rated | | | |
|---|---|---|---|---|
| | Surgeon (%) | Anesthesiologist (%) | Nurse (%) | CRNA (%) |
| Surgeon | 85 | 84 | 88 | 87 |
| Anesthesiologist | 70 | 96 | 89 | 92 |
| Nurse | 48 | 63 | 81 | 68 |
| CRNA | 58 | 75 | 76 | 93 |

*Source*: Borrowed with permission from [24]
Surgeons and Anesthesiologist consistently rated teamwork and collaboration among members of the operating room team higher than their nurse colleagues
*CRNA* certified registered nurse anesthetists

foot choosing to use either name or number according to their preference. The patients had an overall error rate of 11.6 %. Other factors that increase the risk for errors in foot surgery include the fact that patients frequently have disease that affects multiple toes, such as gangrene or rheumatoid arthritis, and the fact that foot pathology is common among diabetics who may not be able to see or feel their feet due to retinopathy and neuropathy [23].

Good teamwork, communication, and redundant systems are the only way to reduce these types of errors. However, as more WSS cases are analyzed it is increasingly clear that "good teamwork" may need to be fostered.

Poor interpersonal dynamics hamper effective teamwork. Too rigid a hierarchy and too steep authority gradients between team members often results in the withholding of critical information and safety concerns. Healthcare organizations are characterized by large authority gradients with physicians generally positioned above the rest of the workforce. This is particularly true within the OR environment.

In 2000, Sexton and colleagues surveyed OR personnel on teamwork climate within the OR. The survey included perceptions about difficulty speaking up, conflict resolution, physician–nurse collaboration, feeling supported by others, asking questions, and the heeding of nurse input [24, 25]. Across all institutions surveyed, surgeons and anesthesiologists perceived that physician–nurse collaboration was much better than nurses did. Among the 60 institutions, more than 80 % of all surgeons rated the quality of communication and collaboration within the OR as high, whereas only 48 % of their nursing colleagues felt the collaboration between nurses and surgeons was high. Similar results were found between nurses and anesthesiologists (see Table 10.5). Nurses and other staff were also less positive about speaking up when having safety concerns. Transforming this "culture" is extremely challenging but there are a number of communication and teamwork strategies that the healthcare industry can adapt from the aviation industry.

The first step is to dampen authority gradients. Methods include techniques such as having the team leader introduce himself, learn the names of other team members, and to explicitly welcome input from all members of the team. To improve communication and information exchanges within groups a number of other tools have been designed to ensure that important information and safety concerns are both heard and acted upon. Two examples are the use of SBAR and CUS words.

SBAR stands for "**S**ituation, **B**ackground, **A**ssessment, and **R**ecommendation." It provides a format for nurses and other team members to structure their communication with physicians in such a way as to capture the latter's attention and to generate an appropriate action. The need for SBAR training grew from the recognition that nurses have been schooled and socialized to report in story format, while physicians have been trained to think and process information in bullet points [26].

The use of CUS words is a tool used to escalate levels of concern by anyone lower on the hierarchy to get the attention of someone higher up. The CUS words are used in escalating order as needed and begin with "I am **C**oncerned about..." then "I am **U**ncomfortable...." and finally, "This is a **S**afety issue!" The key to success is to teach those who are in a position to receive such messages to appreciate the significance of such statements and the need to respond appropriately. Appropriate use of CUS words between the nurse and resident in scenario 12–2 may have prevented the death from the central line placement. Other team training techniques that have been used successfully include the use of checklists, briefings, and debriefings [24, 25, 27].

Institutions that have promoted medical team training programs and the use of checklists, briefings, and debriefings have not only reduced the incidence of surgical errors such as WSS but have also shown a significant reduction in overall surgical mortality as well. Haynes et al. reported a decrease in mortality after initiating a surgery safety checklist involving eight hospitals [28]. Neily and her colleagues demonstrated a dose–response relationship between OR team training and surgical mortality within the Veterans Healthcare Administration System. For each quarter period of team training at a single institution, the risk adjusted mortality rate within that institution decreased 0.5 per 1,000 procedures. Data analysis also showed an almost 50 % greater reduction in mortality rates in the trained VHA institutions when compared to those that had not yet received training [27].

## Conclusion

In Summary, WSS errors are not rare events. Wrong-patient, wrong-side, and wrong-site procedures occur with equal frequency within and outside of the operating room and with the same risk of harm. The Joint Commission created the Universal Protocol as a mandatory safety standard in order to eliminate wrong procedures through the implementation of a preprocedure verification, site marking, and "time-out" process in order to confirm the correct patient, the correct procedure, and the correct side/site prior to the start of any invasive procedure. Up to 70 % of wrong-site procedures can be prevented if the verification and time-out process are performed correctly. In order for the Universal Protocol to be successful there must be 100 % compliance and it must involve the patient and/or family in the process and include active communication between all members of the clinical team.

The remaining 30 % of wrong procedure errors are more difficult to address. Avoidance of these errors requires redundant systems, teamwork, and equal accountability between all members of the operating team. Aggressive education of all employees, both clinical and nonclinical, in the prevention of WSS is essential for a successful prevention program. It must include the education of staff in the risk factors and common errors known to occur at each step along the process. But above all, there must be constant vigilance by all practitioners who participate in invasive procedures both inside and outside the operating room.

## Key Lessons Learned

- There must be a policy and procedure in place at every institution to assure correct patient, correct procedure, and correct site prior to the performing any surgery or invasive procedures.
- Errors in information and communication can occur at multiple steps along the process.
- There must be a verification checklist that ensures that all sources of information have been checked before starting any procedure.
- Ensure that all pertinent radiologic studies and pathology specimens have been reviewed and are consistent with the planned procedure, the medical record, and the patient diagnosis.
- Assure effective communication between all members of the operative or clinical team. Special care should be given when information is transferred during hand-off procedures.
- Include the patient and/or family member in the process at every feasible point.
- Ensure accurate site markings to include right versus left, multiple structures (finger/toes), or levels of the spine. Use the assistance of radiographs, photographs, diagrams, and forms when marking the actual operative site is not feasible.
- Do not allow time pressures to short-cut completion of the verification and time-out process.
- Train the team so that each member feels empowered to raise concerns. Other members must never belittle or dismiss another's inquiry and should halt all procedures until concerns are reconciled.

## References

1. Patterson ES, Roth EM, Woods DD, Chow R, Gomes JO. Handoff strategies in settings with high consequences for failure: lessons for health care operations. Int J Qual Health Care. 2004;16(2):125–32.
2. Facts about the Universal Protocol. 1/2011. Available at http://www.jointcommission.org/assets/1/18/Universal%20Protocol%201%204%20111.PDF. Accessed 13 Jul 2013.

3. Kizer KW, Stegun MB. Serious reportable adverse events in health care. In: Henriksen K, Battles JB, Marks ES, Lewin DI, editors. Advances in patient safety: from research to implementation (Volume 4). Rockville, MD: Agency for Healthcare Research and Quality (US); 2005.
4. Kwaan MR, Studdert DM, Zinner MJ, Gawande AA. Incidence, patterns, and prevention of wrong-site surgery. Arch Surg. 2006;141(4):353–7.
5. Shojania KG, Duncan BW, McDonald KM, Wachter RM, Markowitz AJ. Making health care safer: a critical analysis of patient safety practices. Evid Rep Technol Assess. 2001;2001(43):1–668. i–x.
6. Clarke JR, Johnston J, Finley ED. Getting surgery right. Ann Surg. 2007;246(3):395–403. discussion 395–403.
7. Dunn D. Surgical site verification: A through Z. J Perianesth Nurs. 2006;21(5):317–28.
8. Croteau RJ. Wrong site surgery—the evidence base. Available at http://www.health.ny.gov/professionals/patients/patient_safety/conference/2007/docs/wrong_site_surgery-the_evidence_base.pdf. Accessed 13 Jul 2013.
9. Joint Commission on Accreditation of Healthcare Organizations. A follow-up review of wrong site surgery. Sentinel Event Alert. 2001;(24):1–3.
10. DeVito K. The high costs of wrong-site surgery. Healthc Risk Manage. 2003;8(18).
11. Neily J, Mills PD, Eldridge N, et al. Incorrect surgical procedures within and outside of the operating room. Arch Surg. 2009;144(11):1028–34.
12. Neily J, Mills PD, Eldridge N, Carney BT, Pfeffer D, Turner JR, et al. Incorrect surgical procedures within and outside of the operating room: a follow-up report. Arch Surg. 2011;146(11):1235–9.
13. Hospital settles case of amputation error. New York Times. 1995 May 12; A14.
14. Watters WC. SMAX: early data and practical applications. SpineLine. 2003;4(3):18–20.
15. Wong D, Herndon J, Canale T. An AOA critical issue. Medical errors in orthopaedics: practical pointers for prevention. J Bone Joint Surg Am. 2002;84-A(11):2097–100.
16. WHO Guidelines for Safe Surgery. Safe surgery saves lives. 2009. Available at http://whqlibdoc.who.int/publications/2009/9789241598552_eng.pdf. Accessed 13 Jul 2013.
17. Universal Protocol (Poster). Available at http://www.jointcommission.org/assets/1/18/UP_Poster1.PDF. Accessed 13 Jul 2013.
18. Seiden SC, Barach P. Wrong-side/wrong-site, wrong-procedure, and wrong-patient adverse events: are they preventable? Arch Surg. 2006;141(9):931–9.
19. Ensuring Correct Surgery Directive FAQ. Available at http://www.patientsafety.gov/faq.html#CorrectSurg. Accessed 13 Jul 2013.
20. Gormley GJ, Dempster M, Best R. Right-left discrimination among medical students: questionnaire and psychometric study. BMJ. 2008;337:a2826.
21. Edmonds C, Liguori G, Stanton M. Two cases of a wrong-site peripheral nerve block and a process to prevent this complication. Reg Anesth Pain Med. 2005;30(1):99–103.
22. Lee JS, Curley AW, Smith RA. Prevention of wrong-site tooth extraction: clinical guidelines. J Oral Maxillofac Surg. 2007;65(9):1793–9.
23. Beckingsale TB, Greiss ME. Getting off on the wrong foot doctor-patient miscommunication: a risk for wrong site surgery. Foot Ankle Surg. 2011;17(3):201–2.
24. Makary MA, Sexton JB, Freischlag JA, et al. Operating room teamwork among physicians and nurses: teamwork in the eye of the beholder. J Am Coll Surg. 2006;202(5):746–52.
25. Sexton JB, Makary MA, Tersigni AR, et al. Teamwork in the operating room: frontline perspectives among hospitals and operating room personnel. Anesthesiology. 2006;105(5):877–84.
26. Haig KM, Sutton S, Whittington J. SBAR: a shared mental model for improving communication between clinicians. Jt Comm J Qual Patient Saf. 2006;32(3):167–75.
27. Neily J, Mills PD, Young-Xu Y, et al. Association between implementation of a medical team training program and surgical mortality. JAMA. 2010;304(15):1693–700.
28. Haynes AB, Weiser TG, Berry WR, et al. A surgical safety checklist to reduce morbidity and mortality in a global population. N Engl J Med. 2009;360(5):491–9.

# Chapter 11
# Transfusion-Related Hazards

Barbara Rabin Fastman and Harold S. Kaplan

> *"The greatest follies are often composed, like the largest ropes, of a multitude of strands. Take the cable thread by thread, take all the petty determining motives separately, and you can break them one after the other, and you say, 'That is all there is of it!' Braid them, twist them together; the result is enormous…"*
>
> Victor Hugo

## Introduction

Blood transfusion is a critical component of modern medical care. Concomitant with this, there has been intensive focus on the safety of blood products. Increasingly sophisticated advances in donor screening and blood product testing have markedly reduced the risks of transfusion-transmitted diseases, with the risks of HIV and HCV transmission falling below one in one million transfusions [1, 2]. Worldwide surveillance programs for newly emerging infectious disease threats are expanding, with ongoing efforts targeted at early detection and pathogen inactivation [3, 4]. Three decades of attention primarily centered on the safety of the blood product itself suggests, in part, the societal perception that transfusion risk primarily derives from the potential for disease transmission [5, 6]. Safety efforts such as bacterial culture of platelet products and the selection of male-donated plasma [7] while differing in scope, indicate a focus on the blood product.

B. Rabin Fastman, M.H.A., M.T.(A.S.C.P.)S.C., B.B. (✉) • H.S. Kaplan, M.D.
Mount Sinai School of Medicine, Health Evidence and Policy, One Gustave L, Levy Place, 1077, New York, NY 10029, USA
e-mail: barbara.rabin@mssm.edu; harold.kaplan@mssm.edu

A. Agrawal (ed.), *Patient Safety: A Case-Based Comprehensive Guide*, 161
DOI 10.1007/978-1-4614-7419-7_11, © Springer Science+Business Media New York 2014

**Table 11.1** Steps in the blood and blood product transfusion process

| Donor/blood unit | Recipient/sample |
|---|---|
| • Donor interview/history check | • Recipient evaluation for transfusion need |
| • Donor unit and sample collection | • Recipient sample collection |
| • Donor sample testing (antigens and infectious disease) | • Recipient sample testing |
| | • Cross-match with donor samples |
| • Blood unit labeling | • Release to recipient |
| • Additional testing | • Transport to recipient location |
| • Product manipulation/component processing | • Product administration |
| • Inventory management | • Monitoring of recipient (transfusion reaction/infectious disease) |
| • Distribution/transport to hospital | |
| • Entry into hospital inventory | |
| • Monitor donor on subsequent donations | |

Perhaps reflective of this, until relatively recently there has been less intensity of effort directed at the safe *delivery* of the blood product to the patient, i.e., the transfusion process itself [8]. The transfusion medicine process is lengthy and complex and involves many hands and computers. Originating with screening the blood donor and ending with post-transfusion monitoring of the patient, it involves a blood component(s), as well as samples from the donor, the blood unit, and the patient/recipient. Even at a high level, the process includes numerous steps [9] (Table 11.1).

From reports to the New York State Department of Health, it was calculated that the estimated frequency of a fatal outcome was approximately 1 in 1.8 million transfusions, with as many as one in 14,000 transfusions administered to the wrong patient [10]. Thirteen percent of erroneous transfusions were due to errors in patient sample collection, the pre-analytic phase, while an additional 38% were due to administration to an incorrect recipient, likely a failure of the bedside pre-transfusion verification of patient identity and blood unit information, the post-analytic phase [10]. Recognizing these and other process-related risks, efforts have been increasingly directed at the transfusion process itself, with a focus on reducing error through converging optimization of system and human performance.

All of these endeavors take place within the safety culture of the transfusion service and the hospital. A culture of safety has been characterized as an "informed culture" [11], with event reporting and analysis as a critical component. However, mere *compliance* with mandated reporting, while necessary, is not sufficient. What is required is its *adoption*, that is, events routinely reported within a nonpunitive environment and employed to inform system improvement.

Internal requirements for event reporting systems, external reporting of transfusion-related deaths to the Food and Drug Administration (FDA) and individual state agencies, as well as non-reimbursement by the Centers for Medicare and Medicaid Services (CMS) for preventable adverse events including ABO incompatible transfusion [12] reflect the increased safety concern directed at the transfusion process.

# Case Studies

## Case 1: Mislabeled (Incorrect Patient) Stem Cells Transfused

### Clinical Summary

*Patient A is a 42-year-old female diagnosed with non-Hodgkin lymphoma. Three years after a splenectomy and several rounds of chemotherapy, her lymphoma returned and her physician ordered an autogeneic stem cell transplant. After undergoing induction chemotherapy and stem cell collection on April 1st, she was scheduled to receive her stem cell infusion on April 14th.*

*Patient B is a 63-year-old female newly diagnosed with multiple myeloma. Her physician prescribed induction chemotherapy followed by autogeneic stem cell collection. Patient B's stem cells were also harvested on April 1st and cryopreserved for potential future need. All procedures took place at the same facility: a large academic medical center.*

*On April 12th, the hospital's Transfusion Service/Stem Cell Laboratory received a request to prepare patient A's stem cells for infusion at 10 A.M. on April 14th. On the morning of April 14th, the laboratory technologist removed four canisters from the freezer, each labeled as containing a unit of stem cells for patient A. He thawed the stem cell products, pooled them, and issued the pooled product to the patient's floor. At 2 P.M., while reconciling the morning's paperwork with the tags from the units and the canister labels, he noticed that one of the four canisters that were labeled for patient A actually contained a unit labeled for patient B. He immediately called the floor, but the pooled product had already been transfused (Fig. 11.1).*

### Preliminary Investigation

An investigation began immediately following detection of the error. The initial focus was on the bedside process. How could stem cells labeled for another patient

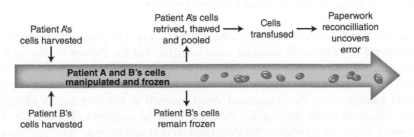

**Fig. 11.1** Case 1 Timeline—Mislabeled stem cells transfused

have been transfused? It was quickly realized that the units, having been pooled and relabeled, no longer retained the information from the bags in which they were frozen. There was no information from the individual units in the pool available at the bedside, so the pre-transfusion bedside check could only confirm that the label on the pooled product matched the patient's information.

While staff on the patient's unit were looking into the product administration phase of this event, the laboratory was working to determine how this mistake could have occurred. The technologist was experienced, having performed this procedure hundreds of times with no ill consequence. Working backwards in time, he recalled that during the thawing process he hadn't checked the tags on the bags themselves against the labels on the canisters from which they were removed. He also did not ask for a "second person check."

## Causal Analysis Method

The method used to perform a root cause analysis in this chapter is known as causal tree or fault tree building. This technique provides a structured, standardized method for uncovering underlying actions, circumstances, and decisions that contributed to an event in question. The tree provides a visual representation of an event which includes all possible causes gathered during the investigation process [13]. The consequent or discovery event is at the top of the tree and is described in terms of the event's consequences—harm, no harm, or a near miss event (an event that could have reached the patient, but was prevented by a barrier). The branches of the causal tree are constructed of precursors which reveal what "set up" the consequent event. Precursors are displayed in both logical and chronological order proceeding across and down the tree. By continuing to ask "why" at each step on each of the tree's branches, all relevant precursors and the root causes of the event are revealed. Root causes are indentified in the bottom boxes of each branch of the tree, and in these examples are coded using the Eindhoven Classification Model—Medical Version (also known as PRISMA) [14]. Causal trees provide a realistic view of how a system is functioning, as well as facilitate the creation of effective and lasting solutions.

## Causal Analysis, Discussion, and Possible Solutions

In the causal tree that was built surrounding Case 1 (Fig. 11.2), the consequent event is described as "Patient A received stem cells labeled for Patient B (one of four units, pooled)." As described, this active error occurred after many preceding, latent events had occurred. Moving down the tree, we can see these latent factors as precursors leading up to the consequent event, as well as the root causes identified along with their codes for classification and trending purposes (Table 11.2). The following issues contributed to this event and were revealed by an investigation and building of the causal tree.

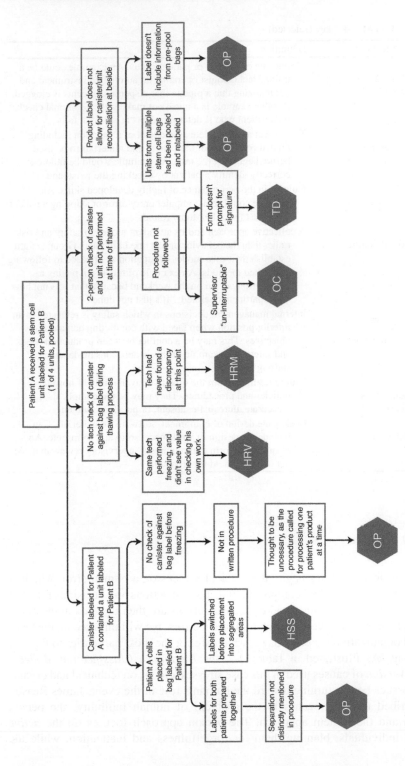

**Fig. 11.2** Case 1 Causal tree analysis—Mislabeled stem cells transfused (refer to Table 11.2 for an explanation of codes)

**Table 11.2** Causal codes key (selected)

| Causal classification | Definition |
|---|---|
| **HRM**<br>Human rule based:<br>  monitoring | Monitoring of process or patient status. An example could be a trained technologist operating an automated instrument and not realizing that a pipette that dispenses reagents is clogged. Another example is a nurse not making the additional checks on a patient who is determined as "at risk" for falls |
| **HRV**<br>Human rule based:<br>  verification | The correct and complete assessment of a situation including related conditions of the patient and materials to be used before beginning the task. An example would be failure to correctly identify a patient by checking the wristband |
| **HSS**<br>Human skill based: slip | Failures in the performance of highly developed skills. An example could be a computer entry error or skipping a patient on the list for phlebotomy rounds |
| **OC**<br>Organizational: culture | A collective approach and its attendant modes to safety and risk rather than the behavior of just one individual. Groups might establish their own modes of function as opposed to following prescribed methods. An example of this is not paging a manager/physician on the weekend because that was not how the department operated; "It's just not done" |
| **OM**<br>Organizational:<br>  management priorities | Internal management decisions in which safety is relegated to an inferior position when faced with conflicting demands or objectives. This may be a conflict between production needs and safety. An example of this is decisions made about staffing levels |
| **OP**<br>Organizational: protocols/<br>  procedures | Failure resulting from the quality or availability of hospital policies and procedures. They may be too complicated, inaccurate, unrealistic, absent, or poorly presented |
| **TD**<br>Technical: design | Inadequate design of equipment, software, or materials. Can include the design of workspace, software, forms, etc. An example is a form that requires a supervisory review that does not contain a field for signature |

## Human Failure

When event investigations begin, they typically focus on a human error, which is only one, and often the final component of a chain of actions and decisions that "set up" the event to occur. Suggested corrective actions are then directed at changing human behavior. But if we *only* look for human errors, we will miss seeing the technical and organizational *system* flaws, which are for the most part easier to fix than are humans. As illustrated in Table 11.2, the *Eindhoven Classification Model—Medical Version* of causes used in this case stresses a focus on technical and organizational issues before turning toward the human's role in the event. James Reason has described two approaches to the problem of human fallibility: the person approach and the system approach. The person approach focuses on the *active* errors of individuals, blaming them for forgetfulness and inattention, while the system approach concentrates on the system's "built-in" *latent* failures, focusing on minimizing or eliminating them, and reinforcing defenses to avert errors or mitigate their effects [15].

In laboratory settings, which are heavily focused on the individual, institutions have traditionally relied upon the "blame and shame" and "blame and (re) train" approaches for staff involved in events. Not surprisingly, these approaches create strong pressure on individuals to cover up mistakes rather than admit to them [16] and do nothing to fix the system flaws that set them up to make the error [17]. Experts in the fields of error and human performance reject these methods [18].

In this case it was quickly realized that the earlier labeling error was undetectable to the nurses at the bedside. The pooled stem cell unit contained the contents from the four canisters and had been labeled with the name that matched three of the canisters and the patient's wristband. The final check for the right blood product and patient at the bedside is typically the "2-person, 3-way check" requiring active verification of paperwork, product label, and patient identity carried out by two qualified staff. In this instance, the checking procedure had been carried out properly. However, this check is often performed improperly, incompletely, or not at all [9]. To account for the human in the process, particularly when distracted or interrupted, bar-code labeling [19], radiofrequency identification tags (RFID) [20], and even palm vein-scanning technology [21] are increasingly being utilized in patient identification.

## Safety Culture

As Reason points out, the quality of a reporting culture is contingent on its response to error [11]. If its routine response is to blame, then reports will be few and far between. However, if blame is limited to behavior which is in reckless disregard of patient safety or is of malicious intent, then it is a component of a "just culture," is supportive of reporting and enhances the opportunity for organizational learning [22].

In this case the technologist did not check his own work, nor did he ask for a second-person review of his work. Although both of these verification steps were part of the written procedure, he was the same technologist who had frozen the cells on the day that they were harvested and believed there was little value in rechecking his own work. He also had never before seen a discrepancy at this stage of the process. Dekker has written about flawed systems and the dangerous complacency resulting from the fact that "Murphy's Law" is wrong, and that "What can go wrong usually goes right," prompting us to then draw the wrong conclusion [23]. In reality, eventually things that usually go right *will* go wrong. We cannot afford to have a safety culture that allows us to become "mindless" about our seemingly flawless processes. "Nothing recedes like success" is an often quoted reminder of this caution [24]. In high reliability organizations (HROs) [25], staff regard success with suspicion and act mindfully, paying close attention to even weak signals in order to detect a problem in its earliest stage.

The technologist did not ask for a second-person check of the canisters against the labeled bags of stem cells inside them because the only other person present in his department was the supervisor, and she had said that she was not to be interrupted that morning. The technologist recognized that he was not following procedure, but believed that it was acceptable. His diligence and vigilance had decreased based on past experience, and he did not consider that an event of low probability could occur.

What allowed this situation to occur? Organizational culture can play a significant role in contributing to error. Organizational culture is characterized by both visible behaviors and the more subtle values and assumptions that underlie them. The cultural focus on individual autonomy, for example, seems to conflict with desired norms of teamwork, problem reporting, and learning [26]. In this case, it was not only acceptable for the supervisor to be unavailable by choice, but the technologist also did not feel comfortable in going against her wishes to interrupt her, even when it was called for by protocol. Westrum has defined culture as "the organisation's pattern of response to the problems and opportunities it encounters" [27]. He states that leaders, by their preoccupations, shape a unit's culture through their symbolic actions and rewards and punishment, and these become the preoccupations of the workforce. The supervisor in this case sent a clear message to staff, setting a culture that allowed, or even encouraged the tech to break with procedure.

Safety culture is not easy to change. A group in the UK performed a literature review covering the processes and outcomes of culture change programs and found little consensus over whether organizational cultures are capable of being shaped by external manipulation to beneficial effect [28]. Key factors that appear to impede culture change are wide and varied. They concluded that while managing culture is increasingly viewed as an essential part of health system reform, transforming cultures that are multidimensional, complex, and often lacking leadership is a huge task. Other studies have shown that culture change is slow and difficult, but possible. Moving toward high reliability, including preoccupation with failure, reluctance to simplify interpretations, sensitivity to operations, commitment to resilience, and deference to expertise can push an organization's culture in the right direction [25].

Developed by the Department of Defense's Patient Safety Program in collaboration with the Agency for Healthcare Research and Quality, TeamSTEPPS is a system that has proven to be one solution in improving safety culture [29]. It is an evidence-based system designed to improve communication among healthcare professionals by integrating teamwork principles into all areas. TeamSTEPPS has been shown to facilitate optimization of information, people, and resources, resolve conflicts and improve information sharing, and eliminate barriers to quality and safety. Communication and teamwork between the transfusion service and the nursing department, for instance, can go a long way in reducing sample collection errors.

## Redundancy

A second-person review of one's work, often performed by a passive visual check, is a common approach in transfusion medicine in both detecting and preventing error. However, passive checks have significant potential for distraction, and dual responsibility does not necessarily enhance human performance. In fact, in a system where two people are responsible for the same task, neither person feels truly responsible. Paradoxically, such safety procedures may provide less, rather than more assurance [30].

Various views exist concerning the value of redundancy. Normal Accident Theorists (NATs) argue that adding redundancy can increase the complexity of a system, and efforts to increase safety through the use of redundant safety devices may backfire, inadvertently making systems fail more often and creating new categories of accidents [31]. Sagan describes the phenomena of "social shirking," another way in which redundancy can backfire. Diffusion of responsibility is a common phenomenon in which individuals or groups reduce their reliability in the belief that others will pick up the slack. In transfusion services, backup systems are often humans that are aware of one another. Awareness of redundant units can decrease system reliability if it leads an individual to shirk off unpleasant duties because it is assumed that someone else will take care of it [32].

On the other hand, High Reliability Theorists (HRTs) believe that duplication and backups are necessary for system safety. Redundancy in High Reliability Organizations (HROs) takes the form of skepticism, in that when an independent effort is made to confirm a report, there are now two observations where there was originally one. Redundancy involves doubts that precautions are sufficient and wariness about claimed levels of competence. HRTs believe that all humans are fallible and that skeptics improve reliability [33].

A slight modification of the traditional two-person check, however, has the potential to resolve this issue. It has been estimated that the average failure rate of error detection by one person passively checking another person's work after the fact is as high as one in ten. But the failure rate in a two-person team with one person performing, and the second person monitoring, and then switching roles, is approximately one in 100,000 trials [34].

## Human Factors

The investigation also showed that the technologist would have been much more prone to ask for the second person check if there was a distinct place for that person's sign-off on the form. Forms and records can and should be designed to effectively control potential mistakes [35]. Had there been a check-box and a place for a signature, the omission would have been apparent, presenting itself in a way that pointed out that the technologist did not follow procedure, and in retrospect perhaps he might not have made the same decision. Computerization of forms and records can also be used as a tool to flag omissions.

The room in which the laboratory procedures were carried out had two separate product processing areas. The investigation revealed that each patient's stem cell products were in their appropriate canisters, having been handled with appropriate segregation. However, one of the units in one of the canisters was then labeled with the incorrect patient information. The labeling of the blood bags was the critical failure step in the process. The technologist had prepared the labels on a desk outside of the cell preparation areas. He believes that the labels were switched before they were placed into their appropriate segregated areas for the labeling process. This failure was compounded by a failure to check the blood bag label against the canister label at the time the blood bag was placed into the container and lack of required documentation of this verification. The potentially detectable error remained undetected.

Knowing the potential risk of confusing products from different patients, the lab had previously put into place the human factors "group or distinguish" rule [36], segregating the areas for each patient's products, but it was not sufficient in this case. Perhaps if the protocol had called for the labels to be prepared in the segregated areas, this error might have been avoided. Other human factor solutions could have made this error visible. If, for instance, each patient's information was printed on a different color label, the error would have been made obvious before the units were frozen and certainly before they were pooled.

## Process

Methods utilized to intercept errors should appear as far upstream in the process as possible. The paperwork/blood bag reconciliation which ultimately made the error visible occurred too late in the process to prevent the incorrectly labeled stem cells from being transfused. It is clear that the additional time to perform this reconciliation in the laboratory before the units are released to the patient is warranted. This way of thinking reminds us once again that patient verification processes may be needed everywhere, not just at the bedside [37].

## Case 2: Red Blood Cells Almost Transfused to Incorrect Patient (Near Miss Event)

### Clinical Summary

*Mr. Sebastian Michaels is a 62-year old who was admitted to the hospital for a Streptococcus pneumonia infection and anemia secondary to chronic lymphocytic*

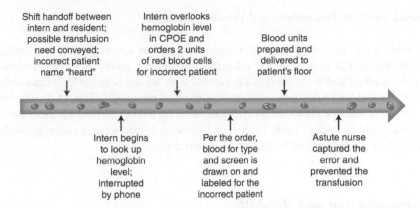

**Fig. 11.3** Case 2 Timeline—Red blood cells almost transfused to incorrect patient

*leukemia. At 11:45 P.M., an intern was rotating onto his shift. He was briefed via verbal handoff by the resident whose 24-h shift was ending, and he began reviewing charts in preparation for his 12 A.M. start. The departing house staff had said to "keep an eye" on Mr. Michaels, as his lab results from earlier in the day showed a hemoglobin level of 7.0. The receiving intern, however, "heard" the name Michael See, another patient on the same unit with whom he was familiar. The intern went to the computer and began to look up Mr. See's latest hemoglobin level, when he was interrupted by his cell phone ringing. The call was brief, and he turned his attention back to the computer. He proceeded to order two units of packed red blood cells for Mr. See without having looked up the hemoglobin value or seen the display of the normal value in the computerized physician order entry (CPOE) system. He then went on with his chart reviews (Fig. 11.3).*

## Preliminary Investigation

Investigation showed that the verbal handoff coupled with the cell phone interruption and compounded by inexperience and lack of sleep all came together in the near-transfusion of an incorrect patient with a normal hemoglobin level. The intern did not specifically look up the patient's hemoglobin level. Although it is presented in a dialog box when ordering blood, one additional click of the mouse constitutes acknowledgement of the value, and the "warning" is in essence overridden. The intern did not remember seeing the dialog box.

From this point on, the error in patient selection became "silent." The blood bank was notified of the order but does not routinely check hemoglobin levels. The laboratory technologist, therefore, did not have any indication that the order was placed on an incorrect patient. The blood units were delivered to the patient's floor, where an astute nurse recognized the error.

**Causal Analysis, Discussion, and Possible Solutions**

In building the causal tree, the consequent event was determined to be "patient almost received a unit of blood intended for another patient." As the question "why?" is asked and we move down the causal tree, we see both the latent and active conditions that led to this near-miss event illustrated pictorially (Fig. 11.4). The bottom boxes of the tree represent the root causes of the event and beneath them are the associated causal codes (Table 11.2). The following issues were revealed through a thorough investigation and building of the causal tree.

## Communication and Handoffs

Communication is a taken-for-granted human activity that is recognized as important once it has failed [38]. Recognition and reporting of the potential for error in communication and the clinical handoff process has exploded in recent years. Handoffs involve the transfer or reassignment of patient care and all associated responsibilities from one person or team to another. In our complex healthcare system, there are many handoffs and associated opportunities for errors. Patients are transferred from unit to unit, from one clinical discipline to another, between procedural areas, and among providers of care. Of the 936 sentinel events reported to the Joint Commission in 2009, and 802 reported in 2010, communication was identified as a root cause in 612 and 661 events, respectively [39]. A study involving interviews with residents about their routine activities and the medical mishaps in which they recently had been involved demonstrated that communication failures were an associated or contributory factor in 91 % of mishaps [40]. Critical, but less often considered or studied issues are handoffs/transfers of patient-related information to and from laboratories.

This hospital did not have a specific policy or procedure regarding verbal handoffs of patient-related information. Further investigation showed that this policy deficit also existed in the blood bank and had been a recognized past source of communication errors concerning blood product orders as well as communication of test results. As a possible solution, many recommend a check list (i.e., of critical patient information) for use in the handoff process [41]. However, in this case a check list might not have prevented the error. There is a large gap that separates hearing from actual *listening*, and seeing from actually *comprehending*. While the departing resident verbalized the name "Mr. Michaels," the receiving intern had another patient, Michael See, on his mind as Mr. See had received a blood transfusion the previous day. Cognitive psychology has shown that as humans, we sometimes see what we expect to see rather than what is actually there. Our perceptions may be biased by our expectations [25]. We can derive from this that we also sometimes hear what we expect to hear, rather than what was actually said.

Many studies have shown that the use of "read-backs," the repeating by the listener of what was said, may be one of the best means to improve verbal communication issues. The College of American Pathology requires that a caregiver is

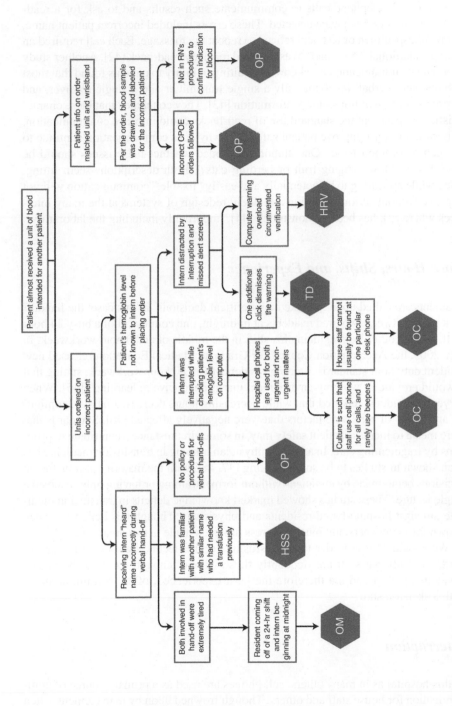

**Fig. 11.4** Case 2 Causal tree analysis—Red blood cells almost transfused to incorrect patient (refer to Table 11.2 for an explanation of codes)

immediately notified of critical laboratory results or "panic values." In a study that looked at 822 telephone calls to communicate such results and to ask for a read-back, 29 errors (3.5 %) were detected. These errors included incorrect patient name, test result, specimen or test, and refusal to repeat the message. Each call required an average additional 2.8 seconds to ask for and receive a read-back [42]. Another study analyzing communication breakdowns resulting in injury to patients found that most failures were verbal, involving only a single transmitter and a single receiver, and misinterpretation of transmitted information [43]. They concluded that these charac-teristics suggest that the standard use of read-backs could improve communication. This would in turn improve patient safety. So why does communication continue to be such a common issue? One author questioned whether these issues should be described as "low hanging fruit or herding cats?" Both descriptions seem fitting, since while appearing to be potentially and easily "fixable," communication will not improve without serious attention and likely redesign of systems at the many inter-faces where risk has been demonstrated [44], importantly including the laboratory.

## Work Hours, Shifts, and Experience

Residents are called upon to make many critical decisions. In this case, the handoff took place between two tired residents at midnight, one coming off of a busy 24 hour shift, and the other just beginning his shift in the middle of an 80 hour workweek. In July 2003, the Accreditation Council for Graduate Medical Education released new resident duty hour standards, limiting residents to an 80 hour workweek, stating that it would provide a "working environment that is conducive to learning" [45]. When surveyed, residents reported that while decreasing hours decreased fatigue, continu-ity and safety of care were factors that were negatively affected [46]. Standards that were meant to increase patient safety may, in some circumstances, contribute to prob-lems by fragmenting care. In addition, physician training in transfusion medicine has been shown in studies to be sorely lacking [47, 48], with the majority of transfusion decisions being made by clinicians without formal training or having only received a single lecture. These studies showed marked knowledge deficits in transfusion medi-cine amongst hospital-based residents and physicians. Though this deficit has been known for many years, not much progress has been made in this area [49].

We should also consider that transfusion service technologists on the "grave-yard," or midnight shift are frequently those without enough tenure to be on the coveted day shift and are therefore the least experienced, often working alone or with a skeleton staff.

## Interruption

In this hospital as in many others, cell phones are used as a primary source of com-munication for house staff and others. Though frowned upon by management, often the same cell phone is used for business and personal, urgent and nonurgent matters, and is frequently a cause of interruption. The omission of a step due to interruption of a well practiced routine, a skill-based error, is not an uncommon active failure [50]. These skill-based errors, or "slips," occur when attention is drawn but then followed by the lack of a subsequent intentional check, with the interrupted automatic activity smoothly resuming but at the wrong place. Skill-based errors are also commonly caused by the process of multitasking, when attention is not truly concentrated on any one task [51]. Particularly in this growing electronic age of cell phones, beepers, alarms, email, texting on-the-go, etc., interruptions are becoming a particularly frequent occurrence in human–computer interactions. Interruption rates are approaching 30 % of all clinical communication [52].

Interruptions and distractions play a major role as a cause of error in the transfusion service. Telephone calls about blood status, buzzing beepers, clinicians dropping in, noisy equipment, and ringing instrumentation alarms do not create an environment conducive to concentration [9]. In the medication process, an innovation to deal with distractions is a "quiet zone" for medication retrieval and preparation where staff may not be interrupted, and the wearing of a red vest or sash signaling "do not disturb" to others. In addition, "quiet zone" signs may be posted on medication room doors and above medication-dispensing machines to remind others of the issue. Perhaps similar methods may be employed during various steps in the transfusion medicine process.

## Conclusion

Due to the possible implications of blood administration, transfusion medicine has been on the forefront of safety innovation in health care. The reduction and elimination of numerous clinical hazards has been demonstrated over many years. Although there is still a heavy reliance on procedural methods in the essentially manual steps constituting the phases of the transfusion chain, recognition of this continued vulnerability has led to increased attention to the transfusion process itself. An encouraging exemplar of this current phase of enhancing the safety of transfusion is the creation of the new role of hospital transfusion safety officer to assist in the effort and advancement of monitoring, identifying, and resolving conditions that may lessen safety.

## Key Lessons Learned

- There has been a shift in transfusion medicine safety from a sole focus on disease transmission and clinical consequence to attention to error-prone process issues.
- In transfusion medicine, we tend to focus on the human error which is only one component of an event. We need to look beyond the human for the contribution of system flaws.

- A culture of safety must be nurtured, including reporting and analysis as a critical component. This culture should cross barriers between the transfusion medicine department and others, including handoff processes.
- Transfusion medicine is highly complex, with many opportunities for redundancy. Normal Accident Theorists believe that redundancy can increase system complexity, while High Reliability Theorists are skeptics, and think that two observations are better than one.
- Designing transfusion medicine processes with human factors in mind will reduce the possibility of errors occurring.
- Causal tree building provides a realistic view of how a system is functioning, as well as facilitates the creation of effective and lasting solutions. It can be particularly useful in transfusion medicine, as it looks at processes beyond the walls of the laboratory.

# References

1. Zou S, Musavi F, Notari E, et al. Prevalence, incidence, and residual risk of major blood-borne infections among apheresis collections to the American Red Cross Blood Services, 2004 through 2008. Transfusion. 2010;50(7):1487–94.
2. O'Brien SF, Yi Q, Fan W, et al. Current incidence and estimated residual risk of transfusion-transmitted infections in donations made to Canadian blood services. Transfusion. 2007;47(2):316–25.
3. Atreya C, Nakhashi H, Mied P, et al. FDA workshop on emerging infectious diseases: evaluating emerging infectious diseases (EIDs) for transfusion safety. Transfusion. 2011;51(8):1855–71.
4. Stramer SL, Hollinger FB, Katz LM, et al. Emerging infectious disease agents and their potential threat to transfusion safety. Transfusion. 2009;49:1S–29.
5. Finucane ML. Public perception of the risk of blood transfusion. Transfusion (Philadelphia, PA). 2000;40(8):1017–22.
6. Slovic P. Perception of risk. Science. 1987;236(4799):280–5.
7. Hashimoto S, Nakajima F, Kamada H, et al. Relationship of donor HLA antibody strength to the development of transfusion-related acute lung injury. Transfusion. 2010;50(12):2582–91.
8. Callum JL, Lin Y, Lima A, et al. Transitioning from 'blood' safety to 'transfusion' safety: addressing the single biggest risk of transfusion. ISBT Sci Ser. 2011;6(1):96–104.
9. Fastman BR, Kaplan HS. Errors in transfusion medicine: have we learned our lesson? Mt Sinai J Med. 2011;78(6):854–64.
10. Linden JV, Wagner K, Voytovich AE, et al. Transfusion errors in New York State: an analysis of 10 years' experience. Transfusion. 2000;40(10):1207–13.
11. Reason J. Managing the risks of organizational accidents. Brookfield, VT: Ashgate; 1997.
12. Lippi G, Guidi GC. Laboratory errors and medicare's new reimbursement rule. Lab Med. 2008;39(1):5–6.
13. Battles JB, Dixon NM, Borotkanics RJ, et al. Sensemaking of patient safety risks and hazards. Health Serv Res. 2006;41(4p2):1555–75.
14. Kaplan H, Battles JB, Van der Schaaf TW, et al. Identification and classification of the causes of events in transfusion medicine. Transfusion. 1998;38(11–12):1071–81.
15. Reason J. Human error: models and management. BMJ. 2000;320(7237):768–70.
16. McIntyre N, Popper K. The critical attitude in medicine: the need for a new ethics. BMJ. 1983;287:1919–23.

17. Trevas D. Building a culture of safety. AABB News: The Magazine for Transfusion and Cellular Therapies Professionals. 2011;13(10):21–4.
18. Leape LL. A systems analysis approach to medical error. J Eval Clin Pract. 1997;3(3):213–22.
19. Murphy MF. Application of bar code technology at the bedside: the Oxford experience. Transfusion. 2007;47:120S–4.
20. Dzik S. Radio frequency identification for prevention of bedside errors. Transfusion. 2007;47:125S–9.
21. Patient-centered care, right in the palm of your hand: a fast, easy, safe registration system makes its debut. News & Views, New York University Langone Medical Center. 2011.
22. Marx D. Whack-a-Mole: the price we pay for expecting perfection: By Your Side Studios; 2009.
23. Dekker S. The field guide to human error investigations. Burlington, VT: Ashgate; 2002.
24. Winchell W. The quotations page. Quotationspage.com, 1897–1972 [cited July 13 2013]. Available from http://www.quotationspage.com/quote/33861.html
25. Weick KE, Sutcliffe KM. Managing the unexpected: resilient performance in an age of uncertainty. 2nd ed. San Francisco, CA: Jossey-Bass; 2007.
26. Carroll JS, Quijada MA. Redirecting traditional professional values to support safety: changing organisational culture in health care. Qual Saf Health Care. 2004;13 Suppl 2:ii16–21.
27. Westrum R. A typology of organisational cultures. Qual Saf Health Care. 2004;13 Suppl 2:ii22–7.
28. Scott T, Mannion R, Davies H, et al. Implementing culture change in health care: theory and practice. Int J Qual Health Care. 2003;15(2):111–8.
29. TeamSTEPPS: National implementation. Rockville, MD: Agency for Healthcare Research and Quality [cited 13 July 2013]. Available from http://teamstepps.ahrq.gov/
30. Linden JV, Kaplan HS. Transfusion errors: causes and effects. Transfus Med Rev. 1994;8(3):169–83.
31. Perrow C. Normal accidents: living with high-risk technologies. Princeton, NJ: Princeton University Press; 1999.
32. Sagan SD. The problem of redundancy problem: why more nuclear security forces may produce less nuclear security. Risk Anal. 2004;24:935–46.
33. Weick KE, Sutcliffe KM, Obstfeld D. Organizing for high reliability: processes of collective mindfulness. In: Sutton R, Staw B, editors. Research in organizational behavior. Stanford, CA: JAI Press; 1999. p. 81–123.
34. Stephenson J. System safety 2000: a practical guide for planning, managing, and conducting system safety programs. New York, NY: Van Nostrand Reinhold Press; 1991.
35. Hinckley CM. Make no mistake!: an outcome-based approach to mistake-proofing portland. Portland, OR: Productivity Press; 2001.
36. Andersson UL. Humanware – practical usability engineering. Victoria, BC: Trafford; 1999.
37. Oops, sorry, wrong patient! a patient verification process is needed everywhere, not just at the bedside. March 10, 2011 [13 July 2013]. Available from http://www.ismp.org/newsletters/acutecare/articles/20110310.asp
38. Dayton E, Henriksen K. Communication failure: basic components, contributing factors, and the call for structure. Jt Comm J Qual Patient Saf. 2007;33(1):34–47.
39. Sentinel event data root causes by event type 2004 -third quarter 2011. The Joint Commission [cited 13 July 2013]. Available from http://www.jointcommission.org/assets/1/18/Root_Causes_Event_Type_2004_2Q2012.pdf
40. Sutcliffe KM, Lewton E, Rosenthal MM. Communication failures: an insidious contributor to medical mishaps. Acad Med. 2004;79(2):186–94.
41. Arora V, Johnson J. A model for building a standardized hand-off protocol. Jt Comm J Qual Patient Saf. 2006;32(11):646–55.
42. Barenfanger J, Sautter RL, Lang DL, et al. Improving patient safety by repeating (read-back) telephone reports of critical information. Am J Clin Pathol. 2004;121(6):801–3.
43. Greenberg CC, Regenbogen SE, Studdert DM, et al. Patterns of communication breakdowns resulting in injury to surgical patients. J Am Coll Surg. 2007;204(4):533–40.

44. Dunn W, Murphy JG. The patient handoff: medicine's formula one moment. Chest. 2008;134(1):9–12.
45. ACGME duty hours frequently asked questions [cited 21 Nov 2012]. Available from http://www.acgme.org/acgmeweb/Portals/0/PDFs/dh-faqs2011.pdf
46. Kort KC, Pavone LA, Jensen E, et al. Resident perceptions of the impact of work-hour restrictions on health care delivery and surgical education: time for transformational change. Surgery. 2004;136(4):861–71.
47. Arinsburg SA, Skerrett DL, Friedman MT, et al. A survey to assess transfusion medicine education needs for clinicians. Transfus Med. 2012;22(1):44–51.
48. O'Brien KL, Champeaux AL, Sundell ZE, et al. Transfusion medicine knowledge in postgraduate year 1 residents. Transfusion. 2010;50(8):1649–53.
49. Eisenstaedt RS, Glanz K, Polansky M. Resident education in transfusion medicine: a multi-institutional needs assessment. Transfusion. 1988;28(6):536–40.
50. Reason J. Human error. Cambridge, UK: Cambridge University Press; 1990.
51. Rabin FB, Kaplan H. Transfusion medicine: the problem with multitasking. In: Wu A, editor. The value of close calls in improving patient safety. Oakbrook Terrace, IL: Joint Commission Resources; 2011.
52. Alvarez G, Coiera E. Interruptive communication patterns in the intensive care unit ward round. Int J Med Inform. 2005;74(10):791–6.

# Chapter 12
# Hospital-Acquired Infections

Ethan D. Fried

*"Not to err at all...is above the wisdom of men; but it is part of a prudent and good man, to learn from his errors and miscarriages, to correct himself for the future."*

Plutarch

## Introduction

Hospital-Acquired Infections (HAIs) are infectious complications of care in health-related institutions including hospitals, nursing homes, rehabilitation facilities, and anywhere else patients are cared for. HAIs typically are the result of a breach of the body's normal barriers to infection in an environment where powerful antibiotics are frequently used. The resultant infection, therefore, is often caused by an organism that is resistant to antibiotic agents and requires even more powerful (and more costly) antibiotics to be used. HAIs account for $6.65 billion of annual healthcare expenditures in the US [1] and in one study of 5 ICUs, extended ICU stays by an average of 5.3 +/− 1.6 days [2]. They are also responsible for up to 99,000 patient deaths every year in US hospitals [3].

In 2008 the Center for Medicare and Medicaid Services (CMS) classified hospital-acquired central line infections, ventilator-associated pneumonias, and surgical wound infections as "never events" and additional payments to hospitals for these complications were stopped. In 2009, hospital-acquired *Clostridium difficile*

E.D. Fried, M.D., M.S., M.A.C.P. (✉)
Department of Internal Medicine, Columbia University College of Physicians and Surgeons, St. Luke's-Roosevelt Hospital, 1111 Amsterdam Avenue, Suite Clark 700, New York, NY 10025, USA
e-mail: EDFried@chpnet.org

A. Agrawal (ed.), *Patient Safety: A Case-Based Comprehensive Guide*,
DOI 10.1007/978-1-4614-7419-7_12, © Springer Science+Business Media New York 2014

colitis was added to this list. Prior to this many institutions viewed HAIs as unpreventable complications; a low percentage of these were tolerated and even expected. By not allowing to be paid for the extra bed-days and other resources needed to treat these infections, healthcare facilities have reinvigorated their efforts to avoid even a single HAI.

## Case Studies

## Case 1

### Clinical Summary

*Four days ago, an 84-year-old man named Charles Frost was sent from the Blessed Virgin nursing home to Central Valley Hospital (CVH) for fever and severe diarrhea. Before the transfer, he was treated for fever related to a deep sacral decubitus ulcer with oral sulfamethoxazole and trimethoprim for 10 days at the nursing home. His white blood cell count went from $10 \times 10^9$ per L, declining to $2 \times 10^9$ per L within 5 days on oral antibiotics. Four days ago he was noted to have voluminous diarrhea which prompted the transfer. On the day that the diarrhea was noted his white blood cell count had soared to $22 \times 10^9$ per L.*

*At the hospital, the patient was diagnosed with presumed Clostridium difficile colitis. He was placed on contact isolation and was admitted to an isolation room at the end of the corridor. Stool samples were collected and sent to the lab facility for C. difficile antigen testing and oral metronidazole was prescribed. After four days of unremitting diarrhea and attempted fluid resuscitation Mr. Frost died in his bed.*

*That very same day, two other elderly patients on the same floor began to have diarrhea. A root cause analysis (RCA) was requested by the hospital's infection control director.*

### Root Cause Analysis

Hospital procedures for cases of *C. difficile* colitis were to place the patient on contact isolation in a private room with a vestibule that had its own sink and antimicrobial soap dispenser. There were only four such rooms in the hospital at each end of the corridors on the third and the fourth floors. A supply of yellow impermeable gowns and gloves in three sizes were to be placed on a rolling table outside the door of the room. The isolation rooms were to be stocked with their own stethoscopes, blood pressure cuffs, and thermometers so that this equipment would not be carried

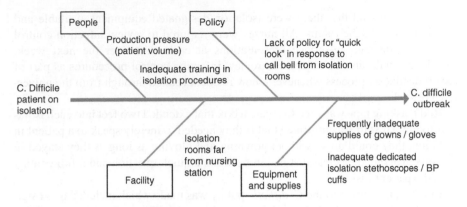

**Fig. 12.1** Stem and leaf diagram for Case 1

from one isolation patient to another. Green placards would be placed on the door of the room indicating that the patient within was on contact isolation and warning visitors to see the charge nurse before entering the room.

In the RCA it was noted that over the last year or so, so many stethoscopes had disappeared and so many blood pressure cuffs had worn out that this equipment was rarely available for the use of each isolation room alone. Each staff member developed different techniques for using equipment that was shared with other patients. Some wiped the equipment down with alcohol wipes. Others placed the chest piece of their own stethoscope into a latex glove to use it for an isolation patient. None of these techniques were considered standard processes nor did the infection control department sanction them.

Central Valley Hospital used an old call bell system for patients to use when calling for help. In many cases when Mr. Frost had pushed his call bell, the nearest nurse simply popped her head into the room to ask what the problem was. If the problem involved some service to be performed the nurse would normally step out and don the protective gear before going back in to assist him. But it was known that sometimes gloves or a gown alone or no barrier equipment at all was used.

The Internal Medicine resident who sat in at the RCA stated that if a team was behind in their rounds and there were no gowns on the cart outside the room that they would go in to the room any way in the interest of time and to be as thorough as possible in following up their work with patients. Furthermore, the resident did not know that alcohol-based hand sanitizer alone was insufficient to eliminate *C. difficile* spores.

In the end, the RCA committee concluded that *C. difficile* had been spread from Mr. Frost to the other two patients because of multiple failures of infection control procedure (Fig. 12.1). As a result it was decided that central sterile technicians would visit each of the four isolation rooms as part of their daily rounds of the nurses' station supply rooms to make sure that they were stocked with enough gowns, gloves, and that there were isolation-designated equipment available and visible in each of the rooms. All nurses were required to review infection control procedures through in-service presentations at each shift over the next week. Residents would also be required to review infection control procedures as part of the credentialing process whenever a new resident rotated through from the university hospital.

Red masking tape was used to create a box that extended two feet into each isolation room. All staff was informed that if they needed to merely speak to a patient in isolation, they could do so without gowning and gloving as long as they stayed in this box. This made a clear distinction between a quick question and a full contact with the patient inside [4].

Finally, an addition to the *C. difficile* policy was made which included a red sign to be placed on the alcohol-based hand sanitizers outside a *C. difficile* patient's room informing all workers and visitors that washing with soap and water at the sink was necessary after contact with the patient inside.

## Case 2

### Clinical Summary

*John Russell was brought by ambulance to the Emergency Department of the St. Joseph's Medical Center after vomiting a large amount of blood in the subway. The patient was a 42-year-old man with a long history of intravenous drug use. Five years ago he had been diagnosed with Hepatitis C and was then lost to follow up. At the time of presentation he was severely dehydrated and peripheral venous access could not be established. A right femoral central venous line was inserted during the emergency upper endoscopy that diagnosed the patient with bleeding esophageal varices. The varices were sclerosed during the procedure. After 12 h in the Emergency Department, 4 units of transfused packed red blood cells and a continuous drip infusion of somatostatin, Mr. Russell was stable enough to be sent to a bed on the regular patient care unit. The patient did well over the first four days of his hospitalization. He was treated with lactulose for his hepatic encephalopathy and his blood counts were monitored twice per day. On hospital day five a student nurse asked about the dressing over the patient's right groin. She also recorded a fever of 39.5 °C. The patient's team of resident physicians pulled the old femoral central line out and sent the tip of the line for culture. Blood cultures were also taken from the patient prior to starting intravenous gentamicin and vancomycin. Within 24 h the blood and the line tip were growing Staphalococcus aureus which was found to be resistant to methicillin by the microbiology lab.*

*The infection control nurse, upon learning of the methicillin-resistant Staphyloccus aureus (MRSA) line tip and positive blood cultures, requested an RCA.*

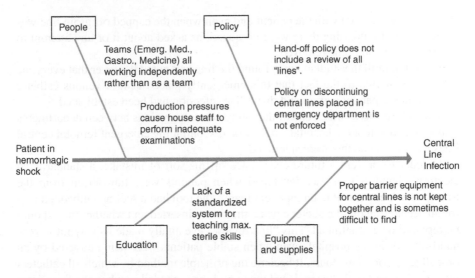

**Fig. 12.2** Stem and leaf diagram for Case 2

## Root Cause Analysis

The RCA in this case revealed several failures of established protocols (Fig. 12.2). The patient was originally brought to the trauma room of the Emergency Department. Owing to his hypotension and the history of hematemesis a reliable intravenous line needed to be rapidly established so that fluid and blood resuscitation could be initiated. As the gastroenterology team was working at the patient's head, the femoral site was chosen for the central line. The patient was admitted to the medical service and was immediately attended to by the "night medicine" team. There was not much for the team to do, however, because the senior resident responsible for screening patients into the Intensive Care Unit was working hard on the patient to stabilize him so that he could be transferred to a regular medical bed instead of a critical care bed. By morning, the night medicine team, which passed the care of the patient to the floor team, had met him only briefly and had examined him in a most rudimentary way. In their hand-off, the night medicine team had failed to mention the femoral line.

Upon taking over the care of this patient, the floor team placed a large bore peripheral line and took patient's history. Since there was already a full history and physical exam written by the night medicine team in the chart, the floor team intern just listened to the heart and lungs and examined the patient's abdomen. It was determined that the team treating Mr. Russell on the medical floor was not even aware that a femoral central line was present. They were giving fluids and medications through the 20 gauge peripheral line in the patient's left forearm which they established after he was transferred to the floor from the emergency department. The floor team intern had not even exposed the groin. Indeed, although shift after shift of nurses worked with the patient and even when the capped off groin line was detected under a dressing there were no questions asked about it or any attempt to determine if it should be removed.

It was not until the student nurse found the line and noted the fever that everyone was finally alerted to the fact that this emergently placed central venous catheter was still in the patient's groin and that a line infection had been established.

The failure was as much one of the faulty communications between departments as it was a breach of the hospital's policy on discontinuing emergent femoral central lines as soon as patients were stabilized.

In order to prevent a future occurrence of this sort of mistake, a standardized electronic handoff tool was developed when patients were transferred from the Emergency Department to an inpatient team. The tool had a section embedded in it for an accounting of all catheters, when they were inserted and whether they should be replaced or discontinued. This "line review" eventually made its way into every hand-off tool in the hospital, even when stable patients were being covered by an on-call team for a few hours. It became the principle method by which all catheters were tracked and pulled or replaced when no longer needed or when the lines were deemed contaminated.

As part of a general review of central line procedures, two additional programs were put into place. central line-associated bacteremia (CLAB) prevention kits were assembled by central sterile services and placed in every nursing unit utility room. These kits contained all of the equipment needed to place sterile central lines including full-length sterile gowns, large fenestrated drapes, chlorhexidine sponges, a suture kit, and clear impermeable dressings. The existence of these kits would eliminate the need for clinicians to search for each of the components and lessen the likelihood of lapses or making due with inadequate equipment.

Finally, the residency program at the university hospital had recently instituted a required simulation exercise for all residents to learn maximal sterile precautions. This was based on a recent report that suggested that this sort of training could prevent many central line infections [5]. Six months later, the "line review," CLAB prevention kits, and simulation training were credited with a 25 % drop in line infections documented by the infection control team.

# Discussion

## Background

Before there was a patient safety movement there was infection control. Indeed it was the 1840s when Oliver Wendell Holmes and Ignaz Semmelweis, before the discovery of germ theory itself, independently recognized that the hands of physicians could transmit some agent from the autopsy table to the womb of expectant mothers during obstetrical examinations that resulted in puerperal sepsis and maternal death. Semmelweis demonstrated that rinsing the hands in a mixture of chlorinated lime dramatically reduced the number of maternal deaths [6]. Soon thereafter we find James Young Hamilton discovering the epidemiology and prevention of surgical site infections and in 1865 Joseph Lister pioneering the use of antisepsis in orthopedic surgery [7].

In 1962, Mortimer and his colleagues demonstrated that the hands of medical personnel transmitted *S. aureus* in a neonatal unit and that hand hygiene with hexachlorophene prevented such transmission [8].

Modest measures in infection control followed in the 1970s but interestingly it was the concern for healthcare workers in the 1980s and the risk of transmission of the Human Immunodeficiency Virus and Viral Hepatitis B from patients to healthcare workers that really got infection control going [9]. Indeed it was the Occupational Safety and Health Administration (OSHA) that required hospitals to protect workers from these pathogens in 1991 rather than any set of regulations that protected patients. By 2003 the Joint Commission for the Accreditation of Healthcare Organizations (TJC) launched an infection control-related sentinel event alert and put institutions on notice that deaths and major morbidity related to nosocomial infections were to be treated as sentinel events and investigated by a team of healthcare professionals and that systemic steps be taken to prevent such events [10].

## Infection Control as a Multidisciplinary Team-Based Enterprise

Thus it can be said that infection control was the first aspect of patient safety that utilized the modern quality improvement team-based approach. Clearly, the success or failure of infection control in hospitals and other healthcare environments rests on the effectiveness of the infection control team. Recommendations made by Haley and Quade et al. in 1980 in a study of the efficacy of nosocomial infection control described the components of the modern infection control program. They reported that an effective team must (1) monitor HAIs and give feedback to workers, (2) institute best practices with regard to sterilization, disinfection, asepsis, and the handling of medical devices, (3) include an infection control nurse and a physician epidemiologist or microbiologist with special skills in infection prevention [11].

As with any quality improvement process, systematic solutions are more effective than individual reminders and staff education projects. The most effective solutions generally work behind the scenes and are so integral to a process as to be harder to perform incorrectly than correctly or they are forced functions which literally do not allow step "B" to be performed until step "A" is completed. These solutions, sometimes called "change concepts" can be as simple as removing a wasteful test from a preprinted laboratory order form. By making physicians write in the name of the test rather than just checking a box, the frequency of ordering this wasteful test is reduced [12]. Designing forced functions into a process is part of the science of human factors engineering. These activities are deliberately "baked-in" to a process to insure that lapses do not occur. The classic example of a forced function outside of medicine is the engineering advance that does not allow a car to be placed in reverse unless the brake pedal is depressed [13]. Antibiotic stewardship is a form of forced function in that one cannot obtain an overused or otherwise risky antibiotic without specifically consulting with an expert who makes sure that the criteria for using it are met.

## Comprehensive Unit-Based Safety Program

In 2003 researchers from Johns Hopkins University led by Peter Pronovost implemented a system in the intensive care units in 127 hospitals in Michigan in what came to be known as the Michigan ICU project. The system was comprised of a Comprehensive Unit-based Safety Program (CUSP) and the Central line-associated bloodstream infection (CLABSI) bundle and it effectively reduced CLABSI rates to zero. In 2008 the Agency for Healthcare Research and Quality began to fund a nationwide expansion of CUSP. Today over 750 hospitals across 44 states and territories utilize the "CUSP: Stop BSI" program and have achieved sustained decreases in CLABSI of 33 % or more. The percentage of participating institutions with zero CLABSI over a 3 month period is now 69 % [14].

The CUSP system is comprised of five steps:

Step one is to educate the staff of a patient care unit on the science of safety training. This training includes an understanding that patient safety is ultimately a product of a system of care and not the effect of any one person. The training also creates a framework of standardized work, checklists, and open discussion about mistakes. This education includes safe design as it applies to both technical work and teamwork. Finally, the training makes the point that teams can only be effective when there is diverse and independent input from all members.

Step two is to identify defects in patient safety from a variety of sources including surveying the staff.

Step three sets up partnerships between unit staff and senior hospital leadership in order to improve communication and provide resources to staff to mitigate identified risks. Staff are held accountable for implementing risk reducing measures.

Step four is to implement a series of rapid cycle projects to reduce risks. Each project triggered by events must identify (1) what happened, (2) why it happened, (3) what was done to reduce risk, and (4) how it is known that risks were actually reduced. The expectation is that staff will complete one project per month.

Step five is to implement unit specific teamwork tools to improve teamwork, communication, and patient safety systems.

## Hand Washing Video Surveillance

Although hand washing was historically the first proven measure to fight HAIs, it remains to this day one of the hardest safeguards to enforce in clinical settings. Healthcare workers' average adherence to hand washing is approximately 40 % based on an oft-cited systematic review [15]. Although biologically inferior to thorough soap and water hand washing, hand disinfection with alcohol-based, self-drying hand rubs, gels, and foams is thought to improve the adherence to any sort of disinfection routine as to make it a superior strategy to reduce the transmission of infection [15]. Even with the ubiquitous placement of alcohol-based hand rub dispensers, the best proven measurement of hand hygiene practices and enforcement is direct and video surveillance [16, 17]. Other metrics of compliance include the consumption of hand hygiene material and dispensers that count the number of actuations. Studies have shown, however, that one of the strongest motivators of hand hygiene adherence is the role modeling of senior clinicians [18].

## Device Utilization Ratio

In 1970, the Center for Disease Control established the National Nosocomial Infections Surveillance System (NNIS) to monitor data on the incidence of nosocomial infections. In 1988 definitions for each type of nosocomial infection were published [19]. Then in 1997 the NNIS created an index to monitor these infections which would allow institutions and health agencies to compare relative rates of device utilization and device-associated infections [20]. This measurement, known as the device utilization ratio (DUR) and the device-associated infection rate, established a standard unit for monitoring hospital-acquired infections due to devices. The DUR compares the number of devices used in a population expressed as percentage of patients in that population. The device-associated infection rate is the number of infections per 1000 device days. The DUR incorporates a time period, a patient population (whole hospital, ICU, NICU, etc.), and the specific device being monitored. For example in a specific ICU in the first week of January there are 12 patients on Monday, 10 on Tuesday, 11 on Wednesday, 9 on Thursday, 11 on Friday, and 12 each on Saturday and Sunday. During that week 4 patients had central lines (defined as any catheter having its termination close to the heart) in place on Monday, 3 on Tuesday, 5 on Wednesday, 3 on Thursday, 3 on Friday, 4 on Saturday, and 2 on Sunday. The DUR therefore is $(4+3+5+3+3+4+2)/(12+10+11+9+11+12+12)=24/77=31$ %. If during that time 3 CLABSI are detected, the CLABSI rate is $3 \times 1,000/24=125$ infections per 1,000 device days.

The DUR can be seen as a surrogate for disease burden or severity with higher percentages implying a sicker population. The device-associated infection rate can be seen a measure of a hospital's success in preventing infections with lower numbers usually associated with better compliance with infection avoiding protocols [21].

## C. difficile Antibiotic-Associated Diarrhea

In the first scenario, stocking of the barrier supplies near the isolation rooms had been an intermittent, added on responsibility of the nursing staff who lacked access to supplies if they were not present in the utility storeroom. Rather than waiting for

**Table 12.1** CLABSI bundle to prevent central line infections

1. Hand hygiene
2. Maximal barrier precautions upon insertion
3. Chlorhexidine skin antisepsis
4. Optimal catheter site selection, with avoidance of the femoral vein for central venous access in adult patients
5. Daily review of line necessity with prompt removal of unnecessary lines

a nurse to note a shortage of supplies and informing the charge nurse who then placed an order with supply, who then brought the supplies to the utility room from where they would be placed at the isolation room by the nurse, the responsibility to monitor and replace isolation supplies was left directly to the supply technicians eliminating several steps in the process. Changes in work flow to make things safer, more efficient, and less wasteful are typical of a relatively new concept in healthcare management called Lean Design [22].

Although it is clear that the education of staff alone is insufficient to insure progress, timely reminders and visually arresting warnings may be helpful. In the case of the "red square" just inside the doorway to isolation rooms and the red labels on alcohol-based sanitizer dispensers, the reminders come just at the time they are needed to make the staff more aware of the steps to avoid cross contamination.

## Central Line-Associated Bloodstream Infection

In the second scenario, a coordinated approach was again employed. In this case, a forced function was built into the electronic handoff procedure that required the user to address all catheters that were placed into the patient. Users would not be able to save or transmit a "signout" in the electronic system unless they addressed these catheters.

At the same time, a team approach needed to be applied to the placement of central lines. Essential components of the CLABSI bundle were the use of hand hygiene prior to beginning the procedure, maximal barrier precautions when placing the line, avoiding the use of the femoral site for insertion, the daily discussion of the ongoing use of the central line, the use of an insertion checklist, and the promotion of a culture of safety for patients [23] (Table 12.1). In addition, the training of residents using simulation with close direct observation and feedback to ensure proper sterile precautions helped to standardize the way they were taught to do the procedure [5].

While the use of the CLABSI bundle and central line insertion checklists clearly reduce infections in short-term use catheters, many medically necessary catheters are designed for long-term use defined as longer than 10 days. To prevent infections associated with these catheters, a strategy to reduce the formation of intraluminal biofilm must be used. Factors that have been proven to reduce the growth of

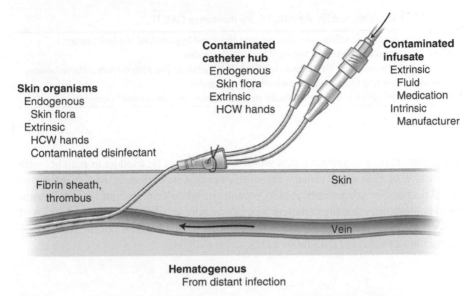

**Skin organisms**
Endogenous
  Skin flora
Extrinsic
  HCW hands
  Contaminated disinfectant

**Contaminated catheter hub**
Endogenous
  Skin flora
Extrinsic
  HCW hands

**Contaminated infusate**
Extrinsic
  Fluid
  Medication
Intrinsic
  Manufacturer

Fibrin sheath, thrombus

Skin

Vein

**Hematogenous**
From distant infection

**Fig. 12.3** Routes for central venous catheter contamination with microorganisms. Potential sources of infection of a percutaneous intravascular device (IVD): the contiguous skin flora, contamination of the catheter hub and lumen, contamination of infusate, and hematogenous colonization of the IVD from distant, unrelated sites of infection. HCW: health care worker (*Source*: Crnich CJ, Maki DG. The promise of novel technology for the prevention of intravascular device-related bloodstream infection. I. Pathogenesis and short-term devices. Clin Infect Dis. 2002 May 1;34(9):1232–42)

intraluminal biofilms include hand hygiene when accessing ports, adequate disinfection of ports, use of split septum rather than mechanical valve needleless connectors, and the replacement of administration sets and add-on devices no more frequently than the manufacturer recommended rate unless contamination occurs [24] (Fig. 12.3).

## Catheter-Associated Urinary Tract Infections

The primary way to prevent catheter-associated urinary tract infections (CAUTIs) seems to be to avoid or limit the use of indwelling urinary catheters. CAUTIs represent up to 40 % of HAIs but their consequences vary greatly. Asymptomatic CAUTIs are rarely associated with adverse outcomes and generally do not require treatment. Bacteriuria and pyuria associated with fever or other urinary tract symptoms, however, can lead to renal failure and sepsis and must be treated accordingly. Strategies for the avoidance of CAUTI include alternatives to indwelling urinary catheters like intermittent catheterization, the use of condom catheters, avoiding bladder irrigation, and the monitoring of bladder with portable ultrasound

**Table 12.2** The bladder bundle: the ABCDE for preventing CAUTI

Adherence to general infection control principles like hand hygiene and sterile insertion
Bladder ultrasound to monitor for the need for catheterization
Condom catheters and other alternatives to indwelling catheters like intermittent catheterization
Do not use indwelling catheters unless absolutely necessary
Early removal of catheters using a reminder or other nurse-initiated removal protocols

equipment. It is recommended that when catheters must be used as in patients with obstructive or functional urinary retention, urinary incontinence in the setting of sacral decubiti or other perineal skin wounds or when urine output must be monitored continuously that the catheters be placed under sterile conditions and removed as soon as possible. As with central intravenous lines a bladder bundle can reduce the number of CAUTI [25] (Table 12.2).

## *Ventilator-Associated Pneumonia*

Healthcare-associated pneumonia (HCAP) and ventilator-associated pneumonia (VAP) account for up to 15 % of HAIs and are associated with as many as 36,000 hospital deaths. Although there is no gold standard for establishing a diagnosis of HCAP, it is generally thought that any pneumonia that develops more than 72 hours into a hospitalization falls into this category. The chief reason for this is that the oropharyngeal flora of hospitalized patients tends to change after 72 hours to gram negative and other resistant organisms. Small amounts of this oropharyngeal flora are thought to be aspirated causing HCAP and VAP. It is important for healthcare providers to recognize that intubation with an endotracheal tube and feeding with a nasogastric tube do not prevent microaspirations of oropharyngeal flora and in the case of the latter, technology may even promote it.

Other factors that may promote HCAP and VAP include supine rather than semi-recumbent positioning in patients who have altered mental status or who are intubated and the use of proton pump inhibitors for acid suppression. As with other HAI, a ventilator bundle or checklist has been shown to reduce morbidity and mortality associated with ventilator-associated pneumonia [26].

The ventilator bundle is designed as much to reduce some of the complications associated with VAP as it is to prevent the aspiration of infectious material. The bundle is listed in Table 12.3. Included in the bundle is the peptic ulcer disease prophylaxis, usually with a long-acting proton pump inhibitor (PPI). While the use of PPI can theoretically allow for gastric bacterial overgrowth normally suppressed by stomach acid and has been associated with the development of HCAP, the incidence of stress-related ulcers in ventilated patient who develop pneumonia is high enough and contributes to so much morbidity and mortality that it is considered safer that

**Table 12.3** The ventilator
bundle to prevent VAPs

| |
|---|
| 1. Head of the bed raised to 30° |
| 2. Daily sedative interruption and assessment of readiness to extubate |
| 3. Peptic ulcer disease prophylaxis |
| 4. Deep venous thrombosis prophylaxis |
| 5. Daily oral decontamination with chlorhexidine |

PPI be used routinely in ventilated patients. Similarly, the rate of deep venous thrombosis (DVT) is higher in patients with VAP and therefore prophylaxis is recommended in the absence of contraindications.

Of all the measures that have been attempted to remove infected material from the oropharynx including regular tooth brushing, subglottic suctioning of the secretions around the cuff of the endotracheal tube, and selective decontamination of the digestive tract, oropharyngeal decontamination with chlorhexidine seems to be the most powerful although the other measures are effective in some studies.

A chapter on hospital-acquired respiratory infections would not be complete without mentioning hospital-acquired tuberculosis, legionella, and aspergillus pneumonia. Each of these pathogens can be controlled with disciplined use of isolation, surveillance, and containment of the offending agents. All suspected tuberculosis patients must be isolated, ideally in rooms equipped with negative pressure and there must be periodic surveillance of the infectious status of healthcare workers with PPD skin tests. Alert institutions will maintain surveillance of the water and air conditioning systems of a hospital to prevent outbreaks of Legionella. Finally, hospitals that undertake renovation projects must use precaution to avoid the airborne spread of Aspergillus, which tends to colonize older construction.

## Surgical Site Infections

Surgical site infections (SSIs) occur in 2–5 % of surgical procedures which amounts to 300,000–500,000 infections each year in the USA. With over 230 million operations occurring annually worldwide, even a 3 % infection rate yields almost 7 million preventable infections [27] each adding more than a week of hospitalization, costing up to $29,000 per patient [28] and increasing surgical mortality by 2- to 11-fold [29].

Many risk factors have been identified that may contribute to SSIs. Recommendations that mitigate these risks can reduce SSI greatly (Table 12.4). Of these, four recommendations stand out. They are appropriate use of prophylactic antibiotics, appropriate hair removal, controlled postoperative glucose control (especially in cardiac surgery), and preventing of postoperative hypothermia (especially in colorectal surgery) [30].

**Table 12.4** Risk factors and recommendations to mitigate the surgical site infections (SSIs) [31]

| Risk factor | Recommendation |
| --- | --- |
| **Patient related** | |
| Glucose control | Control serum levels to below 200 mg/dl |
| Obesity | Adjust dose of prophylactic antimicrobials according to body weight |
| Smoking | Encourage smoking cessation within 30 days of the procedure |
| Immunosuppressive medications | Avoid if possible |
| Nutrition | Do not delay surgery to enhance nutritional support |
| Remote sites of infection | Identify and treat before elective procedures |
| Preoperative hospitalization | Keep as short as possible |
| **Procedure related** | |
| Hair removal | Do not remove unless presence interferes with operation. If necessary remove by clipping and not shaving immediately before surgery |
| Skin preparation | Wash and clean area around surgical site with approved solutions |
| Chlorhexidine nasal and oropharyngeal rinse | No recommendation. Some evidence that nosocomial infections reduced in cardiac surgery |
| Surgical scrub (surgeons hands and forearms) | 2–5 min preoperative scrub with appropriate antiseptic agent is needed |
| Incision site | Appropriate antiseptic agent |
| Antimicrobial prophylaxis | Administer when indicated |
| Timing of prophylaxis | Within 1 h prior to first incision |
| Choice of prophylaxis | Appropriate to surgical procedure |
| Duration of prophylaxis | Stop within 24 h of procedure |
| Surgeon technique | Eradicate dead space |
| Incision time | Minimize |
| Maintaining oxygenation with supplemental $O_2$ | May be important in colorectal procedures |
| Maintain normothermia | Actively warm patient to >36°. Particularly in colorectal surgery |
| **Operating room characteristics** | |
| Ventilation | Follow American Institute of Architects' recommendation |
| Traffic | Minimize |
| Environmental surfaces | Use approved hospital disinfectant to clean visibly soiled or contaminated surfaces and equipment |

# Conclusions

As the preceding cases and the other entities discussed clearly show, HAIs are amongst the most wasteful and destructive of error-related adverse events. In each instance bundled procedures seem to offer some systematic way to avoid these complications. In order for health facilities to adhere to these recommended procedures, that teamwork and discipline are key. Strong leadership that emphasizes the use of bundles and checklists by each and every professional that interacts with the patient has been proven to reduce the burden of these events for patients, institutions, and the cost of health care overall [32].

It is regrettable that one of the most effective means to encourage hospitals to adopt strong infection control measures may be to deny payments for complications like hospital-acquired infections. In 2008 certain hospital-associated infections were included on a list of "never events" as defined by the CMS. These included central line-associated blood stream infections, ventilator-associated pneumonias, surgical wound infections and, later, *C. difficile*-related diarrhea. CMS and other insurers stopped paying hospitals for these complications and the Leapfrog Group recommended that facilities waive any fee they might collect for such events. The effect of this intervention is unknown as most of the new regulations took effect in 2010 [33]. The spotlight on these policies and the public reporting of institutional safety parameters, it is believed, will continue to drive the creation of safer health care.

## Lessons Learned

- Hospital-Acquired Infections are no longer considered acceptable at any level. They are all felt to be a failure of infection control practices and they are no longer paid for by Medicare and many other insurers.
- Many HAIs can be avoided by bundled checklists that usually start with hand hygiene and also include meticulous sterile technique when placing, accessing, and maintaining medical devices.
- The most effective way to prevent device-associated HAIs is to carefully consider when devices need to be used and to remove them as soon as possible.
- As with almost all patient safety initiatives, a systematic and multidisciplinary approach is needed to successfully prevent HAIs which includes administration as well as frontline healthcare providers.

## References

1. Scott RD. The direct medical costs of healthcare-associated infections in U.S. hospitals and the benefits of prevention; 2009. http://www.cdc.gov/ncidod/dhqp/pdf/Scott_CostPaper.pdf. Accessed 6 Feb 2012.
2. Beyersmann J, Gastmeier P, Grundmann H, Bärwolff S, Geffers C, Behnke M, et al. Use of multistate models to assess prolongation of intensive care unit stay due to nosocomial infection. Infect Control Hosp Epidemiol. 2006;27:493–9.
3. Klevens RM, Edwards JR, Horan TC, Gaynes RP, Pollack DA, Cardo DM. Estimating health care-associated infections and deaths in U.S. hospitals, 2002. Public Health Rep. 2007;122:160–6.

4. Frank JN, Behan AZ, Herath PS, Mueller AC, Marhoefer KA. The red square strategy: an innovative method to improve isolation precaution compliance and reduce costs. Am J Infect Control. 2011;39:E208.

5. Khouli H, Jahnes K, Shapiro J, Rose K, Mathew J, Gohil A, et al. Performance of medical residents in sterile techniques during central vein catheterization: randomized trial of efficacy of simulation-based training. Chest. 2011;139:80–7.

6. Bolson M. Hand hygiene. Infect Dis Clin North Am. 2011;25(1):21–43.

7. Lister J. On the antiseptic principle in the practice of surgery. BMJ. 1867;ii.

8. Mortimer Jr EA, Lipsitz PG, Wolinsky E, et al. Transmission of staphylococci to newborns: importance of the hands to personnel. Am J Dis Child. 1962;104:289–95.

9. Occupational exposure to bloodborn pathogen—OSHA. Final rule. Federal Register 1991. 56:64004 Standard 1910.1030.

10. Infection control related sentinel events. Available at http://www.jointcommission.org/assets/1/18/SEA_28.pdf. Accessed 6 Feb 2012.

11. Haley RW, Quade D, Freeman HE, Bennett JV. The SENIC project. Study on the efficacy of nosocomial infection control. Summary of study design. Am J Epidemiol. 1980;111:472–85.

12. Berwick DM. A primer on leading the improvement of systems. BMJ. 1996;312:619.

13. Wachter RM. Understanding patient safety. New York, NY: McGraw Hill; 2008. p. 22.

14. On the CUSP: stop BSI—comprehensive unit-based safety program manual, Agency for Healthcare Research and Quality; April 2009. p. 2.

15. Hospital Infection Control Practices Advisory Committee (HICPAC). Recommendations for preventing the spread of vancomycin resistance. Infect Control Hosp Epidemiol. 1995;16(2): 105–13.

16. Boyce JM, Pittet D. Guideline for hand hygiene in healthcare settings. Recommendations of the Healthcare Infection Control Practices Advisory Committee and the HICPAC/SHEA/APIC/IDSA Hand Hygiene Task Force. MMWR Recomm Rep. 2002;51(RR-16):1–45.

17. Boyce JM. Hand hygiene compliance monitoring: current perspective from the USA. J Hosp Infect. 2008;70 Suppl 1:2–7.

18. Haas JP, Larson EL. Measurement of compliance with hand hygiene. J Hosp Infect. 2007; 66(1):6–14.

19. Duggan JM, Hensley S, Khuder S, Papadimos TJ, Jacobs L. Inverse correlation between level of professional education and rate of handwashing compliance in a teaching hospital. Infect Control Hosp Epidemiol. 2008;29(6):534–8.

20. Garner JS, Jarvis WR, Emori TG, Horan TC, Hughes JM. CDC definitions for nosocomial infections. Am J Infect Control. 1988;16:128–40.

21. National nosocomial infections surveillance (NNIS) report, data summary from October 1986–April 1997, issued May 1997. A report from the NNIS System. Am J Infect Control. 1997;25:477–87.

22. Graben M. Lean hospitals: improving quality, patient safety and employee satisfaction. New York, NY: CRC; 2009. p. 97–117.

23. Pronovost P, Needham D, Berenholtz S, Sinopoli D, Chu H, Cosgrove S, et al. An intervention to decrease catheter related bloodstream infections in the ICU. N Engl J Med. 2006;355(26): 2725–32.

24. The Joint Commission. Preventing central line–associated bloodstream infections: a global challenge, a global perspective. Joint Commission; 2012. p. 56.

25. Saint S, Olmsted RN, Fakih MG, Kowalski CP, Watson SR, Sales AE, et al. Translating health care-associated urinary tract infection prevention research into practice via the bladder bundle. Jt Comm J Qual Patient Saf. 2009;35(9):449–55.

26. How-to guide: prevent ventilator-associated pneumonia. Cambridge, MA: Institute for Healthcare Improvement; 2012. Available at http://www.ihi.org. Accessed 16 Feb 2012.

27. Gawande A. Checklist Manifesto, Metropolitan Books. New York: Henry Holt; 2009. 87.

28. Anderson DJ, Kirkland KB, Kaye KS, et al. Under-resourced hospital infection control and prevention programs: penny wise, pound foolish? Infect Control Hosp Epidemiol. 2007;28(7): 267–73.

29. Kirkland KB, Briggs JP, Trivette SL, et al. The impact of surgical site infections in the 1990s: attributable mortality, excess length of hospitalization and extra costs. Infect Control Hosp Epidemiol. 1999;20(11):725–30.
30. How-to guide: prevent surgical site infections. Cambridge, MA: Institute for Healthcare Improvement; 2012. Available at http://www.ihi.org. Accessed 16 Feb 2012.
31. Mangram AJ, Horan TC, Pearson ML, et al. Guidance for prevention of surgical site infection. Center for Disease Control Hospital Infection Control Practices Advisory Committee. Am J Infect Control. 1999;27(2):97–132.
32. Cook E, Marchaim D, Kaye KS. Building a successful infection prevention program: key components, processes, and economics. Infect Dis Clin North Am. 2011;25(1):1–20.
33. HHS action plan to prevent healthcare-associated infections: incentives and oversight. http://www.hhs.gov/ash/initiatives/hai/executive_summary.pdf. Accessed 29 May 2013.

# Chapter 13
# Hospital Falls

Cynthia J. Brown and Rebecca S. (Suzie) Miltner

> *"As with dying, we recognize erring is something that happens to everyone, without feeling that it is either plausible or desirable that it will happen to us."*
>
> Katherine Schulz

## Introduction

Fall prevention has been the subject of significant attention particularly among community-dwelling older adults. Numerous risk factors for falls have been identified, and national guidelines recommend single and multifactorial interventions that have been demonstrated to reduce falls [1]. In the past decade, numerous studies have also examined fall prevention practices in the hospital setting. Although progress has been made, to date there are no guidelines for fall prevention in the hospital. The most recent American Geriatrics Society (AGS) Fall Prevention Guidelines specifically did not address the hospital setting [1]. Indeed, the definition of a fall has yet to be standardized across all hospitals; however, the definition adopted by the American Nursing Association (ANA) National Database of Nursing Quality

C.J. Brown, M.D., M.S.P.H. (✉)
Geriatric Medicine Section, Department of Medicine/Gerontology, Geriatrics and Palliative Care, Birmingham VA Medical Center, University of Alabama at Birmingham, CH19 Room 201, 1720 2nd Avenue South, Birmingham, AL 35294-2041, USA
e-mail: cynthiabrown@uabmc.edu

R.S. (Suzie) Miltner, Ph.D., R.N.C.-O.B.
School of Nursing, Department of Community Health, Outcomes and Systems, VA Quality Scholars Program, Birmingham VA Medical Center, University of Alabama at Birmingham, NB 328D, 1720 2nd Avenue South, Birmingham, AL 35294, USA
e-mail: rmiltner@uab.edu

Indications (NDNQI), which defines a fall as "an unplanned descent to the floor" is frequently cited [2].

Among all hospitalized patients, inpatient falls have been estimated to range between 2.2 and 12 falls per 1,000 bed days [3]. Rates vary depending on the hospital service and the characteristics of the patient. For example, in one academic center rates for surgical patient were significantly lower at 2.2 falls/1,000 patient–days when compared to medical patients with a rate of 6.8 falls/1,000 patient days. In another prospective study, 53 % of patients who fell were over the age of 65 years [4]. Unfortunately, falls in the hospital are often associated with injury. Approximately 30 % of hospital falls result in minor injuries with up to 15 % leading to serious injuries such as head trauma, fractures, and death. Patients who suffer an injurious fall are more likely to have longer lengths of stay and are at higher risk of admission to a long-term care facility [3, 5]. It is estimated that costs are approximately $4200 higher for patients who sustain a hospital fall [6].

## Falls as a Patient Safety Issue

In 2005, The Joint Commission included falls as a National Patient Safety Goal for the first time [7]. Specifically, hospitals were to reduce the risk of patient harm resulting from falls. Initially, the focus was identification of those at risk, specifically through medication review and nursing interventions. However, the stakes were raised for hospitals in 2008 when falls with injury were declared by the Center for Medicare and Medicaid Services (CMS) to be a "never event." As a hospital-acquired condition, falls with injury during hospitalization were no longer reimbursed at the higher payment for secondary diagnosis [8]. As hospitals grappled with how to reduce their fall and injury rates, researchers were also trying to provide evidence for best practice. There were also some who raised concerns about the unintended consequences of potential fall reduction strategies, specifically those which reduced mobility [9]. Still others have suggested that a culture of patient safety may hold the key to reducing hospital falls and injuries [10].

## Case Studies

### Clinical Summaries

#### Case 1

*Mr. Owen, a 78-year-old male with a history of hypertension, is admitted with an exacerbation of heart failure. He is evaluated by the attending physician and nursing staff who find he has significant shortness of breath with ambulation, and 3+ pitting edema to the knees. He is unable to lay flat in the bed and is begun on oxygen. He is encouraged by the physician and nursing staff to remain in bed and a urinary catheter is placed. Within 24 h his fall risk is assessed per protocol using the Morse Falls Scale [11]. He is found to be at low risk scoring 35/125 points for having a secondary diagnosis and heparin lock. His medications include Furosemide, Metoprolol, Acetylsalicylic acid (ASA), and he is started on Alprazolam for sleep on Day 2. On Day 3 of the hospital stay he is found on the floor by the nursing staff. When asked what happened, he reports he wanted to get the water pitcher on the bedside table, but that it had been pushed out of his reach. When he tried to get up to get it, his legs gave away. He complains of right hip and leg pain and a radiograph reveals a right intertrochanteric fracture.*

## Case 2

*Mrs. McDonald is an 86-year-old woman with early Alzheimer's disease who is admitted with increased agitation and confusion. According to her daughter, Mrs. McDonald lives alone and is able to perform all her own Activities of Daily Living (ADLs) and most Instrumental Activities of Daily Living (IADLs). She stopped driving a year ago, and her daughter does the checkbook and bill paying. Mrs. McDonald's past medical history is significant for atrial fibrillation, hypertension, and osteoarthritis. Her medications include Warfarin, Lisinopril, and Acetaminophen as needed for pain. Her Morse Fall Scale is 50/125 with 15 points each for secondary diagnosis and an assessment of "forgets limitations," and 20 points for heparin lock. Lab work reveals a urinary tract infection as the probable cause of her delirium, and the patient is begun on antibiotics. She is encouraged by the physician and nursing staff to stay in bed. Her daughter spends the night with her and the patient does well overnight. On Day 2 of the hospital stay, just after lunch, Mrs. McDonald gets out of bed without assistance and sustains a fall without injury. She is helped back to bed and is examined by the physician. There are no new orders and she is not reassessed for her fall risk. Mrs. McDonald is encouraged by the nursing staff to use the call button and she nods her head in agreement. However, just before dinner, Mrs. McDonald is again found on the floor, only this time she is noted to have a significant bruise on her left temple area. A CT Scan reveals a subdural hematoma with some midline shift. She is begun on q2 hour neurological checks. Approximately 7 h after her fall, Mrs. McDonald is noted to be unresponsive with shallow respirations. A repeat CT shows significant worsening of her subdural hematoma and she is transferred to the ICU and intubated. She dies the following morning.*

## *Analysis*

In our large, urban medical center, a lot of work had been done to reduce falls including policies for risk assessment, and implementation of technologies suggested to reduce falls such as bed alarms, low beds, and hourly rounding. A root cause analysis (RCA) was completed for each of these individual events, and these two events signaled that there were underlying problems that had not been addressed through the current efforts. Several contributing factors were noted to be systemic issues. For example, Mrs. McDonald's first fall was not directly communicated to the oncoming shift or between resident teams suggesting poor information exchange not only between nursing staff members but also between medical staff members. The organization made the decision to complete an aggregate RCA due to the recurrent problem with over 80 falls per month (3/1,000 patient days).

While RCA is a common tool used to understand the underlying causes of adverse events in healthcare organizations, an expansion of this tool, an aggregate RCA, can help to identify trends and systems issues across similar events [12]. The aggregate RCA can be used in lieu of individual case analysis of adverse events or as a method to analyze high risk processes. Step by step instructions to conduct an aggregate RCA are available [12].

## Aggregate Root Cause Analysis

### Step One: Charter a Team

The first step in an aggregate RCA is to charter a team to gather and analyze all information about all falls that have occurred for a given period of time. An inter-professional team including hospitalists, geriatric specialists, nurses, nursing assistants, physical therapists, risk management, and service line administrators was formed to review the data about all falls from the previous 12-month period obtained from the adverse occurrence reporting system with additional information from the organization's risk management database. Over 900 falls were reported in this large medical center. Ninety percent of falls occurred on inpatient nursing units, so the team decided to focus on this group of patients for further analysis.

### Step Two: Map the Process

In the second step, the team drew a high-level process map of the hospital experience related to falls and falls preventions (Fig. 13.1). When the patient was admitted, a falls risk assessment was completed by the admitting RN. Physicians may do an informal assessment of falls risks during the admission process. A nursing plan of care is developed to address patient nursing needs. A medical plan of care is developed to address the problems that led to the current admission. Theoretically, the nursing and medical plans of care become the largest component of the interdisciplinary plan of care. This plan of care should focus on desired outcomes of care and be implemented, evaluated, and modified as needed throughout the hospitalization. The plan of care should also include interventions to address falls prevention. The team found that the development of nursing and medical plans of care are a

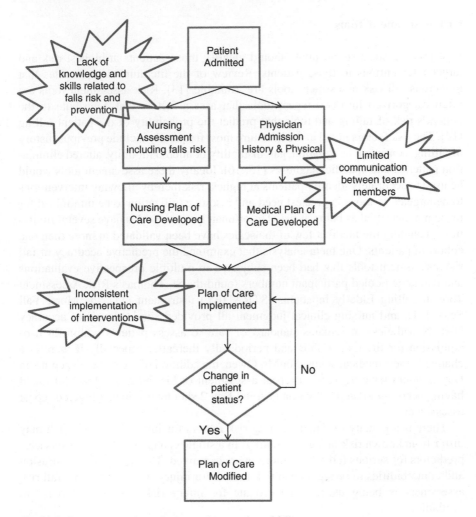

**Fig. 13.1** Process map of assessment and prevention of falls

generally parallel processes with little direct integration of each disciplines' assessed patient needs and plans to address these needs.

## Step Three: Review the General Processes in the System

Next, the team used the high-level process map to review available data on the 900 reported falls, based on each step in the process. Analysis of the data showed several areas for further exploration including assessment of risk, planned interventions, and ongoing communication of patient risk.

## Risk Assessment Tools

Best practice suggests hospitals should identify those patients at highest risk and target interventions to those patients. Review of the literature demonstrates that numerous fall risk assessment tools are available [13]. Risk assessment tools are often categorized into two types: tools that assess factors that contribute to the patient's risk of falling and tools that predict the probability of the patient falling [14]. The risk factors noted in the literature most frequently include previous history of falling, confusion or agitation, gait instability or altered mobility, altered elimination patterns, and specific diagnoses [15, 16]. Ideally these assessment tools would be used to accurately identify patients at highest risk thereby allowing interventions to be targeted to those in greatest need and resource utilization to be minimized for those not identified as being at risk. Unfortunately, these scales have several limitations including the fact that few of the scales have been validated in more than one cohort of patients. One meta-analysis that examined the predictive accuracy of fall risk assessment tools that had been subjected to multiple prospective evaluations and had large pooled participant numbers found the St. Thomas's Risk Assessment Tool in Falling Elderly Inpatients (STRATIFY) instrument [17], the Morse Fall Scale [11], and nursing clinical judgment all provided similar levels of accuracy [18]. Nonetheless, it remains standard of care to assess patients within 24 h of admission for the risk of falls and periodically thereafter especially if there is a change in their medical status. For Mr. Owen, the addition of the benzodiazepine on Day 2 would have triggered a repeat assessment of his fall risk. Mrs. McDonald having been found on the floor on hospital Day 2 should also have triggered repeat assessment.

There is a paucity of data regarding risk factors for injurious falls, which may differ from known risk factors for falling. In a single center study of three services, predictors for serious fall related injury were examined. The study found confusion and comorbidities to be significant risk factors for injury [6]. At this time, fall risk assessment is being used as a surrogate for injury risk until better tools are available.

## Targeted Interventions to Prevent Hospital Falls

### Bed Alarms

Bed and chair alarms are often employed in an attempt to reduce falls. These devices come in a variety of styles including those that are attached to the patient as well as those that are incorporated into the bed. The devices do not prevent a fall. Instead they alert the healthcare providers when a patient is trying to stand or get out of the bed unassisted. To date, the evidence regarding the effectiveness of bed alarms is conflicted. One uncontrolled 12-month before and after study among patients recovering from hip fracture used bed sensors that were linked to a central pager for all patients on the ward to assess the impact on falls. They showed reduced odds of being a faller (average Odds Ratio (OR) 0.55, 95 % CI 0.32–0.94) but no significant reduction in the fall rate [19]. A cluster randomized trial using bed and chair alarms showed no difference in fall rates or the relative risk of being a faller despite good use of the devices on intervention wards [20].

## Low Beds

Beds capable of being lowered to within inches of the floor have also been proposed as both a fall and an injury prevention measure. It has been postulated that if a patient falls from the "low bed" they will be less likely to injure themselves due to the relatively short distance they fall before impacting the ground. In addition, it is harder to get up from the low position, making it more likely staff will have time to intervene when a patient is trying to get out of bed unassisted. A single cluster randomized trial of low height beds that included 22,036 participants found no significant reduction in frequency of patient injuries due to the beds. However, this study also reported no injuries among either the control or the intervention groups [21, 22]. Further work is needed in this area to determine the impact of low height beds to reduce falls and injuries.

## Patient Rounding

Frequent rounding has been proposed as another intervention to reduce falls. Nursing rounds done every 1–2 h has been recommended. In one study of 27 units in 14 hospitals, nurses rounded at 1–2 h intervals with specific actions recommended. They found decreased call bell light usage and increased patient satisfaction. There was a significant reduction in falls with 1 h, but not 2 h rounding [23]. Subsequent studies about rounding have either not used patient falls as an outcome measure or have not found a significant reduction in patient falls associated with rounding [24]. Tucker, et al. found that reduction in the number of falls over time was not sustained with rounding with 4.5, 1.5, and 3.2 falls per 1,000 hospital days during three periods over time [2]. Inconsistent results may be related to weak study designs and the fidelity of the implementation of nursing rounds. For example, Deitrick et al., used ethnographic techniques to examine problems with implementation of hourly rounding. They report that most staff members were unable to verbalize the purpose for hourly rounding [25]. Further work is needed to understand the effect of structured nursing rounds on falls reduction.

## Increased Ambulation

Often patients are encouraged to remain in bed unless assisted to walk for fear of falls. However, there is growing evidence that limiting patient's mobility in an effort to reduce falls may be the wrong approach. There are significant consequences associated with bed rest and low mobility, especially for the older adult, including functional decline and need for new nursing home admission [26]. Thus, a careful weighing of the risks and benefits of increasing patient mobility is warranted. In a recent study exploring the association between ambulatory activity and falls, patient falls were more likely to be associated with cognitive and hospital environmental issues than the actual time spent walking [27]. The Hospital Elder Life Program, which includes scheduled toileting, provision of physical therapy and early mobilization has been shown to not only reduce the incidence of hospital delirium but to also reduce fall rates [28]. Lastly, vonRenteln-Kruse and colleagues tested a structured fall prevention program that included fall risk assessment, assistance with transfers and use of the toilet, provision of ambulatory devices as needed, and early mobilization strategies. There was an 18 % reduction in falls in the intervention group using this protocol [29].

## Multifactorial Interventions

Among community-dwelling older adults, multifactorial interventions which target a variety of risk factors have been very successful in reducing falls [1, 30]. This method has also been utilized in the hospital setting. In a recently published cluster random-ized study at four hospitals, researchers examined the effect of a computerized fall prevention tool kit to reduce falls. Patients were screened using the Morse Fall Scale on admission and specific interventions were identified based on the patient specific risk factors. Dykes, et al. found a reduction in the fall rate (3.15 vs. 4.18 per 1,000 bed–days) with a rate difference of 1.03/1,000 bed–days. In a subgroup analysis, patients who were ≥65 years benefitted the most from the intervention with an adjusted rate difference of 2.08 (95 % CI 0.61–356/1,000 bed–days). The authors noted the number needed to treat to reduce one fall during a typical 3-day hospital stay was 287 [31]. There have been several other studies of multifactorial interven-tions that demonstrated significant reductions in fall rate. Although the tested multi-factorial interventions varied widely, several commonly included components included patient and staff education, post fall review, footwear advice and toileting [32]. In one systematic review and meta-analysis that included only prospective-con-trolled design trials, Coussement and colleagues found no conclusive evidence that acute care hospital fall prevention programs were able to reduce falls [33].

## Step Four: Identify Resources

Our facility dedicated significant personnel and other financial resources for falls pre-vention over the last 5 years as part of improving overall quality, work toward Nursing Magnet Recognition for excellence in professional nursing practice (http://www.nursecredentialing.org/Magnet.aspx), and to address pay for performance penalties. The current falls prevention policy included risk assessment, identification of patients at risk, and use of technology and patient rounding to prevent falls. Falls prevention was incorporated in hospital orientation and ongoing competency assessment programs. Data about falls were reported on quality report cards within the organization. And, most importantly, organizational leaders including the Chief Nursing Officer, Chief Executive Officer, and the Hospital Board were keenly interested and engaged in reducing adverse patient events including patient falls.

## Step Five: Determine Focus of the Aggregate Review

A review of the aggregate data on falls showed that 60 % of the inpatient falls occurred during the evening or night and the majority were related to toileting. In the literature, significant interest has focused on the timing and circumstances of falls and the impact of nurse staffing level. In one study more patients fell in the evening or night and almost 80 % had an unassisted fall. Among patients of all ages, at least 50 % of falls were elimination related. However, for patients who are 65 years of age or older this proportion increased to 83 % [4].

The organization has structural elements in place to support fall prevention including evidence-based policies and procedures, adequate staffing, and staff mix as well as use of technology and other evidence-based interventions. Review of analyses of previous falls showed that staffing was not identified as a contributing factor in the fall.

As the medical center data was similar to other reported data, the team focused on the actual processes of care on the unit level. The team collected additional data from nursing units including interviewing hospitalists, unit managers, staff members, charge nurses, and unit clerks. Some team members completed observations of several nursing units at change of shift and during physician rounds. Examining the processes of care more closely revealed gaps between the structural elements and the actual processes implemented at the unit and shift level. Falls risk assessment, especially after changes in status, were inconsistently reported during nursing hand-offs and even more rarely between nursing and physician staff. Considerable variation was noted in interpretation of the organizational policies between nursing units and between individual nurses. The team focused on this variation in implementation of falls prevention measures.

## Step Six: Determine Root Cause/Contributing Factors

Analysis of the aggregate falls data as well other data suggested three root causes or contributing factors that affected the inconsistency in policy implementation (Fig. 13.2). First, there was inconsistent communication about patients at risk for falls between nursing staff on the units. Worse, there was virtually no communication about patients at risk for falls between nursing staff, physicians, or other

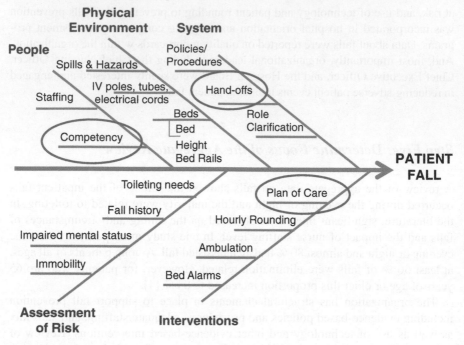

**Fig. 13.2** Fishbone diagram of root causes/contributing factors

members of the healthcare team. No formal document or process was consistently used during nursing change of shift report to ensure consistency of report or to assure risk information was communicated every time. There was more inconsistency on medical surgical units than there was on critical care units that may be related to the number of patients assigned per nurse. In addition, except in most critical care units, nurses did not consistently round with physicians daily. Falls risk was generally not discussed in medical rounds except as follow-up after an adverse event occurred.

Second, several training and competency issues were identified. Medicine and nursing recognize the importance of practitioners who offer care that is supported by evidence, provided in a technically accurate manner and with the humanistic approach that reflects community expectations [34, 35]. Ongoing professional development as well as competence validation should occur in the practice setting [36]. While this organization had annual education related to falls prevention, this education was delivered in a discipline specific, online format with test questions at the end to demonstrate content mastery. Staff frequently bypassed the content and went straight to the test to complete the task. No point of care assessment of content mastery was completed.

Third, there was inconsistent application of the evidence-based policies and procedures in place. Staff attitudes toward certain falls prevention interventions such as hourly rounding were mixed, which is consistent with other studies that suggest that many staff members do not understand the rationale for rounding [25]. In another study, nurses rated rounding benefits for patients, but not for themselves (36.54 vs. 27.83, $p < 0.001$) [37]. In addition, because of the high number of patients identified as being at risk of falls, staff may become inured to the interventions and they become background noise to the numerous other things that must be paid attention to during the shift.

These contributing factors were derived from the aggregate data and team observations of the over 900 falls in this organization. But these were also contributing factors to the two serious events described above. Contrary to evidence, Mr. Owen was instructed by physicians and nursing staff to remain in bed which contributes to debilitation. The structured elements of hourly rounding including placing personal items within reach of the patient were not consistently implemented. And, as previously mentioned, there was no verbal communication of Mrs. McDonald's first fall to oncoming nursing and medical staff, which may have heightened awareness of her risk for a subsequent fall.

The remaining steps (Steps 7 and 8) in the Aggregate RCA process are to further develop the root causes/contributing factors determined in step six, and determine actions to address the root causes, write outcome measures, propose changes to organization leaders for concurrence, and implement the actions [12]. As part of Step Eight, this team determined three actions to address the root causes:

- Units will develop formal processes to communicate falls risk and adverse events to the interprofessional healthcare team. This includes, but is not limited to, nursing change of shift report and nurse and physician communication about the plan of care.
- The organization will develop an interdisciplinary team training program focused on developing staff competencies that will reduce the risk of hospital induced adverse events.
- The organization will develop accountability measures to reduce variation in patient assessment of risk and implementation of interventions designed to reduce risk.

The team developed outcome measures related to each action and communicated their recommendations to hospital leadership. Hospital leaders recognized that these findings could be applied not only to falls but also to multiple issues including hospital acquired pressure ulcers and infections.

# Conclusion

Fall reduction is an ongoing problem in healthcare organizations because of the complexity of the problem. Even with ideal implementation of evidence-based policies and procedures, it is not possible to eliminate all falls in hospitals. Worse, the evidence for the best risk assessment measures and preventive interventions is inconsistent and/or weak. In the absence of strong evidence for interventions, organizations have to look at system/contextual solutions to improve patient safety. Success in reducing falls is dependent on developing unit cultures that exhibit characteristics of high reliability organizations, including creating a state of mindfulness for reliability, sensitivity to operations, reluctance to simplify, preoccupation with failure, deference to expertise not authority, and resilience. There is evidence that programmatic team training can support positive changes in unit culture [38]. Building this reliable culture within an organization requires committed leadership, shared values among team members, and attentiveness to the patient safety risks for hospitalized patients.

## Key Lessons Learned

- Falls are common during hospitalization and associated with adverse outcomes including fractures, head injury, and even death.
- Evidence for the best fall risk assessment measures and preventive interventions is lacking. The most commonly used interventions include bed alarms, low beds, frequent patient rounding by nursing, and increased ambulation. Multifactorial interventions may be more effective than a single intervention.
- Improving communication and teamwork (for example, among nursing staff during shift change and among nursing and physician staff) regarding falls risk assessments and targeted interventions is the key to reducing the risk of falls-related adverse events. Tested interventions may not be enough to reduce falls in hospital systems that do not provide a culture of patient safety.
- The aggregate RCA tool supports process and systems improvement by identifying trends and system issues across groupings of similar events and may be an appropriate tool for patient safety problems like falls, that are high-volume and high-risk.

## References

1. Panel on Prevention of Falls in Older Persons, American Geriatrics Society and British Geriatrics Society. Summary of the updated American Geriatrics Society/British Geriatrics Society clinical practice guideline for prevention of falls in older persons. J Am Geriatr Soc. 2011;59:148–57.
2. Tucker SJ, Bieber PL, Attlesey-Pries JM, Olson ME, Dierkhising RA. Outcomes and challenges in implementing hourly rounds to reduce falls in orthopedic units. Worldviews Evid Based Nurs. 2012;9:18–29.
3. Milisen K, Staelens N, Schwendimann R, De Paepe L, Verhaeghe J, Braes T, et al. Fall prediction in inpatients by bedside nurses using the St. Thomas Risk Assessment Tool in falling elderly inpatients (STRATIFY) instrument: a multicenter study. J Am Geriatr Soc. 2007;55:725–33.

4. Hitcho EB, Krauss MJ, Birge S, Claiborne Dunagan W, Fischer I, Johnson S, et al. Characteristics and circumstances of falls in a hospital setting. A prospective analysis. J Gen Intern Med. 2004;19:732–9.
5. Kannus P, Sievanen H, Palvanen M, Jarvinen T, Parkkari J. Prevention of falls and consequent injuries in elderly people. Lancet. 2005;366:1885–93.
6. Bates DW, Pruess K, Souney P, Platt R. Serious falls in hospitalized patients: correlates and resource utilization. Am J Med. 1995;99:137–43.
7. The Joint Commission. National patient safety goals. Available from http://www.jointcommission.org/assets/1/18/2011-2012_npsg_presentation_final_8-4-11.pdf. Accessed 23 Mar 2012.
8. Department of Health and Human Services, Center for Medicare and Medicaid Services. Hospital-Acquired Conditions (HAC) in Acute Inpatient Prospective Payment System (IPPS) Hospitals. C2010. Available from http://www.cms.gov/Medicare/Medicare-Fee-for-Service-Payment/HospitalAcqCond/Downloads/HACFactsheet.pdf. Accessed 13 May 2013.
9. Inouye SK, Brown CJ, Tinetti ME. Medicare nonpayment, hospital falls, and unintended consequences. New Engl J Med. 2009;360:2390–3.
10. Black AA, Brauer SG, Bell RA, Economides AJ, Haines TP. Insights into the climate of safety towards the prevention of falls among hospital staff. J Clin Nurs. 2011;20:2924–30.
11. Morse JM, Morse RM, Tylko SJ. Development of a scale to identify the fall-prone patient. Can J Aging. 1989;8:366–77.
12. Neily J, Ogrinc G, Mills P, Williams R, Stalhandske E, Bagian J, et al. Using aggregate root cause analysis to improve patient safety. Jt Comm J Qual Saf. 2003;29:434–9.
13. Harrington L, Luquire R, Vish N, Winter M, Wilder C, Houser B, et al. Meta-analysis of fall-risk tools in hospitalized adults. J Nurs Adm. 2010;40:483–8.
14. Morse JM. Nursing research on patient falls in healthcare institutions. Ann Rev Nurs Rev. 1993;11:299–316.
15. Oliver D, Daly F, Martin FC, McMurdo ME. Risk factors and risk assessment tools for falls in hospital in-patients: a systematic review. Age Aging. 2004;33:122–30.
16. Titler MG, Shever LL, Kanak MF, Picone DM, Qin R. Factors associated with falls during hospitalization in an older population. Res Theory Nurs Pract. 2011;25:127–52.
17. Oliver D, Britton M, Seed P, Martin FC, Hopper AH. Development and evaluation of an evidence based risk assessment tool (STRATIFY) to predict which elderly inpatients will fall: case control and cohort studies. BMJ. 1997;315:1049–53.
18. Haines TP, Hill K, Walsh W, Osborne R. Design-related bias in hospital fall risk screening tool predictive accuracy evaluations: systematic review and meta-analysis. J Gerontol Med Sci. 2007;62A:664–72.
19. Sahota O. Vitamin D and inpatient falls. Age Aging. 2009;38:339–40.
20. Shorr RI, Mion LC, Rosenblatt LC, Lynch D, Kessler LA. Ascertainment of patient's falls in hospital using an evaluation service: comparison with incident reports. J Am Geriatr Soc. 2007;55(Supp 4):S195.
21. Haines TP, Bell RAR, Varghese PN. Pragmatic, cluster radomized trial of a policy to introduce low-low beds to hospital wards for the prevention of falls and fall injuries. J Am Geriatr Soc. 2010;58:435–41.
22. Cameron ID, Murray GR, Gillespie LD, Roberstson MC, Hill KD, Cumming RG, et al. Interventions for preventing falls in older people in nursing care facilities and hospitals. Cochrane Database Syst Rev. 2010;(1): Art. No. CD005465.
23. Meade CM, Bursell AL, Ketelsen L. Effects of nursing rounds on patient's call light use, satisfaction, and safety. AJN. 2006;106(9):58–70.
24. Halm MA. Hourly rounds: what does the evidence indicate? Am J Crit Care. 2009;18:581–4.
25. Deitrick LM, Baker K, Paxton H, Flores M, Swavely D. Hourly rounding challenges with implementation of an evidence-based process. J Nurs Care Qual. 2012;27:13–9.
26. Brown CJ, Friedkin RJ, Inouye SK. Prevalence and outcomes of low mobility in hospitalized older patients. J Am Geriatr Soc. 2004;52:1263–70.

27. Fisher SR, Galloway RV, Kuo YF, Graham JE, Ottenbacher KJ, Ostir GV, et al. Pilot study examining the association between ambulatory activity and falls among hospitalized older adults. Arch Phys Med Rehabil. 2011;92:2090–2.
28. Inouye SK, Bogardus ST, Baker DI, Leo-Summers L, Cooney LM. The hospital elder life program: a model of care to prevent functional decline in older hospitalized patients. J Am Geriatr Soc. 2000;48:1697–706.
29. vonRenteln-Kruse W, Krause T. Incidence of in-hospital falls in geriatric patients before and after the introduction of an interdisciplinary team-based fall-prevention intervention. J Am Geriatr Soc. 2007;55:2068–74.
30. Gillespie LD, Robertson MC, Gillespie WJ, Sherrington C, Gates S, Clemson LM, et al. Interventions for preventing falls in older people living in the community. Cochrane Database Syst Rev. 2012;CD007146.pub3.
31. Dykes PC, Carroll DL, Hurley A, Lipsitz S, Benoit A, Chang F, et al. Fall prevention in acute care hospitals: a randomized trial. JAMA. 2010;304:1912–8.
32. Oliver D, Healey F, Haines TP. Preventing falls and fall-related injuries in hospitals. Clin Geriatr Med. 2010;26:645–92.
33. Coussement J, De Paepe L, Schwendimann R, Denbaerynuck K, Dejaeger E, Milisen K. Interventions for preventing falls in acute and chronic care hospitals: a systematic review and met-analysis. J Am Geriatr Soc. 2008;56:29–36.
34. Allen P, Lauchner K, Bridges RA, Franics-Johnson P, McBride SG, Olivarez A. Evaluating continuing competency: a challenge for nursing. J Contin Educ Nurs. 2008;39:81–5.
35. Epstein RM, Hundert EM. Defining and assessing professional competence. JAMA. 2002;287:226–35.
36. Institute of Medicine. Redesigning continuing education in the health professions 2009. Available at http://www.iom.edu/Reports/2009/Redesigning-Continuing-Education-in-the-Health-Professions.aspx. Accessed 23 Mar 2012.
37. Neville K, Lake K, LeMunyon D, Paul D, Whitmore K. Nurses' perceptions of patient rounding. JONA. 2012;42:83–8.
38. Salas E, DiazGranados D, Klein C, Burke CS, Stagl KC, Goodwin GF, et al. Does team training improve team performance? A meta-analysis. Hum Factors. 2008;50:903–33.

# Chapter 14
# Pressure Ulcers

Grace M. Blaney and Monica Santoro

> *"The wise man corrects his own errors by observing those of others."*
>
> Publius Syrus

## Introduction

A pressure ulcer is a localized injury to the skin and/or underlying tissue usually over a bony prominence, as a result of pressure, or pressure in combination with sheer and/or friction [1]. Pressure ulcers typically result from prolonged periods of uninterrupted pressure on the skin, soft tissue, muscle, and bone. Patients who have impaired mobility or sensation and those with diabetes are particularly vulnerable. Each year more than 2.5 million people in the USA develop pressure ulcers [2]. The incidence of pressure ulcers varies considerably by clinical setting and ranges from 0.4 to 38 % in acute care, 2.2 to 23.9 % in long-term care, and 0 to 17 % in home care [3]. The estimated cost for managing a single full-thickness pressure ulcer is as high as $70,000 and the total cost for treatment of pressure ulcers in the USA is estimated at $11 billion per year. In addition to the significant costs and suffering associated with pressure ulcers, they are also associated with increased lengths of stay, morbidity, and mortality [3].

Recognizing that pressure ulcers are of serious concern to the public and healthcare providers, in 2002 the National Quality Forum included Stage III, Stage IV, and

G.M. Blaney, R.N., M.S.N. (✉)
CWOCN, Wound Care RN, Nursing Administration, Winthrop University Hospital, 259 First Street, Mineola, NY 11501, USA

M. Santoro, M.S., B.S.N., C.P.H.Q.
Vice President and Chief Quality Officer, Patient Safety, Quality and Innovation, Winthrop University Hospital, 259 First Street, Mineola, NY 11501, USA

A. Agrawal (ed.), *Patient Safety: A Case-Based Comprehensive Guide*,
DOI 10.1007/978-1-4614-7419-7_14, © Springer Science+Business Media New York 2014

**a**

| Stage | Description | Further Description | Illustration |
|-------|-------------|---------------------|--------------|
| Suspected Deep Tissue Injury | Purple or maroon localized area of discolored intact skin or blood-filled blister due to damage of underlying soft tissue from pressure and /or shear. The area may be preceded by tissue that is painful, firm, mushy, boggy, warmer or cooler as compared to adjacent tissue. | Deep tissue injury may be difficult to detect in individuals with dark skin tones. Evolution may include a thin blister over a dark wound bed. The wound may further evolve and become covered by thin eschar. Evolution may be rapid exposing additional layers of tissue even with optimal treatment. | SUSPECTED DEEP TISSUE INJURY |
| Stage I | Intact skin with non-blanchable redness of a localized area usually over a bony prominence. Darkly pigmented skin may not have visible blanching; its color may differ from the surrounding area. | The area may be painful, firm, soft, warmer or cooler as compared to adjacent tissue. Stage I may be difficult to detect in patients with dark skin tones. May indicate "at risk" persons (a heralding sign of risk). | STAGE 1 |

**b**

| Stage | Description | Further Description | Illustration |
|-------|-------------|---------------------|--------------|
| Stage II | Partial thickness loss of dermis presenting as a shallow open ulcer with a red pink wound bed, without slough. May also present as an intact or open/ruptured serum-filled blister. | Presents as a shiny or dry shallow ulcer without slough or bruising*. This stage should not be used to describe skin tears, tape burns, perineal dermatitis, maceration or excoriation. *Bruising indicates suspected deep tissue injury. | STAGE 2 |
| Stage III | Full thickness tissue loss. Subcutaneous fat may be visible but bone, tendon or muscle are not exposed. Slough may be present but does not obscure the depth of tissue loss. May include undermining and tunneling. | The depth of a Stage III pressure ulcer varies by anatomical location. The bridge of the nose, ear, occiput and malleolus do not have subcutaneous tissue and stage III ulcers can be shallow. In contrast, areas of significant adiposity can develop extremely deep stage III pressure ulcers. Bone/tendon is not visible or directly palpable. | STAGE 3 |

**Fig. 14.1** (**a**, **b**, **c**) Pressure ulcer stages. With permission from National Pressure Ulcer Advisory Panel 2007

c

| Stage | Description | Further Description | Illustration |
|-------|-------------|---------------------|--------------|
| Stage IV | Full thickness tissue loss with exposed bone, tendon or muscle. Slough or eschar may be present on some parts of the wound bed. Often include undermining and tunneling. | The depth of a Stage IV pressure ulcer varies by anatomic location. The bridge of the nose, ear, occiput and malleolus do not have subcutaneous tissue and these ulcers can be shallow. Stage IV ulcers can extend into muscle and/or supporting structures (e.g. fascia, tendon or joint capsule) making osteomyelitis possible. Exposed bone/tendon is visible or directly palpable. | |
| Unstageable | Full thickness tissue loss in which the base of the ulcer is covered by slough (yellow, tan, gray, green or brown) and/or eschar (tan, brown or black) in the wound bed. | Until enough slough and/or eschar is removed to expose the base of the wound, the true depth, and therefore stage, cannot be determined. Stable eschar (dry, adherent, intact without erythema or fluctuance) on the heels serves as "the body's natural cover" and should not be removed. | |

**Fig. 14.1** (continued)

unstageable pressure ulcers acquired after admission to a healthcare setting on the list of Serious Reportable Events (SREs) that are defined as largely preventable [4]. Figure 14.1 demonstrates Pressure Ulcer Staging. In 2008, the Center for Medicare and Medicaid Services (CMS), in response to the Deficit Reduction Act of 2005, defined Stage III and IV pressure ulcers as Hospital-Acquired Conditions (HACs). HACs are not reimbursed at the higher payment for secondary diagnoses if the condition is acquired during hospitalization [5].

Though pressure ulcers are considered an indicator of the quality of care, questions remain as to whether there are situations where they are unavoidable. The National Pressure Ulcer Advisory Panel (NPUAP) convened an international advisory panel consensus conference in 2010 and concluded that *most*, not all, pressure ulcers are avoidable. The panel concluded that there are situations, such as hemodynamic instability exacerbated by movement, that render pressure ulcers unavoidable. The panel noted however that the decision about avoidability is made after the fact, when the processes of care can be evaluated to determine that interventions were consistent with the patient's needs and recognized standards [6].

Comprehensive guidelines on pressure ulcer prevention and treatment are available from a variety of sources. The National Pressure Ulcer Advisory Panel (NPUAP), the European Pressure Ulcer Advisory Panel (EPUAP), and the Wound, Ostomy, and Continence Nurses Society have published systematically developed evidence-based guidelines that address prevention and treatment of pressure ulcers [7, 8]. In addition, the Institute for Healthcare Improvement (IHI) and the Agency for Healthcare Research and Quality (AHRQ) have published comprehensive resource guides that focus on strategies to implement evidence-based practices [2, 9]. Current best practices emphasize the need for a comprehensive, multifaceted, interdisciplinary, systems-based approach implemented in the context of an organizational culture that supports communication and teamwork [2]. Published reports applying these practices describe successful performance improvement initiatives resulting in a decreased incidence in healthcare-acquired pressure ulcers [3, 10, 11].

When a healthcare-acquired pressure ulcer does develop, it is important to use it as an opportunity for learning and improvement through root cause analysis (RCA). RCA provides for the identification of both active and latent failures and for the implementation of improvement strategies that focus on the organizational systems and processes that contributed to the adverse event. In this chapter we describe factors that contribute to the development of pressure ulcers and associated improvement strategies through the lens of two case studies.

## Case Studies

### Case Study 1: Hospital-Acquired Pressure Ulcer in a Nursing Home Patient

#### Clinical Summary

*Mrs. B, an 83-year-old female was transferred from a local nursing home to an acute care hospital on a Friday afternoon with an elevated temperature, abdominal pain, vomiting, and altered mental status. Her past medical history was significant for a cerebrovascular accident (CVA), congestive heart failure (CHF), renal insufficiency, diabetes, peripheral vascular disease (PVD) with resultant bilateral above the knee amputations, and failure to thrive. Other transfer information indicated that Mrs. B was lifted out of bed and spent most of the day in a wheelchair. She required total assistance with hygiene, was incontinent of bladder and bowel, and required total assistance with feeding. Mrs. B is 5'6" and weighs 200 lbs. There was no information in the transfer documentation regarding any skin problems or pressure ulcer prevention methods in use at the nursing home.*

*Mrs. B remained in the Emergency Department (ED) for 18 hours waiting for an inpatient bed. She was maintained 'NPO' pending evaluation of her vomiting and abdominal pain. Because of her altered mental status, the decision was made to have a speech and swallow consult to evaluate her risk for aspiration. The speech and swallow consult was scheduled for Monday. The medical workup revealed pneumonia, urinary tract infection, severe constipation, and dehydration. Mrs. B was started on intravenous fluids, antibiotics, and a bowel cleansing program. Physical exam by the admitting clinician noted the skin to be "intact—no rashes or lesions noted." Medical management of Mrs. B's comorbid conditions was continued.*

*An initial skin assessment was documented by the nurse as "within normal limits" and a Braden Pressure Ulcer Risk Assessment was completed with a score of 12 (a score of 18 or less indicates the patient is at risk). Mrs. B was placed on a pressure redistribution mattress in the ED and on the inpatient unit. Positioning devices and heel offloading devices were implemented. Mrs. B was started on a turning and positioning program and was found to frequently return herself to the right side. Protective skin products were applied. She was maintained on bed rest for the weekend during which time no additional skin assessments were documented. On Monday morning Mrs. B's right ischium was noted to be discolored with a red/slight purple color and it was staged as a "suspected deep tissue injury in evolution" using the staging system of the National Pressure Ulcer Advisory Panel [1]. Eight days later the ulcer had a thin blister over a dark wound bed.*

## Analysis and Discussion

An RCA team was convened including representatives from medicine, nursing, dietary, clinical informatics, and the ED. Nursing representation included staff nurses, nursing assistants, and a wound care nurse. Review of the case revealed that the nurse who conducted the initial assessment identified discoloration over the ischium but did not think it was significant because the skin was intact. The nurse did not document the discoloration, nor did she communicate it to the physician or the oncoming nurse during change of shift report. The admitting physician did not conduct a full skin assessment as part of their physical assessment. Though the hospital had recently implemented a policy for daily skin assessments, the nursing staff over the weekend did not document an assessment. The contributing factors and associated improvement strategies are discussed below and in Fig. 14.2.

## *Policy and Process*

### *Risk Assessment*

The Braden Scale, a validated risk assessment tool was used to assess the patient's risk and the patient was identified as being at risk [12]. The Braden Scale is exhibited in Table 14.1. The Braden scale or other validated tools provide for a standardized and ongoing process to identify patients at risk so that targeted preventive care can be implemented [2]. The tools evaluate factors including mobility, moisture, sensory deficiency, and nutritional status (including dehydration) that impact the patients risk for development of a pressure ulcer. It is important to understand that the tool is only one component of the risk assessment process and should be used in conjunction with an overall clinical assessment. Additional factors that need to be considered in assessing the patients risk for developing a pressure ulcer include but are not limited to an existing pressure ulcer, a prior Stage III or IV pressure ulcer, hypoperfusion states, diabetes, peripheral vascular disease, restraint use, smoking, spinal cord injury, end-of-life/palliative care, and operating room or ED stays. Mrs. B had several of these additional risk factors that needed to be considered in developing and implementing her care plan. Use of a standardized pressure ulcer risk assessment is one of the three critical components incorporated in a pressure ulcer

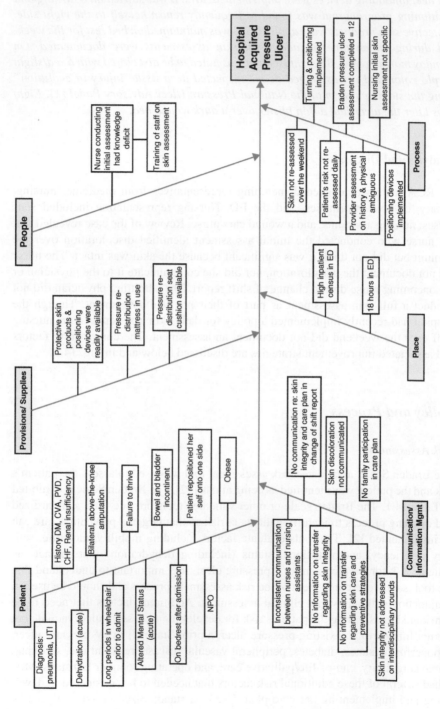

**Fig. 14.2** Cause and effect diagram: Case Study 1

**Table 14.1** Braden scale for predicting pressure sore risk

Patient's Name _____  Evaluator's Name _____  Date of Assessment _____

| | 1 | 2 | 3 | 4 |
|---|---|---|---|---|
| **SENSORY PERCEPTION** ability to respond meaningfully to pressure-related discomfort | **1. Completely Limited** Unresponsive (does not moan, flinch, or grasp) to painful stimuli, due to diminished level of consciousness or sedation. OR limited ability to feel pain over most of body | **2. Very Limited** Responds only to painful stimuli. Cannot communicate discomfort except by moaning or restlessness. OR has a sensory impairment which limits the ability to feel pain or discomfort over ½ of body. | **3. Slightly Limited** Responds to verbal commands, but cannot always communicate discomfort or the need to be turned. OR has some sensory impairment which limits ability to feel pain or discomfort in 1 or 2 extremities. | **4. No Impairment** Responds to verbal commands. Has no sensory deficit which would limit ability to feel or voice pain or discomfort. |
| **MOISTURE** degree to which skin is exposed to moisture | **1. Constantly Moist** Skin is kept moist almost constantly by perspiration, urine, etc. Dampness is detected every time patient is moved or turned. | **2. Very Moist** Skin is often, but not always moist. Linen must be changed at least once a shift. | **3. Occasionally Moist** Skin is occasionally moist, requiring an extra linen change approximately once a day. | **4. Rarely Moist** Skin is usually dry, linen only requires changing at routine intervals. |
| **ACTIVITY** degree of physical activity | **1. Bedfast** Confined to bed. | **2. Chairfast** Ability to walk severely limited or non-existent. Cannot bear own weight and/or must be assisted into chair or wheelchair. | **3. Walks Occasionally** Walks occasionally during day, but for very short distances, with or without assistance. Spends majority of each shift in bed or chair | **4. Walks Frequently** Walks outside room at least twice a day and inside room at least once every two hours during waking hours |
| **MOBILITY** ability to change and control body position | **1. Completely Immobile** Does not make even slight changes in body or extremity position without assistance. | **2. Very Limited** Makes occasional slight changes in body or extremity position but unable to make frequent or significant changes independently. | **3. Slightly Limited** Makes frequent though slight changes in body or extremity position independently. | **4. No Limitation** Makes major and frequent changes in position without assistance. |

(continued)

**Table 14.1** (continued)

| Patient's Name | Evaluator's Name | Date of Assessment |
|---|---|---|

| | 1. Very Poor | 2. Probably Inadequate | 3. Adequate | 4. Excellent |
|---|---|---|---|---|
| **NUTRITION** usual food intake pattern | Never eats a complete meal. Rarely eats more than 1/3 of any food offered. Eats 2 servings or less of protein (meat or dairy products) per day. Takes fluids poorly. Does not take a liquid dietary supplement. OR is NPO and/or maintained on clear liquids or IV's for more than 5 days. | Rarely eats a complete meal and generally eats only about ½ of any food offered. Protein intake includes only 3 servings of meat or dairy products per day. Occasionally will take a dietary supplement. OR receives less than optimum amount of liquid diet or tube feeding. | Eats over half of most meals. Eats a total of 4 servings of protein (meat, dairy products per day. Occasionally will refuse a meal, but will usually take a supplement when offered. OR is on a tube feeding or TPN regimen which probably meets most of nutritional needs. | Eats most of every meal. Never refuses a meal. Usually eats a total of 4 or more servings of meat and dairy products. Occasionally eats between meals. Does not require supplementation. |
| **FRICTION & SHEAR** | **1. Problem** Requires moderate to maximum assistance in moving. Complete lifting without sliding against sheets is impossible. Frequently slides down in bed or chair, requiring frequent repositioning with maximum assistance. Spasticity, contractures or agitation leads to almost constant friction | **2. Potential Problem** Moves feebly or requires minimum assistance. During a move skin probably slides to some extent against sheets, chair, restraints or other devices. Maintains relatively good position in chair or bed most of the time but occasionally slides down. | **3. No Apparent Problem** Moves in bed and in chair independently and has sufficient muscle strength to lift up completely during move. Maintains good position in bed or chair. | |

Total Score

care bundle along with a comprehensive skin assessment and implementation of a care plan to address areas of risk. A care bundle incorporates a group of practices that when implemented together result in better outcomes. It is a way of tying best practices together systematically to address a complex issue such as pressure ulcer prevention and treatment [2].

A comprehensive risk assessment requires updating on regular basis. The patients risk may change rapidly due to changes in their condition that impact mobility, incontinence, or nutrition [9]. In most acute care settings the assessment should be performed on admission, daily, and on transfer or discharge. In some settings it is done every shift, such as in critical care areas. As with the entire care plan, the frequency of risk assessment should be guided by the setting of care and the specific patient's needs [2]. In this case the patient's risk should have been reassessed daily; however, there was no documentation that the reassessment was completed.

*Initial Skin Assessment*

There was a knowledge deficit on the part of the nurse who did the admitting assessment regarding the significance of skin discoloration in the presence of intact skin. This resulted in a lack of documentation and communication to the care team regarding this finding. The team also questioned the thoroughness of the documentation of the initial assessment by both the nurse and physician as a potential contributory cause since the terms "within normal limits" and "intact-no rashes or lesions noted" are vague. One of the challenges in pressure ulcer prevention and treatment is a lack of agreement as to what constitutes a minimal skin assessment. Key to a comprehensive skin assessment are inspection and palpation of the patient's entire skin, looking and touching the skin from head to toe paying particular attention to bony prominences and areas subjected to device-related pressure [2, 9]. It was agreed by the RCA team that the staff required training on a clear definition of the components of a skin assessment including skin temperature, color, moisture level, turgor, and integrity. Identified skin abnormalities should be described by anatomical location, size, and stage (as indicated) [13]. It can be challenging when evaluating intact skin to detect and distinguish Stage I pressure ulcers and suspected deep tissue injuries particularly in dark-skinned individuals [7]. Figure 14.1 demonstrates diagrammatically the differences between a deep tissue injury and Stage I pressure ulcer. Figure 14.3 demonstrates photographic images of a deep tissue injury to a heel as compared to a Stage I pressure ulcer. Careful assessment is required to evaluate these skin changes. Staff should also be educated to identify a blanching response, localized heat, edema, and induration [7]. Retrospective chart review on patients receiving quarterly skin exams revealed that only 50 % of pressure ulcers found on physical exam were documented in the record. In addition, documented pressure ulcers are frequently missing key descriptors such as stage, location, and size [14]. The need for education of all team members (nurses, nursing assistants, and the medical staff) and for resource materials to be readily available at the point of care describing pressure ulcer staging and documentation was identified as an opportunity for improvement. The National

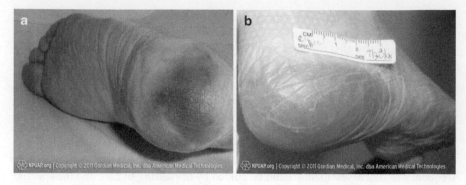

**Fig. 14.3** (**a**) Photograph of deep tissue injury and (**b**) stage I pressure ulcer

Database for Nursing Quality Indicators (NDNQI) has a tutorial for pressure ulcer staging that all staff were required to complete [15]. Sendelbach et al. describes use of a "how to conduct a skin inspection video" as part of a comprehensive system wide initiative that decreased pressure ulcers by 33 % [11]. The same initiative describes modifying the history and physical and nursing forms/electronic templates to prompt complete pressure ulcer documentation. Pressure ulcer prevention algorithms can be integrated into the Electronic Health Record (EHR) with built-in alerts and prompts for skin inspection and prevention strategies. The EHR can also be used to generate real time reports to identify patients at risk as well as to generate referrals (e.g., to wound care nurses). Figure 14.4 demonstrates a screen shot of an electronic template to promote complete assessment and documentation of wounds.

*Reassessment of Skin Integrity*

There was no documentation of reassessment of the skin during the weekend. Skin integrity as well as risk factors can change in a matter of hours in hospitalized patients. Daily skin inspection needs to be a part of routine care for patient's identified at risk [9].

*Turning and Positioning*

Turning and repositioning of patients is an important component in the prevention of pressure ulcers [3, 7]. The nursing staff recalled turning the patient; however, there was inconsistent documentation. The staff identified staffing levels, the acuity of the unit, and the presence of visitors during most of the day as barriers to consistent implementation and documentation. According to the staff interviewed and the documentation, the patient frequently repositioned herself to the right side where the ulcer ultimately developed. The tendency for patients to situate themselves onto an ulcer or

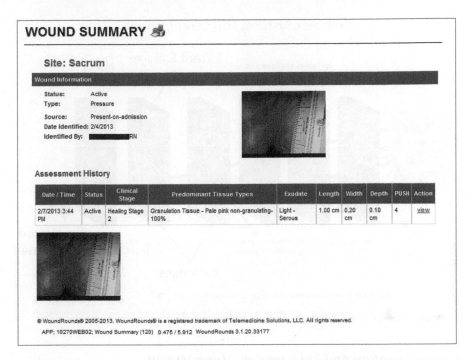

**Fig. 14.4** Screenshot of an electronic system to document wounds and ulcers. With permission from WoundRounds®, a registered trademark of Telemedicine Solutions, LLC

onto tissue at high risk for ulceration has been reported [6]. The team agreed to use tools in the patient's room to remind caregivers to turn and reposition [11]. The degree to which Mrs. B and her family were educated on the importance of repositioning was not documented. The patient and their support system need to be aware of the issues related to nonadherence to the care plan and the potential effects [6].

*Nutrition*

Mrs. B was dehydrated on admission and remained NPO with intravenous hydration due to her clinical condition and the need for a speech and swallow evaluation was delayed due to the weekend. Though not identified as a contributory factor in this case, the evidence indicates that optimizing nutritional status is an appropriate strategy for preventing pressure ulcers [3]. The use of automated alerts to dietary services based on a patient's level of risk in conjunction with a preapproved list of supplements can support timely and reliable implementation of nutrition assessment and planning [11]. The team also identified the potential delays created as a result of speech and swallow evaluations not being conducted 7 days a week. This was referred to the appropriate administrator for follow-up.

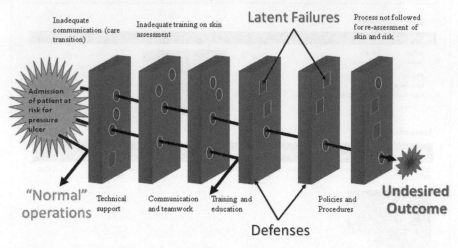

Adapted from: Reason J. Human Error. Cambridge UK; Cambridge University Press; 1990: 208.

**Fig. 14.5** Complex systems and latent failure

## Communication and Information Management

### Continuum of Care (Availability of Pertinent Information)

The lack of documentation from the nursing home related to Mrs. B's skin integrity was identified by the RCA team as potentially the first systems failure that contributed to the development of this wound. A diagram depicting James Reason's theory of how errors occur in complex systems demonstrates the latent systems failures that contributed to this event (see Fig. 14.5) [16].

Effective communication between caregivers is essential during transitions in care. The National Transitions of Care Coalition (NTOCC) is one example of a national program focusing on improving quality of care during transitions. NTOCC has developed a Transitions in Care Compendium that includes "The Care Transition Bundle-Seven Essential Intervention Categories and Crosswalk" [17]. An example of a regional initiative designed to promote collaboration and communication within and throughout the continuum of care related to pressure ulcer assessment, management, and prevention is the New York State Gold STAMP (*S*uccess *T*hrough, *A*ssessment, *M*anagement, and *P*revention) Program to Reduce Pressure Ulcers. Additional information on this program is available at: http://www.health.ny.gov/professionals/nursing_home_administrator/gold_stamp/.

### Communication Among the Care Team

The RCA team identified an opportunity for improvement with communication among all members of the interdisciplinary team. A plan to promote standardized communication between the nursing staff and the nursing assistants was developed. In addition, skin integrity was added as a prompt for discussion on interdisciplinary rounds. Handoff communication of the patient's risk status should be routine, however, one study identified that pressure ulcer risk was not included in handoff communication either verbally or in the written tools [18]. Information about pressure ulcer risk status should be included in all handoff communications and staff should be educated about how to use the risk status in caring for the patient.

## Equipment and Supplies

Mrs. B was on a pressure redistribution surface during her stay in the ED and while on the inpatient unit. The effectiveness of pressure redistribution mattresses is well documented [3]. From a systems perspective, there should be regular reviews of the equipment and supplies available to maximize patient outcomes. It is also important to ensure that supplies are readily available at the point of care [18].

## Patient/Family-Centeredness

The family was present throughout the weekend; however, the RCA team identified that patient and family education was an opportunity for improvement. Information was included in the hospital admission packet and a video was available on the hospital patient education channel but there were not consistent practices to review this information with the patient/family. There are educational materials that have been developed for patients and families to promote involvement in their care plan. One resource is available at: http://www.njha.com/qualityinstitute/pdf/pubrochure/pdf.

## Case Study 2: Occipital and Sacral Pressure Ulcer in a Trauma Patient with Spine Fracture

### Clinical Summary

*Mr. S, a 22-year-old male was admitted to a Level I trauma facility with a gunshot wound to the neck and lower back. He was rushed urgently to the OR to repair major internal abdominal injuries along with an unstable neck and lower thoracic spine fractures. Mr. S lost massive amounts of blood due to intraoperative complications. The neck and thoracic fractures were not stabilized. He was on the OR table for 10 hours. He arrived to the Surgical Intensive Care Unit on multiple medications to maintain his blood pressure. Mr. S was in a firm cervical collar, intubated on a ventilator, sedated, and chemically paralyzed. Due to his hemodynamic instability and an unstable cervical spine, he was not turned.*

*Within 32 hours, Mr. S became hemodynamically stable but developed early signs of pulmonary complications and sepsis with a fever of 104. A specialty bed was obtained for advanced kinetic therapy for aggressive pulmonary treatment which has a firm mattress to help protect the unstable neck and thoracic spine. The bed was set to rotate the patient to 40° angles for pulmonary clearance purposes and the bed was tilted at a 30° angle so his head was elevated higher than his feet.*

*Three days later Mr. S's pulmonary status improved although he remained intubated. He was hemodynamically stable. Mr. S returned to the OR for stabilization of his neck and thoracic spine fractures. Skin inspection postoperatively revealed a purple/maroon color to his sacral/coccyx area suspicious for "deep tissue injury." He also developed eschar to the occiput.*

### Analysis and Discussion

An RCA team was convened and included representatives from nursing, trauma, orthopedics, neurosurgery, wound care, physical therapy, pharmacy, quality, and risk management. The front line staff directly involved in the patient's care were included in the analysis in order to identify barriers to implementation of defined processes and elicit their input on improvement strategies. The team recognized that the patient was critically ill and that the sacral pressure ulcer may have been unavoidable; however, they wanted to review the case to determine if any processes could be improved. The emphasis of all root cause analyses is to focus on the systemic causes of the event and not on the performance of the individual clinicians' involved [19]. The team identified that there were knowledge deficits among the nursing staff on caring for a patient with an unstable spine fracture. In addition, the nursing staff reported receiving conflicting information from the different disciplines.

Contributing to the patient's risk was his hemodynamic instability, the use of vasopressors, sepsis, and prolonged time in surgery. In addition, other factors that may increase the risk of pressure ulcers in critical care patients include: low arterial pressure, prolonged ICU stay, comorbid conditions as previously discussed, as well as sepsis, and the use of vasopressor agents [20]. The patient also developed a pressure ulcer under his cervical collar. Although only 1 % of all pressure ulcers develop on the occipital area, the incidence of pressure ulcers of patients wearing a cervical collar ranges from 23.9 to 44 % [14]. Figure 14.6 shows a photograph of an occipital pressure ulcer that developed under a cervical collar. The contributing factors identified by the RCA team are summarized below and in Figs. 14.7 and 14.8.

### *Communication and Teamwork*

Patients in critical care units have been found to be most at risk in the first week of their ICU stay. This is frequently a time of significant physiologic instability when the staff is managing multiple complex interventions while trying to prevent

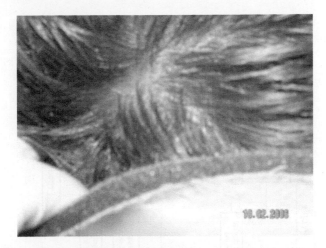

**Fig. 14.6** Photograph of occipital pressure ulcer under a cervical collar. With permission from Black J et al. Medical device related pressure ulcers in hospitalized patients. *International Wound Journal*. Blackwell Publishing Ltd and Medicalhelplines.com Inc. 2010; 7(5)

pressure ulcers. During this time the importance of communication among all disciplines on the care team on strategies to prevent pressure ulcers should not be overlooked [20]. There were questions among the team regarding management of the patient's unstable spinal fractures in the presence of hemodynamic instability and pulmonary complications. This was partially due to inconsistency in the plan of care among the different disciplines. The importance of team-based communication in developing care plans for complex cases such as this was identified as an opportunity for improvement.

## Knowledge Deficit

The RCA team designated a performance improvement team which conducted a literature review on turning and positioning of patients with spinal cord injuries and with cervical collars to implement risk reduction strategies. Standards of care were developed based on the literature review and on literature from the cervical collar manufacturers' [21, 22]. Trauma patients with a length of stay of 2 days or more have a high incidence of skin of breakdown. Head and spinal cord-injured patients as well as patients in cervical collars are particularly at high risk. It is suggested that specific protocols be developed to prevent skin breakdown in these patients. Preventive measures require an interdisciplinary approach that facilitates rapid clearance of cervical spines and aggressive and timely management of patients who must wear cervical collars or other splints including interventions for intermittent removal, replacement, or repositioning [23]. A hospital-wide education program

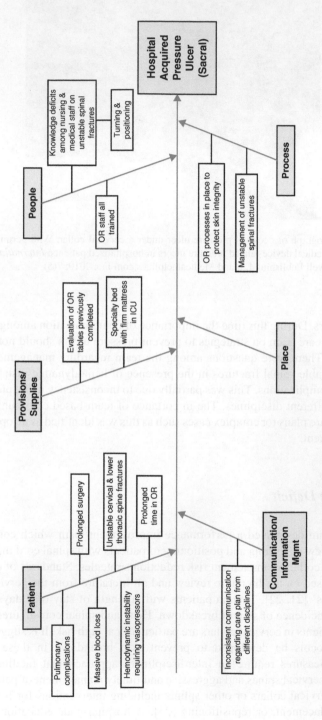

**Fig. 14.7** Cause and effect diagram #1 case study #2

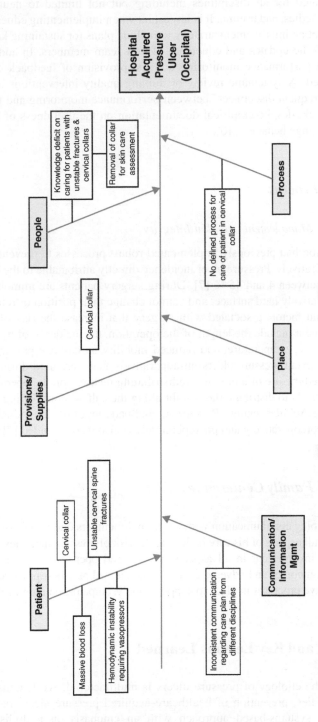

**Fig. 14.8** Cause and effect diagram #2 case study #2

was implemented for all disciplines including but not limited to neurosurgery, nursing, orthopedics, and trauma. It is important when implementing education programs as part of an improvement strategy to include plans for sustaining knowledge including periodic updates and education of new team members. In addition, the importance of performance monitoring and the provision of feedback cannot be overemphasized. A systematic review of nursing quality interventions noted that there was a "frequent disconnect" between performance monitoring and the provision of feedback, despite empirical documentation on the usefulness of audit and feedback to change behavior [10].

## Policy and Process

### Intraoperative Management and Skin Integrity

The organization had previously implemented robust processes to prevent pressure ulcers intraoperatively. Pressure ulcer incidence directly attributable to the operating room ranges between 4 and 45 % [7]. During surgery patients are immobile, positioned on a relatively hard surface, and cannot change their position or relieve pressure. Additional factors associated with surgery that increase the risk of pressure ulcer development include the length of the operation, the incidence of hypotensive episodes, low core temperature, and reduced mobility on day one postoperatively. Strategies to prevent pressure ulcers intraoperatively that were used during Mr. S's surgery included the use of a pressure-redistributing mattress on the operating table and elevating heels to distribute the weight along the calf without putting all of the pressure on the Achilles tendon. Positioning and padding of the individual in a different posture preoperatively and postoperatively is also recommended [7].

## Patient and Family Centeredness

There was ongoing communication with the family members during the course of his stay regarding all aspects of his care including the risk of pressure ulcer development. Patient/family involvement in all aspects of care and open communication about unanticipated outcomes and adverse events is recognized as a best practice and it is important to have processes in place to support timely, empathic communication [24].

## Conclusion and Key Lessons Learned

In summary, the etiology of pressure ulcers is multifactorial. As demonstrated in these case studies, prevention of healthcare-acquired pressure ulcers requires an organizational systems-based approach with an emphasis on multidisciplinary teamwork and continuous improvement. Effective programs include leadership support and medical staff involvement. Essential components of a comprehensive program to prevent pressure ulcers include:

- Implementation of evidence-based processes for risk assessment, skin assessment, and prevention strategies in the care plan
- Monitoring the effectiveness of the pressure ulcer prevention program through defined process and outcome measures
- Robust programs for orientation and ongoing training of staff
- Promoting communication regarding skin integrity and risk factors among the healthcare team
- Use of point-of-care resources for staff on components of a risk and skin assessment and on identification and description of abnormalities
- Use of prompts and available technological support to automate implementation of recommended practices
- Ongoing evaluation of products, equipment, and supplies for pressure redistribution and skin care
- A patient-centered approach that includes patients/family members/caregivers in education regarding risk factors and prevention strategies

# References

1. National Pressure Ulcer Advisory Panel. Updated staging system. c2007. (cited 13 Jul 2013). Available from http://www.npuap.org/pr2.htm
2. Berlowitz D, VanDeusen Lukas C, Parker V, et al. Preventing pressure ulcers in hospitals, a toolkit for improving quality of care. Agency for Healthcare Research and Quality. (cited 13 Jul 2013). Available from http://www.ahrq.gov/research/ltc/pressureulcertoolkit/putoolkit.pdf
3. Reddy M, Sudeep S, Rochon P. Preventing pressure ulcers: a systematic review. JAMA. 2006;296(8):974–84.
4. National Quality Forum. NQF Releases Updated Serious Reportable Events. c2011. (cited 13 Jul 2013). Available from http://www.qualityforum.org/News_And_Resources/Press_Releases/2011/NQF_Releases_Updated_Serious_Reportable_Events.aspx
5. Department of Health and Human Services, Center for Medicare and Medicaid Services. Hospital-Acquired Conditions (HAC) in Acute Inpatient Prospective Payment System (IPPS) Hospitals. c2010. (cited 13 Jul 2013). Available from https://www.cms.gov/HospitalAcqCond/downloads/HACFactsheet.pdf
6. Black J, Edsberg L, Baharestani M, Langemo D, Goldberg M, McNichol L, et al. Pressure ulcers: avoidable or unavoidable? Results of the national pressure ulcer advisory panel consensus conference. Ostomy Wound Manage. 2011;57(2):24–37.
7. National Pressure Ulcer Advisory Panel and European Pressure Ulcer Advisory Panel. Prevention and treatment of pressure ulcers: clinical practice guideline. Washington, DC: National Pressure Ulcer Advisory Panel; 2009.
8. Wound, Ostomy and Continence Nurses Society. Guideline for prevention and management of pressure ulcers. Mount Laurel, NJ: WOCN; 2010.
9. Institute for Healthcare Improvement. How-to Guide: prevent pressure ulcers. c2011. (cited 13 Jul 2013). Available from http://www.ihi.org/knowledge/Pages/Tools/HowtoGuidePrevent PressureUlcers.aspx

10. Soban L, Hempel S, Munjas B, Miles J, Rubenstein L. Preventing pressure ulcers in hospitals: a systematic review of nurse-focused quality improvement interventions. Jt Comm J Qual Patient Saf. 2011;37(6):245–52.

11. Sendelbach S, Zink M, Peterson J. Decreasing pressure ulcers across a healthcare system. JONA. 2011;41(2):84–9.

12. Prevention Plus, LLC. Braden scale for predicting pressure sore risk. (cited 13 Jul 2013). Available from http://www.bradenscale.com/images/bradenscale.pdf

13. Ayello E. Changing systems, changing cultures: reducing pressure ulcers in hospitals. Jt Comm J Qual Patient Saf. 2011;37(3):120–2.

14. Dahlstrom M, Best T, Baker C, Doeing D, Davis A, Doty J, et al. Improving identification and documentation of pressure ulcers at an urban academic hospital. Jt Comm J Qual Patient Saf. 2011;37(3):123–30.

15. National Database of Nursing Quality Indicators. Pressure ulcer training. c2006–2011. (cited 13 Jul 2013). Available from https://www.nursingquality.org/NDNQIPressureUlcerTraining/Default.aspx

16. Reason J. Human error. Melbourne: Cambridge University Press; 1990.

17. Lattimer C. When it comes to transitions in patient care, effective communication can make all the difference. Generations. 2011;35(1):69–72.

18. Jankowski I, Nadzam DM. Identifying gaps, barriers and solutions in implementing pressure ulcer prevention programs. Jt Comm J Qual Patient Saf. 2011;37(6):253–64.

19. Percarpio K, Watts V, Weeks W. The effectiveness of root cause analysis: what does the literature tell us? Jt Comm J Qual Patient Saf. 2008;34(7):391–8.

20. Cox J. Predictors of pressure ulcers in adult critical care patients. Am J Crit Care. 2011;20:364–75.

21. Ackland H, Cooper J, Malham G, Kossmann T. Factors predicting cervical collar-related decubitus ulceration in major trauma patients. Spine J. 2007;32(4):423–8.

22. Webber-Jones J, Thomas C, Bordeaux R. The management and prevention of rigid cervical collar complications. Orthop Nurs. 2002;21(4):19–26.

23. Watts D, Abrahams E, MacMillan C, Sanat J, Silver R, VanGorder S, et al. Insult after injury: pressure ulcers in trauma patients. Orthop Nurs. 1998;17:84–91.

24. National Quality Forum. Safe practices for better healthcare—2009 update: a consensus report. Washington, DC: NQF; 2009. p. 2009.

# Chapter 15
# Diagnostic Error

Satid Thammasitboon, Supat Thammasitboon, and Geeta Singhal

> *"I beseech you, in the bowels of Christ, think it possible you may be mistaken."*
>
> Oliver Cromwell

## Introduction

Diagnostic error is defined as "a diagnosis that was unintentionally delayed (sufficient information was available earlier), wrong (another diagnosis was made before the correct one), or missed (no diagnosis was ever made), as judged from the eventual appreciation of more definitive information" [1]. A diagnostic error may occur due to failure in timely access to care, incorrect interpretation of symptoms, signs, or test results, fault in the differential diagnostic process, or a failure of timely follow-up and specialty referral [2]. It is important to note that not all diagnostic errors result in harm and not all harm related to a diagnostic error is preventable.

S. Thammasitboon, M.D., M.H.P.E. (✉)
Department of Pediatrics, Texas Children's Hospital, Baylor College of Medicine,
6621 Fannin Street, Houston, TX 77030, USA
e-mail: sxthamma@texaschildrens.org

S. Thammasitboon, M.D., M.S.C.R.
Pulmonary Diseases, Critical Care, and Environmental Medicine, Tulane University Health Sciences Center, 1430 Tulane Avenue SL-9, New Orleans, LA 70112, USA
e-mail: sthammas@tulane.edu

G. Singhal, M.D., M.Ed.
Department of Pediatrics, Texas Children's Hospital, 6621 Street Fannin Suite A 210,
Houston, TX 77030, USA
e-mail: grsingha@texaschildrens.org

A. Agrawal (ed.), *Patient Safety: A Case-Based Comprehensive Guide*,
DOI 10.1007/978-1-4614-7419-7_15, © Springer Science+Business Media New York 2014

## Prevalence and Contributory Risk Factors

Diagnostic errors are common, occur in all clinical settings, frequently cause harm to patients, and incur malpractice claims against physicians. Despite their relatively high rate of occurrence, little attention has been paid to them until recently because of difficulty in reliably identifying and analyzing them [3].

The rate of diagnostic error is estimated to be 5–15 % across a wide variety of clinical conditions and settings [4]. Diagnostic error is encountered in every specialty. The estimated rate is <5 % in dermatology, radiology, and pathology, all of which rely on visual interpretation and about 10–15 % in most other fields where data gathering and synthesis play a stronger role [4].

The Harvard Medical Practice Study that preceded the Institute of Medicine report, *To Err is Human,* found that physician errors resulting in adverse events were more likely to be diagnostic than drug related (14 vs. 9 %). Further, diagnostic mishaps comprise the highest proportion to be judged negligent (75 %) and to result in serious disability (47 %) [5].

Diagnostic errors are particularly common in the emergency department (ED) that has been described as a "natural laboratory for the study of error" [6, 7]. In a 2007 study of 122 closed malpractice claims, 65 % involved missed ED diagnoses; of these 48 % were associated with serious harm, and 39 % resulted in death [8]. Typically, missed diagnoses in ED are a result of multiple breakdowns in clinical processes including failure to order an appropriate diagnostic test (58 %), failure to perform an appropriate history and physical exam (42 %), incorrect interpretation of a diagnostic test (37 %), lack of appropriate supervision (30 %), inadequate hand-offs (24 %), and excessive workload (23 %) [8].

Ambulatory care settings are not immune from diagnostic errors. A 2006 analysis of ambulatory care malpractice claims found that of 307 claims, 181 (59 %) involved diagnostic errors [9]. Fifty-nine percent of these errors were associated with serious harm and 30 % resulted in death. Most common process breakdowns leading to harm were failure to order an appropriate diagnostic test (55 %), creating a proper follow-up plan (45 %), obtaining an adequate history, or performing adequate physical examination (42 %). Leading contributing factors for errors were failures in judgment (79 %), vigilance or memory (59 %), knowledge (48 %), and patient-related factors (46 %).

Compared to medication and treatment errors, diagnostic errors cause more harm and are less preventable [4]. They account for more malpractice claims than any other medical mishap and account for twice as many alleged and settled cases as medication error [2, 10]. An analysis of large malpractice carrier data revealed that diagnostic error claims incurred the highest amount of payment: $127 million for diagnostic error versus $123 million for all other categories combined [11].

Based on various studies, commonly misdiagnosed conditions include cancer, infection, fracture, myocardial infarction, pulmonary embolism (PE), neurological conditions, and aneurysms. Since a number of these studies are based on claims data, the results are biased toward more serious and more costly diagnoses. Specifically,

PE and drug reactions are the most frequently missed diagnoses followed by missed lung cancer and colon cancer [2]. A review of 67 patients who died of PE over a 5-year period revealed that the diagnosis of PE was not suspected clinically in 37 (55 %) patients [12]. Estimates show that about 1 in 25 patients with myocardial infarction are sent home, and these patients carry a much higher mortality rate than the patients who are appropriately diagnosed and hospitalized [13].

Diagnostic errors are also of concerning for patients. In a Harris poll conducted by the National Patient Safety Foundation, one in six adults reported being misdiagnosed [14]. Another study found that 55 % of adults in the USA cited "misdiagnosis" as their greatest concern when they see a doctor in an ambulatory clinic [15].

## What Causes Diagnostic Errors

A careful analysis of the literature reveals that similar to other medical errors, diagnostic errors are often multifactorial and are caused by an interplay of system flaws and cognitive defects. This validates the Swiss cheese model of multiple vulnerabilities [16], often one contributing to and compounding the other. In a 2005 study of 100 cases, in 46 % of the cases, both system-related and cognitive factors contributed to diagnostic error. Cases involving only cognitive factors (28 %) or only system-related factors (19 %) were less common, and seven cases were found to reflect "no-fault" without any obvious contributory cause. Combining the pure and the mixed cases, systems factors contributed to the diagnostic error in 65 % of cases [17]. The systems factors leading to diagnostic errors were aligned to other types of medical errors and were most often related to policies and procedures, inefficient processes, and difficulties with teamwork and communication, especially with communication of test results. Of cognitive factors, faulty synthesis of available information was the commonest contributor followed by faulty data gathering such as incomplete history and physical examination. Only rarely did inadequate knowledge of physician cause the diagnostic error [17]. This is clearly in contrast to the prevailing belief that diagnostic errors primarily reflect defective cognition. This also gives hope that appropriate systems and processes can be developed as strategies and solutions to diagnostic errors previously considered not amenable to system fixes.

## Diagnostic Decision Making: A Congnitive Model

Recently, a dual-process model of reasoning with its foundation in cognitive psychology has emerged as a promising framework for understanding how physicians think during a diagnostic process. Based on this model, physicians use two differing modes of cognition—"System 1" and "System 2" [18] (Fig. 15.1). The system 1 mode or, nonanalytical thinking, is the rapid, subconscious, and effortless reasoning

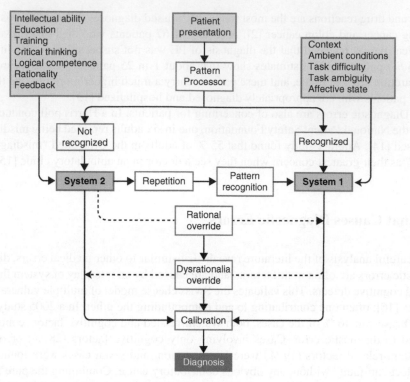

**Fig. 15.1** Dual process model of reasoning applied to diagnosis. Adapted from Croskerry [18]

that is used most of the time in clinical medicine. This is strongly based on pattern recognition—once the combination of clinical features is recognized, physicians automatically go into the System 1 mode of thinking to make a diagnostic decision. However, if the pattern is not recognized or is complex, system 2, or analytical thinking, is activated. The system 2 mode is characterized by slow, deliberate, conscious, and effortful reasoning.

System 1 cognition is based on the physician's clinical competence (experience, fund of knowledge, and intellectual ability) as well as tradition, personal theories, and assumptions that sometimes are not necessarily based on scientific rationale. Although it is a powerful mode for fast and frugal decision making, it can suffer from a number of cognitive biases and, therefore, may lead to a diagnostic error. Table 15.1 describes some of the common cognitive biases that can influence diagnostic decision making. We believe that being aware of such biases will help physicians avoid them in clinical practice [19].

These two systems are not mutually exclusive. It has been suggested that physicians (both novices and experts) use both modes of reasoning to reach final diagnostic decisions in most cases [20]. Early hypotheses generated by System 1 are normally based on experience and theoretical knowledge. Analytical thinking in System 2 may use hypothetico-deduction to deliberately calibrate the early

**Table 15.1** Common cognitive biases [19]

| Failed heuristic/bias | Descriptions |
|---|---|
| *Availability heuristic* | Tendency to accept a diagnosis due to ease in recalling a past similar event or case, rather than based upon statistical prevalence or probability |
| *Representativeness heuristic* | Improper use of pattern recognition to detect representative characteristics (prototype) to diagnose a condition, which can predispose physicians not to consider differential diagnoses |
| *Anchoring* | Tendency to stay with an original diagnosis despite evidence to the contrary |
| *Diagnosis momentum* | The tendency for an opinion or working diagnosis to become almost certain when it is passed from person to person and suppresses further evaluation |
| *Omission* | The tendency toward watchful waiting and reluctance to treat for fear of being held responsible for adverse outcomes, preferring that an event be seen to happen naturally rather than as a result of action taken by a physician |
| *Confirmation* | The tendency to seek out data to confirm one's original idea rather than to seek out or validate disconfirming data |
| *Premature closure* | The tendency to apply closure to the diagnostic process too early on the basis of vivid presenting features that may be convincing for a particular diagnosis, such that the correct diagnosis is not considered |

hypotheses through active search for additional data until a diagnostic threshold is reached. Physicians, however, have to start reasoning from the data to formulate diagnostic hypotheses following the rules or pathophysiological knowledge in complex or rare clinical problems [21]. Studies have shown that using only single mode of reasoning (either analytical or nonanalytical) resulted in lower diagnostic performance when compared with using a combined approach [22]. Physicians should learn and develop the strategic use of the dual process model of reasoning. Some practical cognitive strategies are suggested later in this chapter.

## Steps in Diagnostic Decision Making

Making a clinical diagnosis is a multi-step iterative process that requires listening, collecting data regarding symptoms, performing focused examinations, ordering appropriate tests, synthesizing data, and analyzing results. These steps have been summarized in the Diagnostic Error Evaluation and Research (DEER) taxonomy. This provides another useful framework of classifying the causes of diagnostic errors based on where in the process the error occurred [2].

In an analysis of 583 self-reported diagnostic errors using the DEER chart audit tool, it was found that laboratory and radiology testing (including test ordering, performance, and physician processing) accounted for the largest proportion of errors (44 %), followed by physician assessment (32 %) (including hypothesis generation, weighing or prioritizing, and recognizing urgency or complications). In terms of identifying the specific process failure that occurred, failure or delay in considering the diagnosis accounted for the largest number of diagnostic failures (19 %), followed by failure or delay in ordering needed tests and erroneous laboratory or radiology reading of tests in almost equal frequency (11 %) [2] (Table 15.2).

**Table 15.2** Diagnostic Error Evaluation and Research (DEER) Taxonomy

| Diagnostic process | Failures |
| --- | --- |
| **1. Access/presentation** | A. Failure/delay in presentation |
|  | B. Failure/denied care access |
| **2. History** | A. Failure/delay in eliciting a critical piece in history |
|  | B. Inaccurate/misinterpreted critical piece in history |
|  | C. Suboptimal weighing of a critical piece of history |
|  | D. Failure/delay to follow up on a critical piece of history |
| **3. Physical examination** | A. Failure/delay in eliciting a critical physical exam finding |
|  | B. Inaccurate/misinterpreted critical physical exam finding |
|  | C. Suboptimal weighing a critical physical exam finding |
|  | D. Failure/delay to follow up on a critical physical exam finding |
| **4. Tests** | **Ordering** |
|  | A. Failure/delay in ordering needed test(s) |
|  | B. Failure/delay in performing ordered test(s) |
|  | C. Suboptimal test sequencing |
|  | D. Ordering of wrong test(s) |
|  | E. Test(s) ordered wrong way |
|  | **Performance** |
|  | F. Sample mix-up/mislabel |
|  | G. Technical errors/poor processing of specimen/test |
|  | H. Erroneous laboratory/radiology reading of test |
|  | I. Failed/delay transmission of result to physician |
|  | **Clinical processing** |
|  | J. Failed/delayed follow-up action in test result |
|  | K. Erroneous physician interpretation of test |
| **5. Assessment** | **Hypothesis generation** |
|  | A. Failure/delay in considering correct diagnosis |
|  | B. Suboptimal weighing/prioritizing |
|  | C. Too much weight to lower probability/priority diagnosis |
|  | **Recognizing urgency/complications** |
|  | D. Failure/delay to recognize/weigh urgency |
|  | E. Failure/delay to recognize/weigh complications |
| **6. Referral/consultation** | A. Failure in ordering referral |
|  | B. Failure/delay obtaining/scheduling ordered referral |
|  | C. Error in diagnostic consultation |
|  | D. Failure/delay communication/follow-up consultation |
| **7. Follow up** | A. Failure to refer patient to close/safe setting/monitoring |
|  | B. Failure in timely follow-up/rechecking of patient |

Adapted with permission from Gordon Schiff, MD
*Source*: Schiff et al. [2]

# Case Studies

## Case 1: Clinical Summary: Missed Diagnosis of Appendicitis

A 3-year-old female with acute myeloid leukemia presented with acute febrile neutropenia. Broad spectrum antibiotics were initiated secondary to concerns for sepsis but were discontinued after 5 days of negative cultures. The patient was currently undergoing intensive chemotherapy. She subsequently developed abdominal pain, the severity of which was difficult to interpret due to the child's trepidation around physicians. During the next few days, she developed low-grade fevers. The oncologist suspected typhlitis as the cause of abdominal pain in this febrile neutropenic patient. The pediatric surgeons concurred with the working diagnosis and recommended serial abdominal exams. After 3 more days of persistent symptoms, an abdominal computed tomography (CT) was obtained. Findings were consistent with severe acute appendicitis with perforation without evidence of typhlitis. The patient was taken to the operating room for an appendectomy.

## Case 1: Analysis and Discussion

The fundamental error in this case involved a misdiagnosis of typhilitis and this led to a further delay in making the correct definitive diagnosis of appendicitis.

The diagnostic decision making in this case demonstrates a number of cognitive biases that contributed to the delayed diagnosis of this patient. The oncologists diagnosed typhlitis based on the *availability heuristic*. The pattern of right lower quadrant pain in a neutropenic patient was quickly thought to be typhlitis. This *premature closure* set in motion a *diagnostic momentum* by which a particular diagnosis became established with inadequate evidence. Because the case had been framed for typhlitis, the surgical team followed the diagnostic momentum with *anchoring and confirmation biases* by acting upon an incomplete investigation. They were reluctant to order any potentially unnecessary abdominal imaging because of concerns of subjecting the patient to unnecessary radiation in light of an already working diagnosis. The surgeons then exhibited *omission bias* through watchful waiting. A complete evaluation, in retrospect, should have included an abdominal CT scan to rule out appendicitis, which is considered a surgical emergency in a 3-year-old child with right lower-quadrant pain.

### Physician (Over) Confidence in Diagnostic Abilities

The *worst-case scenario* rule should have been applied in this case to diagnose in a timely fashion and avoid the complications of perforated appendicitis. Overconfidence in diagnostic abilities is a common but under-recognized contributor to diagnostic errors. One critical element is physicians' miscalibration of their diagnostic ability. In a 2005 study, medical students, residents, and attending physicians were given 36 diagnostically challenging cases, each with a definitive correct diagnosis. The study found that students were overconfident in 25 % of cases in which their confidence and correctness was not aligned, residents were overconfident in 41 % of cases, and attendings in 36 % [23]. In case 1, the whole team of physicians was overly confident about the working diagnosis and failed the attempt to rule out the commonly missed surgical condition, acute appendicitis.

The diagnosis of acute appendicitis based on clinical findings is often challenging in children. One important issue is that in the absence of a systematic feedback loop informing physicians regarding the accuracy of the diagnosis made, physicians may never learn of the diagnostic error as misdiagnosed patients may also simply leave the practice seeking care elsewhere. Enhancing feedback to physicians regarding diagnoses and errors would increase calibration and reduce overconfidence regarding their own diagnostic error rate [4].

## Case 2: Clinical Summary: Hyperkalemia Wrongly Attributed to Hemolysis

*A 4-month-old infant with shock was transferred to the emergency department (ED) from a community hospital. He presented with cardiopulmonary arrest. The ED team successfully resuscitated the patient and transferred him to the pediatric intensive care unit (PICU) with presumed diagnosis of septic shock. He developed another arrest soon after the PICU admission. Upon review of the transfer documentation, the patient was found to have serum potassium level of 8.5 mEq/L at the outside hospital and 9.0 mEq/L at the ED. He stabilized after the hyperkalemia was addressed. The abnormal potassium level was ascribed by the referring physician to be due to a hemolyzed specimen. He responsibly repeated the test but did not include the presumably spurious laboratory data in the transfer communication. The patient was later found to have congenital adrenal hyperplasia.*

## Case 2: Analysis and Discussion

The fundamental error in this case involved non-standardized detection and reporting of hemolyzed samples, wrong assumption made by the referring physician that the hyperkalemia was due to a hemolyzed specimen and failure to consider other possibilities and breakdown in information management.

## The "Swiss Cheese" Model of Multiple Vulnerabilities

Case 2 demonstrates a cascade of errors caused by a combination of cognitive and system-related factors. It is a medical mishap that fits Reason's Swiss cheese model. Although the diagnosis was delayed mainly due to system-related errors, cognitive biases were a contributing factor. The dilemma involving the issue of hemolyzed specimens is a common challenge particularly in pediatric patients. *Omission bias* led physicians to repeat testing rather than acting upon it despite a reading of a life-threatening level of potassium. A tendency toward inaction (do no harm) is a common approach in medicine because of a belief that the physician is more likely to be blamed for taking action than for inaction when a negative outcome occurs. Some physicians may become desensitized by frequent erroneous results and get into the habit of disputing them rather than considering the risks and benefits of watchful waiting. Because a hemolyzed specimen is so common in pediatric practice, some physicians may develop the logical fallacy of assuming or suspecting that every report of hyperkalemia involves a hemolyzed sample. Inadequate knowledge sometimes forms the basis of cognitive errors. The physician at the outside hospital did not recognize the strikingly abnormal laboratory data as an ominous sign of congenital adrenal hyperplasia. The potassium level of 9 mEq/L was reported to the ED after the patient had already been transferred to the PICU.

## Errors Associated with Laboratory Testing

Given technological advances in laboratory testing, the risk of errors has decreased significantly within the processes occurring in the laboratory. Still, recent studies from various clinical settings such as primary care, internal medicine, and ED attest that the rates of errors in test request and result interpretation are unacceptably high and cause diagnostic errors [9, 24]. An attempt must be made to improve physician knowledge about laboratory tests and the correct interpretation of test results. The sources for point of care knowledge should be readily available. The narrative interpretation and interpretive comments in the test reports should be provided [25]. At the systems level, explicit guidelines to improve coordinated care between laboratory specialists and physicians must be developed and implemented. In cases of specimen with hemolysis, laboratory personnel should always ask for a new sample. If the specimen is found to be hemolyzed, it cannot simply be rejected especially in case of critical reading of hyperkalemia, but the laboratory should alert the physician so that any in vivo hemolysis or hyperkalmia can be ruled out.

## System-Related Errors

The frequency of system-related factors varies with the types of errors. Based on Graber et al. [17], delayed diagnosis had more system-related errors (89 %), whereas wrong diagnosis had more cognitive errors (92 %). In one of the largest physician-reported cases of diagnostic errors using DEER taxonomy to localize the breakdowns in the diagnostic process, the testing process had the greatest number of reported process failure. Failures in the physician assessment process, cognitive errors, were slightly fewer [2]. Common system-related factors that contribute to diagnostic errors include those related to specimen identification, test tracking, reporting of abnormal and critical test results, and transitions in care. The breakdown in information management, including inefficient processes, poor communication, and coordination of care, are among the most common causes [17, 26]. In looking back at the case, poor handoff and communication at transition of care from the referring physician to the hospital contributed to the error. Critical information was lost during inter-facility transfer and was not conveyed in a timely enough manner to prevent the patient's second cardiopulmonary arrest in the PICU.

## Remedies for Diagnostic Errors

### Strategies to Enhance Diagnostic Decision-Making

Based on existing knowledge in cognitive psychology and medical decision making, physicians should learn about sources of cognitive errors and familiarize themselves with different cognitive approaches to making better decisions. Further research is required to investigate if these proposed cognitive strategies can actually optimize diagnostic decision making and decrease errors. Suggested cognitive strategies to avoid biases and errors include the following [27]:

1. Decrease reliance on memory: To improve diagnostic accuracy, physicians should force themselves to use memory aids such as mnemonics, flash cards, or computer applications with algorithms and checklists to reduce cognitive load for diagnostic thinking.
2. Enhance metacognitive skills to promote System 2 processes: Metacognition is "thinking about one's own thinking." Physicians should take time to actively reflect and regulate their own thinking and affective process. Forced generation of a comprehensive list for differential diagnosis and routine use of a "diagnostic pause" (Fig. 15.2) to check one's diagnostic thinking are examples of metacognitive strategies [28].

**Diagnostic Pause**

- Does the diagnosis make sense?
- Are these clinical findings what I expected?
- Did I have any cognitive or affective biases?
- Did I omit any serious/life-threatening conditions?
- Am I comfortable with this decision?
- Does the patient agree with me?

If the patient is not comfortable, or does not believe your diagnosis, reevaluate the situation

Fig. 15.2 Diagnostic pause: a tool to foster metacognition. Adapted from Quirk [28]

3. Develop cognitive forcing strategies: Individual physicians should develop generic and specific strategies to monitor and override predictable cognitive biases in particular clinical situations (e.g., diagnose anxiety disorder only by careful exclusion, always investigate for multiple drugs in case of suicidal ingestion, be extra cautious prior to making the conclusion when reliable history is not available, etc.) [19].
4. Use group decision making: In doubtful situations, collective wisdom likely produces an optimal solution. If a patient care conference is not practical, sharing one's decision making with another colleague to reflect on diagnostic thinking is still valuable (i.e., thinking out loud with feedback).
5. Personal accountability: People generally put more effort into decision making when they know that they will be held accountable. Personal accountability using timely constructive feedback will lead to better calibration of future decisions [23].

Addressing issues at the system level can assist physicians with cognitive aspects of diagnostic error. An organization may consider the following system-level strategies to reduce cognitive errors:

1. Provide resources for diagnostic decision support system (DDSS): There are commercially available Web-based DDSS applications that provide comprehensive list of differential diagnoses and accurate estimates of disease probability. Isabel (http://www.isabelhealthcare.com) is an example of available applications that uses patient's demographics and clinical features to produce a list of possible diagnoses. These resources could also include point-of-care general medical knowledge references [4, 27].
2. Provide resources and encourage use of clinical guidelines and clinical algorithms: These resources enhance physician adherence with evidence-based medical practice which helps prevent cognitive biases inherent to human judgment and reduce errors.
3. Incorporate forced use of checklists into the diagnostic process: Diagnostic checklists can be used to prevent reliance on memory and overconfidence for error-prone diagnoses (e.g., chest pain, dizziness). Strategic use of checklists can: (1) guide physicians to optimize their cognitive approach, (2) remind physicians to consider a complete list of possible diagnoses for a given clinical problem, and (3) remind physicians of common pitfalls or biases when diagnosing certain diseases (cognitive forcing strategies) [29].

## Strategies to Reduce System-Related Errors

While progress has been made in understanding systems causes of diagnostic errors, studies evaluating the effectiveness of these system-level interventions are lacking. The important first step should focus on changing the perception of diagnostic errors from "errors in judgment," "errors in thinking," or "physician mistakes" to

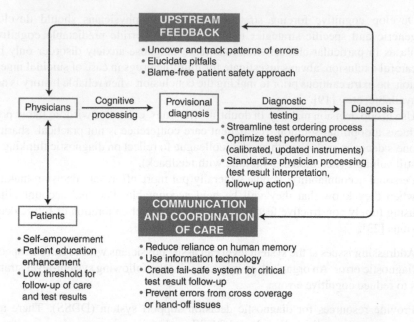

**UPSTREAM FEEDBACK**
• Uncover and track patterns of errors
• Elucidate pitfalls
• Blame-free patient safety approach

Physicians — Cognitive processing → Provisional diagnosis — Diagnostic testing → Diagnosis
• Streamline test ordering process
• Optimize test performance (calibrated, updated instruments)
• Standardize physician processing (test result interpretation, follow-up action)

**COMMUNICATION AND COORDINATION OF CARE**

Patients
• Self-empowerment
• Patient education enhancement
• Low threshold for follow-up of care and test results

• Reduce reliance on human memory
• Use information technology
• Create fail-safe system for critical test result follow-up
• Prevent errors from cross coverage or hand-off issues

**Fig. 15.3** Closed-loop feedback in the diagnostic process

errors related to cognitive processing, communication, and system design [27]. This Just Culture oriented patient safety approach would allow scientific studies of diagnostic errors to find effective strategies to minimize them. Figure 15.3 summarizes multifaceted patient safety approach to improve the diagnostic process. System strategies to enhance communication and coordination of care may include the following:

1. Optimize the use of electronic health records to facilitate transfer of patient information across clinical settings. The system needs to ensure that the required follow-up action is completed to close the information loop. The system can generate automatic messages reporting test results to physicians and patients and schedule follow-up in timely and reliable fashion [4, 30].
2. Enhance the laboratory-clinical interface by developing coordinated care between laboratory specialists and physicians in all steps of diagnostic testing. This approach could minimize inappropriate test requests and misleading interpretation of laboratory data [31].
3. Ensure that specialty expertise is available when needed. An attempt should be made to identify and resolve any potential barriers that compromise effective communication and coordination among all clinical services [27].
4. Encourage and educate patients to be active participants in every step of the diagnostic process. The patient is a crucial part of the "safety net" for system errors. Patients should offer the complete story, remind physicians to consider

other possibilities, disagree with their physicians, ask for clarification, and request timely follow-up [32].

5. Develop a system for reliable follow-up. Timely follow-up is crucial for high-risk diagnoses or symptoms for which a diagnosis has not been made (e.g., patients should not think that "no news is good news") [4].

6. Establish pathways for reliable, blame-free upstream feedback to physicians in cases of misdiagnosis-related harm. This can be achieved by developing chart audit protocols to look for changed diagnoses (e.g., comparing ED diagnoses to subsequent diagnoses, reviewing all readmissions, morbidity and mortality conferences, sentinel event analysis, etc.) [4, 27].

## Conclusion and Key Lessons

Diagnostic errors are common, costly, and can result in adverse consequences for patients, families, and healthcare professionals. Physicians should educate themselves about sources of errors, analyze different processes involved in the diagnostic process, and healthcare organizations should implement explicit strategies to minimize cognitive- and systems-related factors leading to diagnostic errors.

### Key Lessons

1. Diagnostic errors place serious financial burden on the healthcare system and can be devastating for affected patients, families, and physicians.
2. Causes of diagnostic errors are often multifactorial; cognitive processing errors and system design flaws are contributory factors.
3. Physicians should familiarize themselves with the science of diagnostic decision making and common biases that can affect their decisions.
4. While many available system-level strategies can be implemented to reduce diagnostic errors, further research is required to prove their effectiveness.

## References

1. Graber M. Diagnostic errors in medicine: a case of neglect. Jt Comm J Qual Patient Saf. 2005;31(2):106–13.
2. Schiff GD, Hasan O, Kim S, et al. Diagnostic error in medicine: analysis of 583 physician-reported errors. Arch Intern Med. 2009;169(20):1881–7.
3. Wachter RM. Why diagnostic errors don't get any respect—and what can be done about them. Health Aff. 2010;29(9):1605–10.
4. Berner ES, Graber ML. Overconfidence as a cause of diagnostic error in medicine. Am J Med. 2008;121(5 Suppl):S2–23.

5. Leape LL, Brennan TA, Laird N, et al. The nature of adverse events in hospitalized patients. Results of the Harvard medical practice study II. N Eng J Med. 1991;324(6):377–84.

6. Bogner MS. Human error in medicine. Hillsdale, NJ: Lawrence Erlbaum Associates; 1994.

7. Croskerry P, Sinclair D. Emergency medicine: a practice prone to error? Can J Emerg Med. 2001;3:271–6.

8. Kachlia A, Ghandi TK, Puopolo AL, et al. Missed and delayed diagnoses in the emergency department: a study of closed malpractice claims from 4 liability insurers. Ann Emerg Med. 2007;49(2):196–205.

9. Gandhi TK, Kachalia A, Thomas EJ, et al. Missed and delayed diagnoses in the ambulatory setting: a study of closed malpractice claims. Ann Intern Med. 2006;125:488–96.

10. Lippman H, Davenport J. Sued for misdiagnosis? It could happen to you. J Fam Pract. 2010;59(9):498–508.

11. Hanscom R. CRICO/RMF community targets diagnostic error. CRICO/RMF insight. Available at http://www.rmf.harvard.edu/education-interventions/crico-rmf-insight/ archives/092007/ art1.htm. Accessed 10 Feb 2012.

12. Pidenda LA, Hathwar VS, Grand BJ. Clinical suspicion of fatal pulmonary embolism. Chest. 2001;120:791–5.

13. Wachter RM. Understanding patient safety. New York, NY: McGraw-Hill Medical; 2008.

14. Golodner L. How the public percieves patient safety. Newsletter of the National Patient Safety Foundation 2004;1997:1–6.

15. Isabel Health Care. Misdiagnosis is an overlooked and growing patient safety issue and core mission of Isabel Healthcare. 20 Mar 2006. Available at http://www.isabelhealthcare.com/pdf/ USsurveyrelease-Final.pdf. Accessed 13 Jul 2013.

16. Reason J. Human error: models and management. BMJ. 2000;320:768–70.

17. Graber ML, Franklin N, Gordon R. Diagnostic error in internal medicine. Arch Intern Med. 2005;165:1493–9.

18. Croskerry P. Context is everything or how could I have been that stupid? Healthc Q. 2009;12:e171–6.

19. Croskerry P. Achieving quality in clinical decision making: cognitive strategies and detection of bias. Acad Emerg Med. 2002;9:1184–204.

20. Eva KW. What every teacher needs to know about clinical reasoning. Med Educ. 2005;39:98–106.

21. Balla JI, Heneghan C, Glasziou P, Thompson M, Balla ME. A model for reflection for good clinical practice. J Eval Clin Pract. 2009;15:964–9.

22. Ark TK, Brooks LR, Eva KW. The best of both worlds: adoption of a combined (analytical and non-analytical) reasoning strategy improves diagnostic accuracy relative to either strategy in isolation. Proceedings of the annual meeting of the Association of American Medical Colleges, 5–10 Nov 2004, Boston.

23. Friedman CP, Gatti GG, Franz TM, et al. Do physicians know when their diagnoses are correct? Implications for decision support and error reduction. J Gen Intern Med. 2005;20(4): 334–9.

24. Wahls TL, Cram PM. The frequency of missed test results and associated treatment delays in a highly computerized health system. BMC Fam Pract. 2007;8:32.

25. Plebani M. Interpretive commenting: a tool for improving the laboratory-clinical interface. Clin Chim Acta. 2009;404:405–51.

26. Singh H, Graber M. Reducing diagnostic error through medical home–based primary care reform. JAMA. 2010;304(4):463–4.

27. Pennsylvania Patient Safety Authority. Diagnostic error in acute care. Pa Patient Saf Advis. 2010;7(3):76–86. Available at http://patientsafetyauthority.org/ADVISORIES/Advisory Library/2010/Sep7(3)/Pages/76.aspx. Accessed 13 Jul 2013.

28. Quirk ME. Intuition and metacognition in medical education: keys to developing expertise. New York, NY: Springer; 2006.

29. Ely JW, Graber ML, Croskery P. Checklists to reduce diagnostic errors. Acad Med. 2011;86:307–13.
30. Schiff GD, Bates DW. Can electronic clinical documentation help prevent diagnostic errors? N Engl J Med. 2010;362:1066–9.
31. Plebani M, Laposata M, Lundberg GD. The brain-to-brain loop concept for laboratory testing 40 years after its introduction. Am J Clin Pathol. 2011;136(6):829–33.
32. Schiff GD, Kim S, Abrams R. Diagnosing diagnostic error: lessons from a multi-institutional collaborative project. In: Henriksen K, Battles JB, Marks ES, et al., editors. Advances in patient safety: from research to implementation. Rockville, MD: Agency for Healthcare Research and Quality; 2005. p. 255–78. AHRQ pub No. 05-0021-2.

# Part III
# Special Considerations

# Chapter 16
# Patient Safety in Pediatrics

Erin Stucky Fisher

*"We are built to make mistakes, coded for error."*

Lewis Thomas

## Introduction

Medical errors and patient harm events that occur in pediatric patients differ from those of adults, due to different physical characteristics, developmental issues, and the dependent/legal/vulnerable state of the child [1, 2]. Although error and harm due to medications [3–6] are most prevalently cited, diagnostic errors, patient misidentification [7], communication failures, and lack of information system customization are some of the other frequent problems associated with pediatric safety events [8, 9]. It is also important to keep in mind that the definition of a pediatric patient is not always limited by age; young adults with chronic and/or unusual diseases are often cared for in the pediatric healthcare setting [10]. Healthcare safety failures for children are many and include lack of proper equipment (e.g., adult-sized oxygen saturation monitor probes causing erroneous results), over or misuse of technology (e.g., radiation dosing for computed tomography higher than necessary to produce adequate image), lack of awareness of age-specific norms (e.g., vital sign changes misinterpreted, resulting in either excessive or conversely no action taken), and failure to anticipate environmental influences (e.g., hypothermia due to cold rooms or lack of bundling resulting in physiologic stress) [2, 3, 7, 9].

E.S. Fisher, M.D., M.H.M. (✉)
University of California San Diego, San Diego, CA, USA

Rady Children's Hospital San Diego, 3020 Children's Way,
MC 5064, San Diego, CA 92123, USA
e-mail: estucky@rchsd.org

A. Agrawal (ed.), *Patient Safety: A Case-Based Comprehensive Guide*,     249
DOI 10.1007/978-1-4614-7419-7_16, © Springer Science+Business Media New York 2014

To date, reports on the epidemiology of pediatric safety events have been focused primarily on the hospital setting [2]. Medication errors are not surprisingly the most commonly cited safety event (5–50 % of errors) and include a combination of calculation, formulation, dispensing, and administration errors [2, 5]. There is potentially greater risk of error commitment in the medication process for pediatric patients than for adult patients due to weight-based prescribing needs, dynamic age and disease-state physiologic and developmental changes, and medication delivery issues that are unique to children [2, 5]. In addition patient misidentification, delays in care, miscommunication, intravenous access problems, and other incidents have also been reported, some at rates of up to 10 % [3, 7, 9]. Although ambulatory reports are fewer, one multi-center study similarly demonstrated that medication errors occurred most commonly (32 %); however, administrative (documentation) and diagnostic errors were also often reported (22 and 15 %, respectively) [6]. Importantly, communication was deemed a contributing factor in 67 % of all reported events [6]. In all settings, there are challenges to obtaining accurate and timely error reports and to implementing durable solutions. Despite these unique challenges, the approach toward identification, resolution, and abatement of pediatric harm follows the same tenets of healthcare safety mentioned elsewhere in this book.

A number of case examples could serve to instruct on pediatric error and harm. As noted, although most information has come from inpatient reports, ambulatory errors are of great importance as well but largely underreported [2]. Several entities have worked to call attention to pediatric errors and system solutions, including the Institute for Healthcare Improvement (IHI) (High Alert Medications in Pediatrics) [11], the American Academy of Pediatrics (AAP) (Patient Safety Policy statement) [2], The Joint Commission (TJC) (various resources) [12], and the Agency for Healthcare Research and Quality (AHRQ) (Patient Safety Indicators) [13]. These groups suggest both technology-based solutions such as pediatric-specific electronic health record and computerized decision support systems as well as some very basic changes, such as mandatory weight recording in kilograms, that highlight the stark contrast in work yet to be done in pediatric healthcare safety. While the case examples below cannot address all aspects of pediatric error and harm, they call out some of the issues unique to children that deserve attention.

## Case Studies

### Case 1A: Delayed Diagnosis Leading to Orchiectomy in a 9-Month-Old Infant

#### Clinical Summary

*A.B. is a 9-month-old, previously healthy, term male seen at a community emergency department (ED) with parental concern for crying and fussiness for several hours. On arrival vital signs were noted to be stable except for an elevated heart rate thought to be due to crying. Examination was normal except for left-sided scrotal swelling. Over the next 4½ h, the ED physician obtained a scrotal sonogram which was read as nondiagnostic for torsion; the on-call pediatrician was called to admit the patient for pain management and further evaluation. Upon assessment in the ED, the pediatrician called for urgent transfer to the local children's hospital and immediate urologic consult. The child was met in the children's ED by the urologist and taken directly to the operating room where left orchiectomy was performed due to a necrotic testis.*

## Case 1B: Missed Diagnosis of Inflammatory Bowel Disease in an Adolescent

### Clinical Summary

*L.M. is a 16-year-old boy with inflammatory bowel disease (IBD) admitted to a large community hospital for upper arm cellulitis thought to be due to an abrasion that occurred when he fell (helmeted) from his bicycle 2 days prior to admission. The cellulitis improved with treatment. On the day of discharge the patient had a bloody stool and abdominal pain which was recorded by the nurse. A resident assessed the patient when the parent was at work. The patient stated he "was fine" and wanted to go home; he was discharged. One day later, the patient was admitted to the children's hospital for a severe IBD flare.*

### Analysis: Case 1A and Case 1B

These two cases highlight the added vigilance needed when caring for pediatric patients of varied ages. What happened in each case? The first case underscores the need for age and/or disease state- specific criteria for pediatric assessments in community settings as is recommended by the Emergency Medical Services for Children and the American Academy of Pediatrics [14]. Delay in obtaining and interpreting radiological images and delay in transfer to a facility where definitive treatment can be rendered are not uncommon at sites where personnel and facilities do not frequently care for infants. The second case highlights the need to recognize the impact of unrelated acute medical needs on underlying chronic disease states, to assess and account for clinical changes in the face of patient denial, and to balance adolescent autonomy with family engagement when rendering medical decisions. Adolescents are a special challenge, particularly those with chronic disease who may hesitate to complain, do not want to stay in the hospital, or fail to advocate for themselves when they have issues they would like raised. What can be done to prevent recurrence of these failures? Protocols should be written for pediatric consultation and testing that acknowledge skill sets available for rendering services to children of different ages and underlying disease states. Patient- and family-centered care (PFCC) principles [15] and a team approach toward care for adolescents should be fostered. Pediatricians should have a presence on relevant hospital committees and should participate in case reviews of pediatric-aged patient events that occur at any site in the facility. While these short cases focus briefly on the importance of advocacy in community settings, the children's hospital case below offers detail on a review process and solution planning that can translate to any setting.

## Case 2: Pediatric Patient Harm Due to Multiple System Failures

### Clinical Summary

*H.M. is a 5-year-old female, ex-28-week gestation preterm with chronic lung disease (CLD), developmental delay, status post-gastrostomy tube (GT) with fundoplication in infancy and GT closure 1 year ago, history of oral aversion, admitted with CLD exacerbation. During the hospitalization she was diagnosed with atypical pneumonia, started on macrolide therapy, given increased dose of intravenous (IV) steroids, and her home medications were changed to IV form due to severe respiratory distress and both metabolic and respiratory acidosis noted on blood gas analysis. She was improving with treatment by hospital day (HD) 5 but the following day the Code Blue Team was called for respiratory failure. She spent three days intubated in the intensive care unit and was eventually discharged home on HD 12.*

*What happened? The critical event unfolded over approximately 36 h (see Table 16.1). When the Code Blue Team was called, the child had no respiratory drive and had low blood pressure (75/40). After she was intubated it was clear she had pulmonary edema but despite adequate ventilator support, she required significant cardiovascular medication infusions to maintain her blood pressure. She was restarted on her IV steroids at the same dose she had received on admission (2 mg/kg every 8 h). Over the next several hours her blood pressure was under much better control and she was weaned off the cardiovascular medications the following day. It was noted that she had not been placed on oral steroids on HD 5 after her IV steroids were stopped. She had been on 1 mg/kg/day as an outpatient for the week prior to admission due to her increasing respiratory symptoms and had been on every other day steroids for the past several months for her CLD.*

### Root Cause Analysis

What was the next step? A Root Cause Analysis (RCA) led to the discovery of multiple failures and proposed solutions. The RCA process includes asking "why" and "how," offers solutions, and expects actions based on these proposed solutions. Questions on normal policy/procedures, process disruptions, human factors, training, individual performance, equipment, environment, information technology, as well as solution planning are included. The commonly used TJC RCA template [16] goes further to identify organizational leadership investment in promoting the culture of safety and assuring systems are in place to recognize and report errors.

**Table 16.1**  Case 2: Relevant timeline

| Hospital day (HD) and time | Event | Note |
|---|---|---|
| *HD#5* | | |
| 09:30 | Bedside clinical rounds performed; patient is off oxygen with stable baseline respiratory effort and vital signs. Heart rate (HR) 74, blood pressure (BP) 108/65, respiratory rate (RR) 22, oxygen saturation 94 %. Plan made to stop the intravenous (IV) steroids and change to oral steroids | 1 |
| 11:00 | Nurse calls intern for orders. Intern discontinues the IV steroids. No order for oral steroids is placed | 2 |
| 16: 30 | Mother arrives at the hospital and notes her daughter "looks tired." Nurse encourages mother to get her daughter to nap | 3 |
| 19:10 | Father arrives at bedside for the night; mother goes home to care for siblings. Father is updated on the plans of the day | 4 |
| 19:32 | Night nurse calls intern with concern that the patient has had poor oral intake all day. Intern orders IV fluids at maintenance rate | 5 |
| *HD#6* | | |
| 02:35 | Night nurse is taking vital signs, notes HR elevated to 110, patient asleep. Father is sleeping at the bedside. Nurse calls intern about elevated HR. Intern believes this is due to inadequate fluids and orders a 20 mL/kg bolus of normal saline and increases the rate of the IV fluids to 1.5 times maintenance | 6 |
| 06:35 | Mother arrives and father leaves for work, stating things "were fine" overnight | 7 |
| 07:15 | Mother calls nurse with concern about her daughter's breathing and says she is more "clingy." Nurse reassures mother | 8 |
| 09:30 | Bedside clinical rounds are performed. The monitor alarms while the patient is fussy with the exam. Mother restates her concerns and is told the patient will be monitored carefully | 9 |
| 11:12 | Mother calls the nurse to watch her daughter's breathing. Intern is called for "needing oxygen—saturation dips." Orders given for oxygen to keep oxygen saturations greater than 95 %. Charge nurse notified (RR 38, oxygen saturation is 89–90 %, HR 118, BP 89/54) | 10 |
| 13:10 | Nurse records respiratory rate at 33; oxygen saturation on 1 L is 88–90 %. She notes breathing a bit more labored but patient is "calmer". Nurse increases the oxygen to 2 L per minute. Intern notified "turning up the oxygen" | 11 |
| 14:11 | Mother calls the nurse, stating she is concerned that her daughter does not want to eat and is "tired." Nurse reassures mother | 12 |
| 15:08 | Nurse calls intern because the monitor is alarming for HR. Intern is told the patient is sleeping, on 3 L oxygen and that the saturations have been "off" and "not picking up well." The nurse has called for a new monitor saturation probe | 13 |
| 15:22 | Nurse enters room to change probe and finds patient cyanotic and pale, with RR of ~6. Code Blue is called | 14 |

How did this particular event happen? In this case, the hospital staff did not follow established *policies and procedures* (Table 16.2). The hospital's "Ask More" Policy directs staff to notify the Charge Nurse if urgent patient care changes have not been resolved with usual conversation and interventions and to continue to pursue resolution of the concern by elevating the issue to the covering physician and others including the Chief of Staff. The Charge Nurse stated she was told by the nurse that "the patient is a little worse but the resident has been called" and inferred that the issue was being resolved. Documentation of communications between the nurse and intern was unclear or missing. While there was a notation that the nurse notified the intern of changes in the patient's condition, detail on what changes were reported was not documented and the notation indicated only "MD aware." The intern failed to examine H.M. and notify the supervising resident or attending physician of the concerns as he thought his management plan had resolved the problem.

*Human factors* overlaid these procedural failures. Critical thinking was not evident a number of times. The intern did not order resumption and arguably a taper of oral steroids on HD5 as the IV steroids were discontinued, and further, on HD6 the medication list was not reviewed by the team as this could have alerted them to the omission of the steroid. The nurse stated she was distracted and did not document her work on HD6 until late morning, so early morning events and vital signs were not available for the rounding team. The rounding team and in particular the intern separately likely committed one of a variety of cognitive errors: anchoring (fixation on initial features of a case and not adjusting for later information); availability bias (focusing on what readily comes to mind as the source of the problem); and posterior probability (undue influence by what has happened with the patient or similar patients in the past) [17, 18]. Failure to recognize shock, in this case due initially to sudden discontinuation of steroids, is not uncommon in children [19]. The interpretation of heart rate elevation due to inadequate volume status instead of assessing for all causes of tachycardia resulted in excessive IV fluid administration in this fluid-sensitive CLD patient and ultimately led to pulmonary edema. Hypoxia was interpreted as a "normal" variation seen in CLD patients; however, these patients rely on hypoxia for respiratory drive [20]. Administration of oxygen to this patient, without addressing respiratory support needs, removed the drive and caused the respiratory rate to drop. The nurse interpreted patient "calm" as overall improvement. As much of the tachypnea was an attempt to compensate for metabolic acidosis from shock, the inability to ventilate caused a precipitous drop in pH and resulted in cardiorespiratory failure. Children are at greater risk for respiratory failure than adults due to anatomic issues (such as limited cartilage support of airway, small airway diameter, larger and more horizontally placed epiglottis, and narrow subglottis), limited gas exchange (fewer and smaller alveoli and fewer collateral channels for ventilation between alveoli) and immature respiratory drive (underdeveloped central respiratory control and respiratory muscles and compliant chest wall) [21]. CLD patients on steroids and diuretics not only have limited reserve but also develop tolerance for chronic hypoxia and hypercapnia. Often symptom changes are subtle (tiring or decreased appetite) with a dramatic worsening and more classic signs of respiratory failure then occurring within minutes [20, 21].

**Table 16.2** CASE 2: Root cause analysis (RCA) (only applicable issues listed)

| Patient: H.M. | | | MRN: 1234567 | |
|---|---|---|---|---|

**Participants**

Attending physician; Quality Management Medical Director; Pediatric Residency Associate Program Director; Patient Safety Officer; Risk Management/Quality Management nurse specialist; Nursing Unit Director; bedside nurse; Unit Charge Nurse; Pediatric Chief Resident; participant pediatric resident; Quality Management Nurse Coordinator

| Issue type | Issue | Root cause | Actions and solutions | Discussion Involved party ( ) Associated Timeline Note Number from Table 16.1 [ ] |
|---|---|---|---|---|
| Policy/procedures | Normal policy/ procedures followed? | X | Re-education | No. "Ask More" Policy not followed (nurse, intern) [11–13] |
| Policy/procedures | Any missteps in the process? | X | Re-education; "Ask More" Policy change | Yes. Verbal and written communication not clear (nurse, intern, Charge Nurse) [6, 9–11, 13] |
| Policy/procedures | Other concerns? | X | Re-education | Yes. Failure to examine and communicate (intern) [5, 6, 10, 13] |
| Human factors | Relevant human factors? | X | Rounds change; Pediatric Early Warning System (PEWS) | Yes. Failure of critical thinking skills; communication; distraction (nurse, intern, resident, attending physician) [2, 6, 8–13] |
| Performance factors | Did performance meet expectations? | | Training | No. (intern, nurse) [2, 5, 6, 8–13] |
| Recurrence risk | Could this event happen to other patients? In other areas? | | Dissemination | Yes |

**Solutions Planned**

List here details on actions and solutions. Include pilots, dissemination plan, and assessment of outcome of changes made

| Solution | For Whom? | Responsible Party |
|---|---|---|
| **Re-education:** Provide re-education on: Documentation; communication; "Ask More" Policy; use of Situation-Background-Assessment-Recommendation (SBAR) tool; CLD patient risks | Nurse, intern | Unit Nursing Educator; Pediatric Chief Resident; Pediatric Residency Associate Program Director; attending physician |

(continued)

**Table 16.2** (continued)

| Solution | For Whom? | Responsible Party |
|---|---|---|
| **Policy change:**<br>Revise "Ask More" Policy to require Charge Nurse bedside assessment for any patient about whom s(he) is called. Assessment to include review of documentation and care plans | Nursing | Quality Management Department with Nursing and Medical Staff leadership |
| **Rounds change:**<br>Pilot medication review and order writing on rounds for resident patients (all units). Pharmacist to participate when available. | Residents, nurses, pharmacy | Nursing Unit Directors, Pediatric Chief Resident; Pediatric Residency Associate Program Director; Pharmacy Director |
| **New education and orientation:**<br>1. Add SBAR, PEWS, Rapid Response Team (RRT) and Code Blue Team scenario to hospital staff annual education<br>2. Revise family hospital orientation to emphasize family-initiated RRT | Hospital staff, families | Human Resources; Hospital Education Department; Customer Service; Nursing Unit Directors |
| **Dissemination:**<br>1. Re-distribute SBAR tool, revised "Ask More" Policy, revised family hospital orientation, and notification of addition to annual hospital staff education to all clinicians<br>2. Give participant family feedback on plans and actions taken | Medical staff, hospital staff, residents, family | Nursing Unit Directors; Associate Pediatric Chief Resident; Pediatric Residency Associate Program Director; Risk Management/Quality Management nurse specialist |
| **System intervention:**<br>Pilot PEWS program on this Nursing unit | All on unit | Nursing Unit Director |
| **Training:**<br>1. Successfully complete communication education that includes role play<br>2. Successfully participate in mock scenarios that include use of PEWS and RRT | Nurse, intern | Unit Nursing Educator; Pediatric Residency Associate Program Director |
| **Assessment of Changes:**<br>Track PEWS and rounds outcomes at 30 and 60 days. Disseminate these practices across all units within 90-120 days (pending pilot results) | Medical staff, hospital staff | Quality Management Department |

Communication failed numerous times. While the intern did admit hearing the words "tachypnea" and "desats [sic]," the level of concern was not apparent in the tone used by the nurse on the phone and the importance of these did not register with the intern. The hospital's communication tool using the situation–background–assessment–recommendation (SBAR) [22] format was not used. The intern was not asked to reassess the patient and thus assumed H.M. had improved with increasing the oxygen level. The mother was concerned but was repeatedly told her child had CLD so "the breathing can get better and worse again like this." Ignoring parental concerns, in particular related to a patient with chronic disease, is not uncommon but leads to errors and decreased family satisfaction [15, 23].

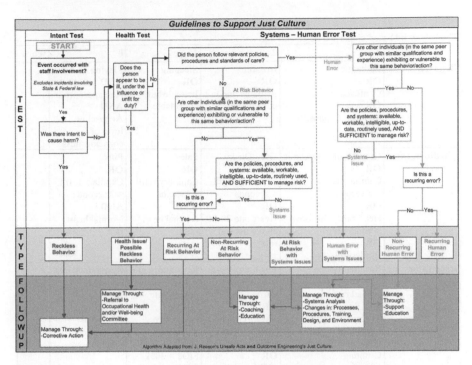

**Fig. 16.1** Just Culture algorithm from Rady Children's Hospital San Diego. With permission from Dr. Glenn Billman, Quality Management and Patient Safety, Rady Children's Hospital San Diego

Other considerations such as staffing, resource availability, environment of care, information technology, leadership, presence of proactive error surveillance systems, and culture of safety were not found lacking. The event was deemed at high risk for recurrence, as the failure points were not unique to the patient, personnel, or environment. Despite this, it was also agreed that nurse and intern *performance expectations* were not met as noted above.

What can be done to prevent this from happening again? Solution planning used quality improvement tools such as failure mode effects analysis (FMEA) and cause-and-effect diagram [24], available facility rapid response team activation data, and the organization's Just Culture algorithm (Fig. 16.1). Just Culture acknowledges that humans are fallible and provides an atmosphere of trust in which people are encouraged to report errors while individuals are still held accountable for risky or unacceptable behavior [25]. Key issues identified in this case were as the following: lack of clarity of roles within the "Ask More" policy, limited team discussions about what clinical changes warrant notification of more experienced clinicians, lack of awareness of high risk populations' more subtle signs of deterioration, difficulty in interpreting level of parental or nursing concerns, and over-reliance on judgment and experience despite concerning objective data such as vital signs. FMEA scores for each of these failures were rated high, each with low likelihood of ability to be detected and high likelihoods of recurrence and risk for future patient harm.

| Pediatric Early Warning Score | | | | | |
|---|---|---|---|---|---|
| Vital signs are based on normal ranges for age | | | | | |
| | 3 | 2 | 1 | 0 | score |
| **Behavior** | Lethargic OR Confused OR Reduced response to pain | Irritable OR Agitated and not consolable | Sleeping OR Irritable but consolable | Playing or Alert AND Age appropriate AND At baseline level of consciousness | |
| **Cardiovascular** | Grey/Mottled/Cyanotic OR Capillary refill ≥ 5 seconds OR Heart rate ≥ 30 above normal rate OR Bradycardia | Grey/Dusky OR Capillary refill 4 seconds OR Heart rate 20-30 above normal rate | Pale OR Capillary refill 3 seconds AND Normal heart rate | Pink OR Capillary refill ≤ 2 seconds AND Normal heart rate | |
| **Respiratory** | Respiratory rate 5 below or ≥ 40 above normal rate OR Moderate to severe retractions OR Grunting OR FiO2 ≥ 50% OR > 8Liters/min oxygen | Respiratory rate ≥ 20 above normal rate OR Mild Retractions OR FiO2 ≥ 40% OR ≥ 5Liters/min nasal cannula or ≥ 6Liters/min mask oxygen | Respiratory rate ≥ 10 above normal rate OR Accessory muscle use OR FiO2 ≥ 30% OR ≥ 2Liters/min oxygen | Respiratory rate normal AND No retractions AND No oxygen requirement | |

**Extra points:** Nebulizer treatments 4 or more per hour =2 points
Persistent vomiting after surgery =2 points
Frequency of nursing assessments are noted below. Use the PEWS algorithm.

| 0-2 | Green: Every 4-6 hours | 3 | Yellow: Every 2 hours | 4 | Orange: Every 1 hour | ≥ 5 | Red: Every 30 minutes |

**Fig. 16.2** Pediatric Early Warning Score action algorithm from Rady Children's Hospital San Diego. With permission from Dr. Glenn Billman, Quality Management and Patient Safety, Rady Children's Hospital San Diego

Using the Just Culture algorithm (Fig. 16.1), the nurse and intern's actions in this event were best described as consistent with "at risk behavior with systems issues," which resulted in targeted training. Of solutions implemented (Table 16.2), the revision to the parent orientation on rapid response team (RRT) use and piloting of a new Pediatric Early Warning Score (PEWS) required the most investment of resources and cultural sensitivity. The PEWS tool, first described in the UK and since modified by others, rates the cardiac, respiratory, and behavior (neurologic) status of a patient [26]. The rating in each category is associated with a point value that is combined to yield a composite score (Fig. 16.2). The real power of

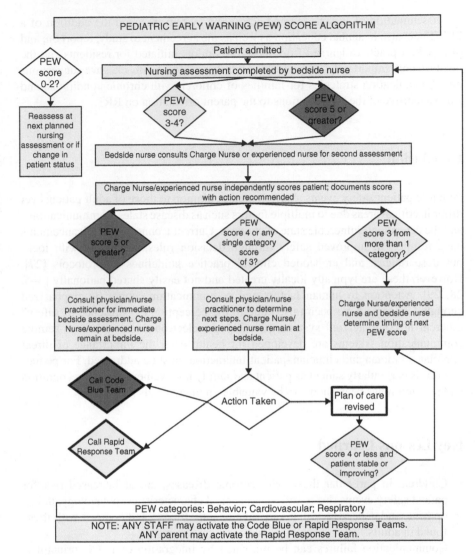

**Fig. 16.3** Pediatric Early Warning Score (PEW) Score Algorithm

the PEWS tool, however, comes from the associated action algorithm (Fig. 16.3), which prescribes specific tasks based on the patient's composite score. Staff's concerns regarding overuse of the RRT system and also of over-reliance on the PEWS system for patient assessment were abated through engagement in development of the parent orientation materials and the PEWS algorithm, respectively.

Dissemination of lessons learned across the system included the addition of a PEWS scenario to annual education as well as the agreement to study, report on, and diffuse best practices learned from the PEWS pilots initiated for residents and the involved unit. Importantly, the participant family received feedback, gave suggestions on communication strategies for families of children with chronic conditions, and was supportive of the modifications to the parent orientation on RRT.

## Conclusion

Pediatric patient safety events share elements common to those of adult patients yet differ in critical areas due to multiple factors such as disease states, communications, and the dependent/vulnerable state of children. Current technological advancements have resulted in improved safety through decision rules, order sets with lock-out dose ranges, and embedded clinical practice guidelines in protocols [27]. However, these are typically locally created and not easily shared nationally [3–5, 28, 29]. Attention to human factors and communication cannot be emphasized enough. Concerns have been raised due to the perception of "presence of safety" inherent in computerized systems [30, 31]. As electronic and moreover remote communication systems are developed for health care, the importance of direct clinician–clinician and clinician–patient interaction must be addressed. For pediatrics this is particularly salient as patient and family involvement in error recognition and resolution has been shown to be valuable on many levels [32].

## Key Lessons Learned

- Children, in particular those with chronic diseases, are at increased risk for patient safety events due to different physical characteristics, physiology, development, and dependency that vary significantly by age and contrast with those found in adults.
- Communication failures can be mitigated by integration of PFCC principles, clearly written policies, constructive education, Just Culture, and use of appropriate technological support.
- In all healthcare settings, advocacy for, initiation of and engagement in pediatric safety initiatives is essential to ensure safe healthcare delivery for children. This is particularly poignant in settings where children are cared for less frequently and/or pediatric expertise is limited.
- Pediatric patient safety events should be reviewed in an interdisciplinary manner. System and human factors solutions should be disseminated across the facility wherever possible, with targeted education, training, and coaching applied as appropriate.

# References

1. National Research Council. To err is human: building a safer health system. Washington, DC: Institute of Medicine National Academy Press; 1999.
2. Miller MR, Takata G, Stucky ER, et al. Principles of pediatric patient safety: reducing harm due to medical care. Pediatrics. 2011;127(6):1199–210.
3. Miller MR, Robinson KA, Lubomski LH, Rinke ML, Pronovost PJ. Medication errors in pediatric care: a systematic review of epidemiology and an evaluation of evidence supporting reduction strategy recommendations. Qual Saf Health Care. 2007;16:116–26.
4. Kaushal R, Jaggi T, Walsh K, Fortescue EB, Bates DW. Pediatric medication errors: what do we know? What gaps remain? Ambul Pediatr. 2004;4:73–81.
5. Kaushal R, Bates DW, Landrigan C, McKenna KJ, Clapp MD, Federico F, et al. Medication errors and adverse drug events in pediatric inpatients. J Am Med Assoc. 2001;285:2114–20.
6. Mohr JJ, Lannon CM, Thoma KA, et al. Learning from errors in ambulatory pediatrics. In: Henriksen K, Battles JB, Marks ES, et al., editors. Advances in patient safety: from research to implementation. Washington, DC: Agency for Healthcare Research and Quality; 2005. p. 355–68.
7. Suresh G, Horbar JD, Plsek P, Gray J, Edwards WH, et al. Voluntary anonymous reporting of medical errors for neonatal intensive care. Pediatrics. 2004;113:1609–18.
8. Upperman JS, Staley P, Friend K, Neches W, Kazimer D, et al. The impact of hospitalwide computerized physician order entry on medical errors in a pediatric hospital. J Pediatr Surg. 2005;40(1):57–9.
9. MacLennan AI, Smith AF. An analysis of critical incidents relevant to pediatric anesthesia reported to the UK National Reporting and Learning System, 2006–2008. Pediatr Anesth. 2011;21:841–7.
10. Berry JG, Hall DE, Kuo DZ, Cohen E, Agrawal R, et al. Hospital utilization and characteristics of patients experiencing recurrent readmissions within children's hospitals. J Am Med Assoc. 2011;305(7):682–90.
11. Institute for Healthcare Improvement. Five million lives campaign how to guide: pediatric supplement. Available at http://www.ihi.org/knowledge/Pages/Tools/HowtoGuidePrevent HarmfromHighAlertMedicationsPediatricSupplement.aspx. Accessed 02.10.2012.
12. The Joint Commission. Pediatric patient safety resources. Available at http://www.jcrinc.com/Search/. Accessed 02.10.2012.
13. Agency for Healthcare Research and Quality. Patient safety indicators. Available at http://www.qualityindicators.ahrq.gov/. Accessed 02.10.2012.
14. American Academy of Pediatrics Committee on Pediatric Emergency Medicine, American College of Emergency Physicians Pediatric Committee, Emergency Nurses Association Pediatric Committee. Joint Policy Statement—guidelines for care of children in the emergency department. Pediatrics. 2009;124(4):1233–43.
15. Committee on Hospital Care and The Institute for Patient-Family Centered Care. Patient- and family-centered care and the pediatrician's role. Pediatrics. 2012;129(2):394–404.
16. The Joint Commission. Root cause analysis framework. Available at http://www.joint commission.org/Framework_for_Conducting_a_Root_Cause_Analysis_and_Action_Plan/. Accessed 02.10.2012.
17. Croskerry P. Achieving quality in clinical decision making: cognitive strategies and detection of bias. Acad Emerg Med. 2002;9:1184–204.
18. Croskerry P. The importance of cognitive errors in diagnosis and strategies to minimize them. Acad Med. 2003;78(8):775–80.
19. Brierley J, Carcillo JA, Choong K, Cornell T, DeCaen A, et al. Clinical practice parameters for hemodynamic support of pediatric and neonatal septic shock: 2007 update from the American College of Critical Care Medicine. Crit Care Med. 2009;37(2):666–88.

20. Thoracic Society of Australia and New Zealand, Fitzgerald DA, Massie RJ, Nixon GM, Jaffe A, et al. Infants with chronic neonatal lung disease: recommendations for the use of home oxygen therapy. Med J Aust. 2008;189(10):578–82.

21. Stucky ER, Haddad GG. Respiratory failure. In: Elzouki AY, editor-in-chief; Haddad GG, section editor.Textbook of clinical pediatrics. 2nd ed. New York, NY: Springer Science+Business Media; 2012. p. 2141–8, Part 13.

22. Insitute for Healthcare Improvement. SBAR technique for communication: a situational briefing model. Developed by Michael Leonard. Available at http://www.ihi.org/knowledge/Pages/Tools/SBARTechniqueforCommunicationASituationalBriefingModel.aspx. Accessed 02.10.2012.

23. Holm KE, Patterson JM, Gurney JG. Parental involvement and family-centered care in the diagnostic and treatment phases of childhood cancer: results from a qualitative study. J Pediatr Oncol Nurs. 2003;20(6):301–13.

24. Institute for Healthcare Improvement Tools. Available at http://www.ihi.org/search/pages/results.aspx?k=tools. Accessed 02.10.2012.

25. Reason J. Human errors: models and management. BMJ. 2000;320:768–70.

26. Akre M, Finkelstein M, Erikson M, Liu M, Vanderbilt L, et al. Sensitivity of the pediatric early warning score to identify patient deterioration. Pediatrics. 2010;125(4):e763–9.

27. Farrar K, Caldwell NA, Robertson J, Roberts W, Power B, et al. Use of structured paediatric-prescribing screens to reduce the risk of medication errors in the care of children. Br J Healthc Comput Info Manage. 2003;20(4):25–7.

28. Cabana MD, Rand CS, Powe NR, Wu AW, Wilson H, et al. Why don't physicians follow clinical practice guidelines? A framework for improvement. J Am Med Assoc. 1999;282(5):1458–65.

29. Pronovost PJ, Goeschel CA, Olsen KL, Pham JC, Miller MR, et al. Reducing health care hazards: lessons from the commercial aviation safety team. Health Aff. 2009;28(3):479–89. (Published online 7 April 2009; 10.1377/hlthaff.28.3.w479). Accessed 02.10.2012.

30. Bomba D, Land T. The feasibility of implementing an electronic prescribing decision support system: a case study of an Australian public hospital. Aust Health Rev. 2006;30(3):380–8.

31. Love JS, Wright A, Simon SR, Jenter CA, Soran CS, Volk LA, et al. Are physicians' perceptions of healthcare quality and practice satisfaction affected by errors associated with electronic health record use? J Am Med Inform Assoc. 2012;19(4):610–4.

32. Daniels JP, King AD, Cochrane DD, Carr R, Shaw NT, et al. A human factors and survey methodology-based design of a web-based adverse event reporting system for families. Int J Med Inform. 2010;79(5):339–48.

# Chapter 17
# Patient Safety in Radiology

Alan Kantor and Stephen Waite

> *"Every great mistake has a halfway moment, a split second*
> *when it can be recalled and perhaps remedied."*
>
> Pearl S. Buck

## Introduction

Technical advances over the last few decades have fostered the development of imaging modalities that provide rapid and more accurate patient evaluation than previously possible. This has led to a rapid growth in the use of imaging, particularly Computed Tomography (CT) scans, that have nearly doubled the US population's exposure to ionizing radiation. The principle behind radiologic safety is that practitioners of radiology employ proper techniques and have the necessary skills to 'obtain image quality consistent with the medical imaging task' while 'minimizing radiation dose' [1].

As a core service to most clinical areas of medicine and surgery, suboptimal radiology processes of care can contribute to preventable medical errors leading to patient harm. Reader variability in interpretation, inappropriate recommendations, mislabeled imaging, inappropriate protocols with resultant unnecessary ionizing

A. Kantor, M.D. (✉)
Department of Radiology, Kings County Hospital Center, 451 Clarkson Avenue,
Brooklyn, NY 11203, USA
e-mail: alan.kantor@nychhc.org

S. Waite, M.D.
Department of Radiology, SUNY – Downstate Medical Center, 450 Clarkson Avenue,
Brooklyn, NY 11203, USA
e-mail: Stephen.Waite@nychhc.org

A. Agrawal (ed.), *Patient Safety: A Case-Based Comprehensive Guide*,
DOI 10.1007/978-1-4614-7419-7_17, © Springer Science+Business Media New York 2014

radiation, and communication errors are some of the potential errors in radiology that can harm patients [2, 3].

Although diagnostic imaging has been found to be an effective tool, it is important to remember that ionizing radiation is not without risks such as cancer, burns, and other injuries. Indeed X-rays have been officially classified as a carcinogen. Steps are, therefore, needed to eliminate avoidable exposure to radiation [4].

Patient safety issues such as ensuring proper patient identification and medication reconciliation, critical for patient safety in imaging departments, are reviewed in detail in previous chapters and not specifically addressed in this chapter. Interventional Radiology, a procedure-oriented radiology subspecialty, follows the Universal protocol for prevention of wrong site, wrong procedure, and wrong person surgery with preprocedure verification, site marking, and a time out. This protocol is identical to that discussed in the chapter on patient safety in surgical specialties and likewise not further addressed in this chapter.

The necessity of keeping radiation dose to patients as low as possible is called the ALARA, or "As Low As Reasonably Achievable," principle [5]. This principle is more important than ever given the rapid growth of radiologic procedures, especially "high technology" and relatively high radiation modalities such as CT scanning. A study of Medicare nonmanaged care enrollees demonstrated that on average, between 1998 and 2001, utilization per Medicare enrollee increased 16 % per year for magnetic resonance imaging (MRI) and 7–15 % for CT, ultrasound (US), interventional radiology, and nuclear medicine. In contrast, general radiography increased only 1 % per year [6]. Between 2000 and 2007, imaging was the most rapidly growing of all physician services in the Medicare population. In the emergency department (ED) setting, growth has occurred in radiography, CT, and ultrasound, but the growth in CT has been found to be considerably more rapid than the other modalities. The share of all ED imaging attributable to CT increased from 14.2 % in 2000 to 29.0 % in 2008 [7]. Indeed, the explosion of imaging has nearly doubled the US population's total exposure to ionizing radiation over the past two decades [4]. It is, therefore, critically imperative for radiologists to take an active role in implementing practices that promote patient safety.

## Case Studies

In this chapter, we discuss two case studies: the first illustrating inappropriate imaging secondary to poor communication and the second demonstrating the potential for patient harm resulting from ignorance regarding the risks of ionizing radiation. Detailed root cause analyses (RCA) of both cases are presented along with suggested risk reduction strategies and corrective actions.

## Case Study 1

### Description and Analysis

*RJ is a 70-year-old man with a history of angina pectoralis who presented to radiology for a routine CT scan of the chest without intravenous contrast for preoperative evaluation before CABG.*

The first job of the radiologist when confronted with an imaging requisition is to determinate the appropriateness of the requested diagnostic procedure or test. Medical appropriateness can be defined as "the indication to perform a medical procedure is appropriate when the expected health benefit (e.g., increased life expectancy, relief of pain, reduction in anxiety, improved functional capacity) exceeds the expected negative consequences (e.g., mortality, morbidity, anxiety of anticipating the procedure, pain produced by the procedure) by a sufficiently wide margin that the procedure is worth doing" [8]. The American College of Radiology (ACR) has developed evidence-based guidelines to help physicians request the "most appropriate imaging or treatment decision for a specific clinical condition" [9].

For example, if the clinicians requested a noncontrast study to exclude pulmonary embolism, this would be an inappropriate indication given that the suspected condition cannot be excluded without administration of intravenous contrast material. The patient would be subjected to radiation (not to mention anxiety, cost, and inconvenience) without the clinical question being answered. In the case above, noncontrast CT scan of the chest is recognized as a useful method to detect the location of aortic calcification for surgeons to determine an operative cannulation site [10]. This study would, therefore, seem appropriate although the ACR has not commented on this specific indication.

### The Case Continued

*CT scan of the chest without contrast was performed and revealed no evidence of aortic calcification; however, incidental note was made of a well-defined noncalcified round peripheral 4 mm nodule in the very posterior basilar right lower lobe. The interpreting radiologist recommended "3-month-follow up imaging."*

The finding of a lung nodule in this case can best be described as an "incidentaloma." One definition of an incidentaloma is "…a finding in a radiological study totally unrelated to the clinician's reason for requesting the radiological examination, that is, a finding that is incidentally noted…" [11]. These incidental findings (IFs) have been increasingly observed as diagnostic imaging has become able to detect smaller and subtler structures. Although this unsought information can be beneficial to patients (such as in the case of unsuspected tumors), frequently this information is detrimental [12]. The IFs often generate uncertainty and anxiety among both physicians and patients in addition to the financial cost of resultant procedures and follow-up imaging [13]. One study [14] reported that 40 % of 1,426 research imaging examinations demonstrated at least one IF. CT scan of the abdomen/pelvis generated more IFs than CT scan of the chest. Further complicating the problem is that guidelines for handling these IFs have not been available until recently leading to disparate recommendations regarding workup of the same lesions by different radiologists [12, 14].

In order to better standardize the workup of an IF, the ACR published an approach to management of incidentalomas noted on CT scans of the abdomen and pelvis [15]. These guidelines are expected to reduce risk to patients from unnecessary examinations, limit the cost of managing IFs to patients and the healthcare system, achieve consistency in managing and reporting of IF, and provide guidance to radiologists concerned about the risk of litigation should an IF prove to be clinically relevant.

Similarly in 2005, the Fleischner Thoracic Radiology Society formed evidence-based guidelines for the management of incidental nodules. Much of the evidence they used to formulate guidelines came from lung cancer screenings trials where nodules were meticulously followed. Their recommendations were a departure from the prevailing standard of care, which required monitoring of all nodules until they demonstrated 2-year stability [16]. Four years after the publication of these guidelines, 79 % of a poll of 834 radiologists reported awareness of the guidelines. However, compliance to the guidelines' recommendations has been less promising with only 34.7–60.8 % of radiologists reporting conformance [17]. The etiology of this lack of compliance is uncertain though there are incentives for radiologists to favor further follow-up examinations. A follow-up examination that confirms a stable lesion is unlikely to be questioned but failure to recommend appropriate additional tests is a typical malpractice allegation. Evidence-based guidelines such as the Fleischner criteria are expected to decrease variability in practice and become the standard of care. Radiologists would likely enjoy a substantial measure of legal protection by virtue of having followed recognized management guidelines [18].

Based on available data that <1 % of nodules <5 mm in diameter in patients without a history of cancer demonstrate malignant behavior, the Fleischer society recommendation for nodules less than or equal to 4 mm in size is no follow-up in a low-risk patient or 12-month-follow-up in a high-risk patient (history of smoking or other known risk factors) to assess for growth [16].

The interpreting radiologists' recommendation that this nodule be followed up in 3 months is, therefore, too aggressive and exposes the patient to unnecessary radiation. Even if this 4 mm nodule doubled in volume yearly, in a 3-month period it would only become approximately 5 mm in diameter. This 1 mm of growth in diameter is not reliably measured on CT examination [16, 19]. This nodule should ideally either be ignored or if the patient is high risk, yearly follow up is recommended.

A more significant error in analysis is the fact that the radiologist failed to compare this examination to a previous CT scan of the abdomen and pelvis performed 5 years prior in 2006. The lung bases are seen usually on CT scans of the abdomen and pelvis and if the radiologist compared the current CT scan of the chest with the prior abdominal imaging, he would have noted that the nodule was present and unchanged. One study demonstrated that 39.1 % of abdomen CTs had noncaclfied nodules at the lung bases [20], and it is critically important that in the workup of nodules, old studies are analyzed. Two-year stability is considered good evidence of benignity [16] especially for solid nodules. It is generally accepted practice in the radiology and legal communities that radiologists have a duty to compare current with previously obtained radiographs [21].

The appropriate recommendation would, therefore, have been that this nodule was unchanged for 5 years and is almost invariably benign; therefore, no follow-up imaging was warranted. Assuming that the old films were available for comparison, this radiologist was not compliant with the standard of care.

## The Case Continued

*Three months later the patient returned for follow-up imaging. As ordered by refer-ring clinician, a high resolution CT scan of the chest (HRCT) was performed in order to evaluate nodule stability. As per HRCT protocol, the patient was scanned in not only supine but also prone position. Expiratory images were also performed.*

Upon returning to the institution for follow-up imaging, this patient underwent further inappropriate imaging. In addition to instructing physicians on the appropri-ate studies to order, it is the radiologists' job to apply the appropriate protocol to provide the optimal quality image with the lowest possible radiation dose [22]. Individualized protocoling of studies further insures that the ordering clinicians' question is fully answered by the examination performed [23]. Appropriate proto-coling also includes the elimination of unnecessary sequences.

In a retrospective analysis, 52.2 % of 500 patients had an unindicated series of scans, most often-delayed phase imaging, after contrast administration. This is yet another source of excess radiation that has not garnered much public attention. It is estimated that should these unnecessary studies be eliminated there would be 63 % decreased radiation exposure [24].

The HRCT protocol is used for diagnosis and assessment of interstitial lung disease and its expiratory and prone series are not necessary in the evaluation or follow up of solitary pulmonary nodules. In regard to protocoling this examination and determining appropriateness, the radiologist neglected his position as a "gatekeeper."

A "gatekeeper" can be defined as a person who is positioned between an organi-zation and the individuals who wish to utilize the resources within that organization. As radiologists are involved in the diagnostic workup of patients; they are posi-tioned to assume the role of gatekeepers by facilitating the appropriate allocation of imaging resources. Unfortunately, many radiologists are uncomfortable acting in this capacity. As the radiologist is not the primary caregiver, many radiologists feel at a disadvantage when discussing imaging options or negotiating urgency when clinicians demand examinations. Furthermore, there is concern regarding potential tension with their referral base [25].

Considering the rapid technological advances in radiology, regular clinico-radiographic meetings/lectures greatly enhance the clinicians' ability to order appropriate studies. Our radiology department has regular meetings with clinical staff and continually educates clinicians regarding appropriate imaging modalities.

Computerized decision support systems can also help facilitate the appropriate allocation of resources [26, 27]. These support systems have been found to be helpful in limiting imaging to evidence-based applications by providing real time appropriateness information to the providers ordering imaging. These systems can also be restrictive and not allow clinicians to order examinations without accepted indications [28]. These systems not only improve utilization but also decrease the amount of time radiologists spend contacting clinicians regarding inappropriate studies thereby improving efficiency [27]. Studies have demonstrated that these systems indeed decrease inappropriate utilization of advanced imaging tests [28].

Indeed, as opposed to a HRCT or even routine CT scan of the chest, this patient would have been best served getting a low dose CT scan. Reducing the tube current-time level substantially reduces the radiation dose received by the patient compared to the standard dose. A low dose protocol is especially useful in the workup of patients with lung nodules because the cumulative radiation dose of repetitive follow-up CT examinations can be reduced considerably without a significant difference in sensitivity [29].

Lastly, radiologists should be specific regarding recommendations to clinicians. Since the interpreting radiologist was not clear regarding the type of follow-up imaging recommended either in the report or via direct communication, the ordering resident ordered an HRCT thinking that this would be the best imaging modality for his patient. On a basic level, radiologists are responsible for communicating significant results directly to clinicians. Communication failures are an increasing proportion of medical malpractice payments. The total indemnity payment for US claims regarding communication errors in medicine increased from $21.7 million in 1991 to $91.0 million in 2010. The most common contributing factor in cases associated with communication failure was a failure of clinicians to communicate with patients followed by failure of communication of results with referring clinicians [30]. An increasing cause of malpractice litigation involves the failure on the part of the radiologist to communicate significant abnormal radiologic findings directly to referring clinicians with secondary serious patient injury [31]. Clinicians often report that they rely on the radiologist to provide guidance regarding patient management [32] and most clinicians like specific recommendations [33]. Given that radiologists are the defacto experts on incidentalomas, direct communication would have been especially warranted in our case.

One solution to close the communication gap is the use of critical test result management (CTRM) software. These products communicate critical test results from radiologists to ordering clinicians, allowing for reliable caregiver communication workflow and expedited patient care. Real-time performance measurements can be measured allowing the assessment of performance goals and targets for turnaround time and compliance. At our institution, the CTRM alerts the ordering clinician that an important result is pending, sends alerts until the message is retrieved, escalates the notification according to preset rules, and sends verification to the reporting radiologist when the message has been retrieved.

**Case Continued**

*The same radiologist read the patient's follow-up imaging and reported that there was no change in nodule size or density and no evidence of significant interstitial lung disease. Short-term follow up to establish 2-year stability was recommended. At time of this writing, RJ is awaiting further follow-up imaging.*

Although no measurable damage occurred to this patient, it is clear that this case presents multiple opportunities for physician and system level improvements in order to optimize patient care.

Radiologists are largely invisible to their patients; indeed 80–90 % of radiologists do not meet their patients [34]. In the absence of effective communications strategies, customer management initiatives, and added value, radiologists are in danger of being perceived as an invisible underappreciated technical commodity. By communicating results and specific recommendations to referring clinicians, radiologists are providing value that uniquely contributes to the management of the patient [35].

Although important to educate the individual radiologist in regard to errors in judgment, it is equally important to examine the system under which they are operating given the possibility that many of these problems are systemic. In order to monitor and improve upon departmental performance, institutions have become more quantitative. To this end some departments have initiated the use of departmental, institutional, and individual performance based "score" cards for individual practitioners. These scorecards assess different measures such as clinical services, education, research, professionalism, and communication. Performance indicators such as communication with referring physicians and participation in continuing medical education can be assessed and monitored in this fashion. A deliberate and organized approach is needed in order to improve upon the goal of high quality patient care and develop a departmental culture of quality improvement [36, 37].

## Case Study 2

### Clinical Summary

*A 28-year-old pregnant female, estimated 26 weeks gestational age, presented to the emergency room (ER) with a 1-day history of lower abdominal pain and nausea. On physical exam, the uterine size was appropriate for dates and there was mild lower abdominal tenderness. CBC revealed an elevated WBC count of 11.7 K. An obstetrical US showed a normal appearing intrauterine gestation and placenta. The examining physician made a provisional diagnosis of acute appendicitis and recommended a CT scan of the abdomen and pelvis for further evaluation. He informed the patient of a small risk of fetal malformations and some increased risk of the child developing cancer later in life. He also discussed the benefits of having the examination and discovering appendicitis early. The patient, worried about her unborn child, declined the examination and chose a wait and see approach. Her symptoms waxed*

**Table 17.1** Estimated conceptus doses from radiographic and fluoroscopic examinations

| Examination | Typical conceptus dose (mGy) |
|---|---|
| Cervical spine (AP, lateral) | <0.001 |
| Extremities | <0.001 |
| Chest (PA, lateral) | <0.002 |
| Thoracic spine (AP, lateral) | <0.003 |
| Abdomen (AP) | |
| 21-cm Patient thickness | 1 |
| 33-cm Patient thickness | 3 |
| Lumbar spine (AP, lateral) | 1 |
| Limited IVP[a] | 6 |
| Small bowel study[b] | 7 |
| Double contrast barium enema study[c] | 7 |

*Note*: *AP* anteroposterior projection, *PA* posteroanterior projection
[a]Limited IVP assumed to include four abdominopelvic images. Patient thickness 21 cm assumed
[b]A small bowel study is assumed to include a 6-min fluoroscopic examination with the acquisition of 20 digital spot images
[c]A double contrast barium enema study is assumed to include a 4-min fluoroscopic examination with the acquisition of 12 digital spot images (adapted with permission, from reference)

*and waned over the next few hours, but eventually progressed and her WBC count increase to 13.6 K. The patient was taken to the operating room where a perforated appendix was discovered. A 3-day-postoperative hospital stay was uneventful.*

## Analysis

The decision to perform an imaging study utilizing ionizing radiation for a pregnant or potentially pregnant patient requires an appropriate risk/benefit analysis, including the benefits and risks to the unborn fetus/embryo. Surveys have documented a lack of awareness by both clinicians and radiologists of the radiation dose associated with common imaging procedures [38]. A survey of ER physicians and radiologists documented that the majority were unaware of the cancer risks associated with imaging studies that expose the patient to ionizing radiation, particularly CT scans [39]. In contrast, other studies demonstrate that a substantial number of obstetricians and family physicians perceive the risk to be far greater than should be a realistic concern [40].

The risk to the fetus can be divided into two categories: deterministic/teratogenic effects and stochastic effects such as the increased incidence of malignancy associated with radiation exposure. The incidence of fetal malformations secondary to radiation exposure is dependent upon gestational age at the time of exposure and the dose to the fetus. The occurrence of these effects increases with higher exposure but have a defined lower threshold, below which they are not observed. Tables 17.1 and 17.2 show the approximate dose administered for common imaging procedures [41]. Table 17.3 lists the risks of teratogenic effects at different stages of gestation and the threshold below which these effects are not seen [42]. A CT scan of the abdomen and pelvis, which has the highest fetal dose of common noninvasive imaging procedures, subjects the fetus to approximately 25 mGy which is well below the

**Table 17.2** Estimated conceptus dose from single CT acquisition

| Examination | Dose level | Typical conceptus dose (mGy) |
|---|---|---|
| Extra-abdominal | | |
| Head CT | Standard | 0 |
| Chest CT | | |
| Routine | Standard | 0.2 |
| Pulmonary Embolus | Standard | 0.2 |
| CT angiography of coronary arteries | Standard | 0.1 |
| Abdominal | | |
| Abdomen, routine | Standard | 4 |
| Abdomen/Pelvis routine | Standard | 25 |
| CT angiography of Aorta (chest through pelvis) | Standard | 34 |
| Abdomen/Pelvis (stone protocol)[a] | Reduced | 10 |

[a]Anatomic coverage is the same as for routine abdominopelvic CT, but the tube current is decreased and the pitch is increased because standard image quality is not necessary for detection of high contrast stones (adapted with permission, from reference)

**Table 17.3** Summary of suspected in utero induced deterministic radiation effects

| Menstrual or gestational age | Conception age | <50 mGy | 50–100 mGy | >100 mGy |
|---|---|---|---|---|
| 0–2 weeks (0–14 days) | Prior to conception | None | None | None |
| 3rd and 4th weeks (15–28 days) | 1st–2nd weeks (1–14 days) | None | Probably none | Possible spontaneous abortion |
| 5th–10th weeks (29–70 days) | 3rd–8th weeks (15–56 days) | None | Potential effects are scientifically uncertain and probably too subtle to be clinically detectable | Possible malformations increasing in likelihood as dose increases |
| 11th–17th weeks (71–119 days) | 9th–15th weeks (57–105 days) | None | Potential effects are scientifically uncertain and probably too subtle to be clinically detectable | Increased risks of deficits in IQ or mental retardation that increase in frequency and severity with increasing dose |
| 18th–27th weeks (120–189 days) | 16th–25th weeks (106–175 days) | None | None | IQ deficits not detectable at diagnostic doses |
| >27 weeks (>189 days) | >25 weeks (>175 days) | None | None | None applicable to diagnostic medicine |

level of 100–150 mGy where there should be any concern for fetal abnormality. Studies that do not directly image the fetus including plain films and CT scans of the head, chest, and extremities expose the fetus to far less radiation. The 1977 report of the National Council of Radiation protection and Measurements (NCRP)

[43] states "The risk (of abnormality) is considered to be negligible at 5 rad (50 mGy) or less when compared to the other risks of pregnancy, and the risk of malformations is significantly increased above control levels only at doses above 15 rad (150 mGy). Therefore, the exposure of the fetus to radiation arising from diagnostic procedures would rarely be cause, by itself, for terminating a pregnancy." This sentiment is echoed by the American college of Obstetrics and Gynecology (ACOG) "Women should be counseled that X-ray exposure from a single diagnostic procedure does not result in harmful fetal effects. Specifically, exposure to less than 5 rad (50 mGy) has not been associated with an increase in fetal anomalies or pregnancy loss [44]."

Models estimating risk of developing malignancy after exposure to low levels of ionizing radiation (low levels defined as <100 mSv) are controversial, with most supporting a linear-no-threshold (LNT) model. According to this model, even the smallest dose of radiation has the potential to increase the risk on malignancy in humans [45]. It should be noted that average annual background radiation is 3 mSv/year.[1] Although precise quantification is impossible, estimates for increased lifetime risk of dying from malignancy range from 0.35 % (estimated from 14 %/1,000 mSv for neonates) [46] to 1.0 % lifetime risk of developing malignancy [42] for a CT scan of the abdomen and pelvis delivering a fetal dose of 25 mSv. It should be noted that in many instances, low dose protocols and newer CT scanners with advanced iterative reconstruction imaging algorithms can provide diagnostic quality images at significantly lower radiation dose than those listed.

In our scenario, the clinician's decision to inform the patient of the potential risks and benefits of the examination overstated the risk of "mutation" and presented the risk of future cancers in manner that caused the patient undue anxiety and a delay in performing the study. Instead of simply stating there is a small chance this examination will increase your child's risk of getting cancer, the statement could have been rephrased "you child will have nearly the same chance of living a healthy life as any other child under similar medical circumstances, because the actual risk that your child might develop cancer is very small" [42]. A study by Larson et al. [47] demonstrated that a brief informational handout providing patients with information of the potential increased risk of cancer secondary to pediatric CT scans, increased patient understanding, and increased their level of concern. It did not, however, cause parents to refuse studies recommended by their doctor.

The clinician was correct in his assessment that the rapid diagnosis of acute appendicitis is critical especially in pregnant patients. Several studies have documented that perforated appendicitis is associated with increased rates of fetal mortality compared to uncomplicated appendicitis with rates of fetal loss ranging from 6 to 37 % [48]. He did not, however, consider the diagnostic capabilities of other imaging modalities such as US and MRI that do not expose the patient to ionizing radiation. The reported usefulness of ultrasound in evaluating acute appendicitis is variable

---

[1] Milligray (mGy) a measurement of the absorbed radiation dose and millisieverts (mSv) a measure of the biologically effective dose are equivalent for the medical X-ray examinations discussed in this chapter.

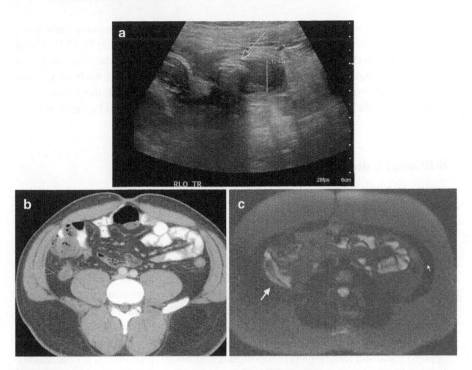

**Fig. 17.1** (**a**) Ultrasound of the right lower quadrant demonstrating a dilated appendix (12.3 mm) with an appendicolith in the lumen (*arrow*). (**b**) CT scan of the abdomen and pelvis at the level of the iliac crests demonstrating a dilated appendix (*arrowhead*) with surrounding inflammatory change. (**c**) MRI of the pelvis, using T2* weighted fat suppression technique demonstrating a mildly dilated appendix with surrounding edema

with report of up to 92 % in the ability to evaluate the appendix in pregnancy [49, 50]. This difficulty increases with increasing uterine size and increasing patient size [51]. For patients early in pregnancy for whom the appendix cannot be adequately assessed with US, and for patients in later stages of pregnancy, MRI is now considered a viable alternative (Fig. 17.1a–c). The negative laporatomy rate and perforation rate, commonly accepted indicators for clinically suspected acute appendicitis, are comparable to the published rates for CT scan [52]. In addition, both CT scan and MRI can also evaluate for other causes of acute right lower quadrant pain such as ovarian torsion, hemorrhagic cyst, hydronephrosis, and pyelonephritis. MRI can also be performed without the use of oral or gadolinium-based intravenous contrast media. It should be noted that the majority of academic medical centers prefer CT to MRI for imaging abdominal complaints in pregnant women, especially in the second and third trimesters [53]. Therefore, if MRI is unavailable or cannot be performed (i.e., patient is claustrophobic), a CT scan using a low dose protocol (i.e., techniques decreasing tube current and increasing pitch) is still recommended and safe. Radiology departments, in consultation with radiation physicists and referring clinicians, should develop algorithms for imaging of pregnant patients for common clinical indications including entities such as appendicitis, pulmonary embolism, and urolithiasis that are routinely imaged with computed tomography [54]. These algorithms should consider, when possible, the use of imaging modalities that do not use ionizing radiation. If modalities that use ionizing radiation such as CT are required or recommended, protocols that limit the radiation dose to the patient without diminishing the diagnostic accuracy of the examination should be employed.

## Additional Considerations

### Pediatric Radiation Safety

The Alliance for Radiation Safety in Pediatric Imaging, a coalition of Imaging societies and organizations, began as a subcommittee of the Society of Pediatric Radiology. Their goal through the Image Gently Campaign is to change imaging practices by increased awareness of the importance of decreasing radiation doses when imaging children in addition to providing practical means of dose reduction. The Campaign website (http://www.imagegently.org) has portals for parents and patients explaining risks and benefits of imaging studies as well as portals for referring physicians and radiologists to algorithms and imaging protocols for decreasing patient radiation exposure. This effort has spawned the Image Wisely Campaign providing similar services and information for adult patient populations.

### Communication

Communication between radiologists and clinicians has traditionally been synonymous with reporting of results. The Joint Commission's National Patient Safety Goal #2 "improve the effectiveness of communication among caregivers" also centers on the reporting of critical results of tests and diagnostic procedures in a timely manner [55]. Through the cases presented, we posit that the need for this communication starts at the time of ordering and continues through the report resulting phase. Information on radiation dose from various medical imaging studies and associated risk (although sometimes inaccurate, confusing, and misleading) is readily available on the internet, including applications available for mobile electronic devices (Fig. 17.2). The most direct way to provide this information to clinicians is through a decision support system. As alluded to earlier, a physician requesting an examination should be provided with information regarding appropriateness of the study requested as well as a list of equally or more appropriate examinations to address the clinical concern. The recommendations are often accompanied by relevant references in the medical literature, and the recommendations are customizable to the diagnostic algorithms and clinical guidelines of individual institutions.

Studies have concluded that most physicians believe that informed consent for communicating the risk of radiation-induced cancer should be obtained from patients

**Fig. 17.2** Screenshot of an application for mobile cellular devices demonstrating approximate radiation dose (patient) for radiographic studies. (Courtesy of Tidal Pool Software)

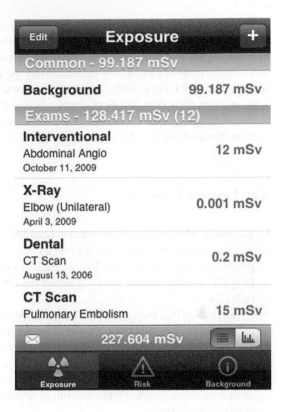

undergoing radiation-based imaging [56]. This information, however, must be presented in a positive manner that informs patients not only of the minimal risks associated with the examinations but also the risks of not having an appropriately performed, medically indicated study [57, 58]. One author has suggested that ALARA, which only addresses the risk of ionizing radiation with no perceived benefit, be replaced with a mantra for medical imaging that includes a benefit/risk ratio. He suggested a replacement such as AHARA which says that the benefit/risk ratio for medical applications using ionizing radiation be As High As Reasonably Achievable [59].

## MRI Safety

We noted in this chapter that MRI does not utilize ionizing radiation and indeed, there is no documented health risks associated with exposure to the magnetic fields generated by presently employed MRI scanners. They are, however, powerful magnets attracting any iron containing ferromagnetic object with the ability to create projectiles from objects the size of hairpins to oxygen tanks with potentially harmful consequences. Rapidly changing magnetic fields can also induce electrical voltages and currents in non-ferromagnetic conductive materials including some implants, pumps, and pacemakers.

MRI facilities must, therefore, carefully screen patients and personnel for external objects such as jewelry, dentures, personal electronic devices, and hearing aid and internal objects such as shrapnel, pacemakers, surgical/aneurysm clips, implants, and metallic prosthesis that may be affected by the magnet. These policies must also address non-MRI personnel ranging from housekeepers to Emergency Service workers such as police and firefighters who may have need to enter the scanner room. Most MRI suites are designed as a small maze with checkpoints to help prevent inadvertent entry to the scanner room.

# Key Lessons Learned

* Technical advances over the last few decades have fostered the development of imaging modalities that provide rapid and more accurate patient evaluation than previously possible. This has led to a rapid growth in the use of imaging, particularly CT scans that have nearly doubled the US population exposure to ionizing radiation.
* Concern over this increased level of radiation exposure, and the financial cost of these studies, has mandated a more comprehensive evaluation of their appropriate use
* Implementation of computerized decision support systems using evidence-based guidelines such as the American College of Radiology's appropriateness criteria should be considered. Direct consultation with imaging specialists should be available at all times.
* Radiology departments need to implement imaging protocols tailored to specific indications and patient populations that limit radiation exposure to a point As Low As Reasonably Achievable (ALARA).
* Reports of Imaging studies should be structured, and include a review of available, pertinent prior imaging studies. Recommendations for follow up should include the rationale for the recommendation, the specific examination suggested and time of follow up.
* Electronic critical results management system are helpful in ensuring that reports of imaging studies are delivered to an actionable clinician in a timely manner.
* Physicians should involve patients in the decision-making process. Information concerning the benefits and risks of proposed imaging studies should be presented to patients clearly and honestly without creating unnecessary anxiety.

# References

1. American College of Radiology. ACR practice guideline for diagnostic reference levels in medical X-ray imaging. Available at http://www.acr.org/Quality-Safety/Radiology-Safety/Radiation-Safety. Accessed 13 May 2013.
2. Kohn LT, Corrigan JM, Donaldson MS, editors. To err is human: building a safer health system. Washington, DC: National Academy Press; 2000.

3. Johnson CD, Krecke KN, Miranda R, Roberts CC, Denham C. Quality initiatives: developing a radiology quality and safety program: a primer. Radiographics. 2009;29(4):951–9.
4. The Joint Commission. Radiation risks of diagnostic imaging. Sentinel Event Alert. 2011;24(47):1–4.
5. Coursey C, Frush D. What radiologists should know. Appl Radiol. 2008;37(3):22–9.
6. Bhargavan M, Sunshine JH. Utilization of radiology services in the United States: levels and trends in modalities, regions, and populations. Radiology. 2005;234(3):824–32.
7. Rao VM, Levin DC, Parker L, Frangos AJ, Sunshine JH. Trends in utilization rates of the various imaging modalities in emergency departments: nationwide Medicare data from 2000 to 2008. J Am Coll Radiol. 2011;8(10):706–9.
8. Sistrom CL. The appropriateness of imaging: a comprehensive conceptual framework. Radiology. 2009;251(3):637–49.
9. American College of Radiology. 'ACR Appropriateness Criteria'. Available at http://www.acr.org/ac. Accessed 13 Jul 2013.
10. Nishi H, Mitsuno M, Tanaka H, Ryomoto M, Fukui S, Miyamoto Y. Who needs preoperative routine chest computed tomography for prevention of stroke in cardiac surgery? Interact Cardiovasc Thorac Surg. 2010;11(1):30–3.
11. Berlin L. The incidentaloma: a medicolegal dilemma. Radiol Clin North Am. 2011;49(2):245–55.
12. Megibow AJ. Preface imaging of incidentalomas. Radiol Clin North Am. 2011;49(2):xi–xii.
13. Casarella WJ. A patient's viewpoint on a current controversy. Radiology. 2002;224(3):927.
14. Orme NM, Fletcher JG, Siddiki HA, Harmsen WS, O'Byrne MM, Port JD, et al. Incidental findings in imaging research: evaluating incidence, benefit, and burden. Arch Intern Med. 2010;170(17):1525–32.
15. Berland LL, Silverman SG, Gore RM, Mayo-Smith WW, Megibow AJ, Yee J, et al. Managing incidental findings on abdominal CT: white paper of the ACR incidental findings committee. J Am Coll Radiol. 2010;7(10):754–73.
16. MacMahon H, Austin JH, Gamsu G, Herold CJ, Jett JR, Naidich DP, et al. Guidelines for management of small pulmonary nodules detected on CT scans: a statement from the Fleischner Society. Radiology. 2005;237(2):395–400.
17. Eisenberg RL, Bankier AA, Boiselle PM. Compliance with Fleischner Society guidelines for management of small lung nodules: a survey of 834 radiologists. Radiology. 2010;255(1):218–24.
18. MacMahon H. Compliance with Fleischner Society guidelines for management of lung nodules: lessons and opportunities. Radiology. 2010;255(1):14–5.
19. Erasmus JJ, McAdams HP, Connolly JE. Solitary pulmonary nodules: part II. Evaluation of the indeterminate nodule. Radiographics. 2000;20(1):59–66.
20. Rinaldi MF, Bartalena T, Giannelli G, Rinaldi G, Sverzellati N, Canini R, et al. Incidental lung nodules on CT examinations of the abdomen: prevalence and reporting rates in the PACS era. Eur J Radiol. 2010;74(3):e84–8.
21. Berlin L. Must new radiographs be compared with all previous radiographs, or only with the most recently obtained radiographs? Am J Roentgenol. 2000;174(3):611–5.
22. Tsapaki V, Rehani M, Saini S. Radiation safety in abdominal computed tomography. Semin Ultrasound CT MR. 2010;31(1):29–38.
23. Bassignani MJ, Dierolf DA, Roberts DL, Lee S. Paperless protocoling of CT and MRI requests at an outpatient imaging center. J Digit Imaging. 2010;23(2):203–10.
24. Guite KM, Hinshaw JL, Ranallo FN et al. Ionizing Radiation in Abdominal CT: Unindicated Multiphase scans are an important source of medically unnecessary exposure. Journal of the American College of Radiology. 2011;8(11):756–761.
25. You JJ, Levinson W, Laupacis A. Attitudes of family physicians, specialists and radiologists about the use of computed tomography and magnetic resonance imaging in Ontario. Healthc Policy. 2009;5(1):54–65.
26. Mohd Ramli N, Faridah Y. The boiling frog syndrome: a radiologist's perspective. Biomed Imaging Interv J. 2010;6(4):e36.
27. Knechtges P, Carlos R. The evolving role of the radiologist within the health care system. J Am Coll Radiol. 2007;4(9):626–35.

28. Blackmore CC, Mecklenburg RS, Kaplan GS. Effectiveness of clinical decision support in controlling inappropriate imaging. J Am Coll Radiol. 2011;8(1):19–25.
29. Christe A, Torrente JC, Lin M, Yen A, Hallett R, Roychoudhury K, et al. CT screening and follow-up of lung nodules: effects of tube current-time setting and nodule size and density on detectability and of tube current-time setting on apparent size. Am J Roentgenol. 2011;197(3):623–30.
30. Gale BD, Bissett-Siegel DP, Davidson SJ, Juran DC. Failure to notify reportable test results: significance in medical malpractice. J Am Coll Radiol. 2011;8(11):776–9.
31. Berlin L. Failure of radiologic communication: an increasing cause of malpractice litigation and harm to patients. Appl Radiol. 2010;39(1–2):17–23.
32. Berlin L. Relying on the radiologist. Am J Roentgenol. 2002;179(1):43–6.
33. Doğan N, Varlibaş ZN, Erpolat OP. Radiological report: expectations of clinicians. Diagn Interv Radiol. 2010;16(3):179–85.
34. Glazer GM, Ruiz-Wibbelsmann JA. The invisible radiologist. Radiology. 2011;258(1):18–22.
35. Silver MA. The invisible radiologist meets the new math, climate change, and business 101. Radiol Bus J. 2009. Available at http://www.imagingbiz.com/articles/rbj/the-invisible-radiologist-meets-the-new-math-climate-change-and-business-10. Accessed 13 Jul 2013.
36. Donnelly LF, Gessner KE, Dickerson JM, Koch BL, Towbin AJ, Lehkamp TW, et al. Quality initiatives: department scorecard: a tool to help drive imaging care delivery performance. Radiographics. 2010;30(7):2029–38.
37. Abujudeh HH, Kaewlai R, Asfaw BA, Thrall JH. Quality initiatives: key performance indicators for measuring and improving radiology department performance. Radiographics. 2010;30(3):571–80.
38. Shiralkar S, Rennie A, Snow M, Galland RB, Lewis MH, Gower-Thomas K. Doctor's knowledge of radiation exposure: questionnaire study. BMJ. 2003;3327:371–2.
39. Lee CI, Haims AH, Monico EP, Brink JA, Forman HP. Diagnostic CT scans: assessment of patient, physician, and radiologist awareness of radiation dose and possible risks. Radiology. 2004;231:393–8.
40. Ratnapalan S, Bona N, Chandra K, Koren G. Physicians' perceptions of teratogenic risk associated with radiography and CT during early pregnancy. Am J Roentgenol. 2004;182:1107–9.
41. McCollough CH, Schueler BA, Atwell TD, Braun NN, Regner DM, Brown DL, et al. Raddiation exposure and pregnancy: when should we be concerned? Radiographics. 2007;27:909–17.
42. American College of Radiology. ACR practice guideline for imaging pregnant or potentially pregnant adolescents and women with ionizing radiation. Reston, VA: American College of Radiology; 2008.
43. National Council on Radiation Protection and Measurements. Medical radiation exposure of pregnant and potentially pregnant women. NCRP report no. 54. Bethesda, MD: National Council on Radiation Protection and Measurements; 1977.
44. ACOG Committee on Obstetric Practice. Guidelines for diagnostic imaging during pregnancy. ACOG Committee opinion no. 299, September 2004 (replaces no. 158, September 1995). Obstet Gynecol. 2004;104:647–51.
45. National Research Council. Health risks from exposure to low levels of ionizing radiation: Beir VII Phase 2. Washington, DC: The National Academies Press; 2006.
46. Brenner DJ, Elliston CD, Hall EJ, Berdon WE. Estimated risks of radiation-induced fatal cancer from pediatric CT. Am J Roentgenol. 2001;176:289–96.
47. Larson DB, Rader SB, Forman HP, Fenton LZ. Informing parents about CT radiation exposure in children: it's OK to tell them. Am J Roentgenol. 2007;189:271–5.
48. Long SS, Long C, Lai H, Macura KJ. Imaging strategies for right lower quadrant pain in pregnancy. Am J Roentgenol. 2011;196:4–12.
49. Pedrosa I, Zeikus EA, Levine D, Rofsky NM. MR imaging of acute right lower quadrant pain in pregnant and nonpregnant patients. Radiographics. 2007;27:721–53.
50. Cobben LP, Groot I, Haans L, Blickman JG, Puylaert J. MRI for clinically suspected appendicitis during pregnancy. Am J Roentgenol. 2004;183:671–5.

51. Pedrosa I, Lafornara M, Panharipande PV, Goldsmith JD, Rofsky NM. Pregnant patients suspected of having appendicitis: effect of MR imaging on negative laparotomy rate and appendiceal perforation rate. Radiology. 2009;250(3):749–57.
52. Wieseler KM, Bhargava P, Kanal KM, Vaidya S, Stewart BK, Dighe MK. Imaging in pregnant patients: examination appropriateness. Radiographics. 2010;30:1215–33.
53. Jaffe TA, Miller CM, Merkle EM. Practice patterns in imaging of the pregnant patient with abdominal pain: a survey of Academic Centers. Am J Roentgenol. 2007;189:1128–34.
54. Shital SJ, Reede DL, Katz DS, Subramaniam R, Amorosa JK. Imaging of the pregnant patient for nonobstetric conditions: algorithms and radiation dose considerations. Radiographics. 2007;27:1705–22.
55. The Joint Commission. National patient safety goals 2011. Oakbrook Terrace, IL: The Joint Commission; 2011.
56. Karsli T, Kalra MK, Self JL, Grosenfeld JA, Butler S, Simoneaux S. What physicians think about the need for informed consent for communicating the risk of cancer from low-dose radiation. Pediatr Radiol. 2009;39:917–25.
57. Dauer LT, Thornton RH, Hay JL, Balter RB, Williamson MJ, St Germain J. Fears, feelings, and facts: interactively communicating benefits and risks of medical radiation with patients. Am J Roentgenol. 2011;196:756–61.
58. McCollough CH, Guimaraes L, Fletcher JG. In defense of body CT. Am J Roentgenol. 2009;193:28–39.
59. Wagner LK. Toward a holostic approach in the presentation of benefits and risks of medical radiation. Health Phys. 2011;101(5):566–71.

# Chapter 18
# Patient Safety in Anesthesia

Brian Bush and Rebecca S. Twersky

*"We often discover what will do, by finding out what will not do; and probably he who never made a mistake never made a discovery."*

Samuel Smiles

## Introduction

The field of anesthesiology has had a long relationship with issues related to patient safety. Early practitioners recognized that the administration of anesthetic agents was fraught with danger for patients, and some of the initial large-scale studies aimed at examining rates of morbidity and mortality in medical practice focused on surgery and anesthesia [1]. For the period spanning the 1950s through the 1970s, estimates of mortality caused by anesthesia care itself attributed one or two deaths to every 10,000 patient encounters. It was not until the late 1970s, however, that the sources of human error and mechanical malfunction leading to patient injury were analyzed in depth. In 1978, Cooper et al. employed the critical incident analysis technique developed in the aviation industry to examine the etiology of human errors in anesthesia mishaps. He and others later expanded on this work to suggest how hospital systems could be improved to minimize risks to patients [2].

B. Bush, B.A.
College of Medicine, SUNY Downstate, 450 Clarkson Avenue, Box 947, Brooklyn,
NY 11203-2098, USA
e-mail: bbush526@gmail.com

R.S. Twersky, M.D., M.P.H. (✉)
Department of Anesthesia, SUNY Downstate Medical Center, 450 Clarkson Avenue,
Box #6, Brooklyn, NY 11203-2098, USA
e-mail: Rebecca.Twersky@downstate.edu

A. Agrawal (ed.), *Patient Safety: A Case-Based Comprehensive Guide*, 281
DOI 10.1007/978-1-4614-7419-7_18, © Springer Science+Business Media New York 2014

Faced with mounting costs of professional liability insurance in the mid 1980s, the American Society of Anesthesiologists (ASA) became the first major professional society to champion the cause of patient safety. In 1985, the Anesthesia Patient Safety Foundation (APSF) and the ASA Closed Claims Project were created. The APSF was charged with raising awareness of patient safety issues and creating programs to address problems identified [3]. The Closed Claims Project was designed to collect and analyze data from closed insurance claims to identify sources of patient injury [4]. In 1984, Harvard Medical School voluntarily imposed standards for patient monitoring during the administration of anesthesia at all of its teaching hospitals, which were used as a model for more comprehensive standards adopted by the ASA in 1986 [5]. Importantly, these standards required some means of continuously monitoring ventilation and circulation, which had become more feasible with the introduction of new technologies such as capnography and pulse oximetry.

With this emphasis on patient safety, clear improvements were seen in the following decades. Analysis of closed claims has revealed a significant drop in death and brain damage as a cause for legal action against anesthesiologists [6]. The ASA continues to seek improvements in patient safety. In the past 5 years, dozens of standards, guidelines, and statements have been published with intent of improving outcomes [7]. Moving forward, the specialty of anesthesiology will continue to maintain its position as a leader in patient safety and improvements in care.

# Case Studies

## Case 1: The Impaired Anesthesiologist

### Timeline

*2:30 a.m.:* An alarm on a pulse oximetry sensor alerts the nursing staff in the Post Anesthesia Care Unit (PACU) to a patient in distress.

*2:33 a.m.:* After assessing the patient and recognizing respiratory distress, the nursing staff administers oxygen and pages the senior resident on call.

*2:38 a.m.:* The resident does not respond to the page, and the patient's oxygen saturation levels are continuing to range from 78 to 86 %. The senior anesthesia resident is paged overhead to the PACU. Following no response, the nurse pages the junior anesthesia resident.

*2:40 a.m.:* The junior anesthesia resident on call reports to the PACU and finds the patient disoriented and making poor respiratory effort. After inquiring as to the whereabouts of the senior resident, the junior resident decides to intubate the patient on his own. The intubation is performed successfully, and subsequent pulse oximetry and arterial blood gas measurements confirm the stabilization of the patient.

*3:15 a.m.: The junior resident locates the senior resident in the call room. The senior resident is sleeping and is difficult to arouse. Upon awakening, the senior resident is groggy and incoherent. There are empty vials of fentanyl and used syringes on the floor next to the bed.*

*7:00 a.m.: The junior resident notifies the Operating Room (OR) director of his senior resident's behavior, and the senior resident is confronted about suspected substance abuse. The resident confesses to injecting himself with fentanyl he had collected during cases the prior day.*

*7:30 a.m.: The program director is made aware of the situation. A urine sample is requested, and plans are made to immediately suspend the resident and arrange for substance abuse treatment.*

While in treatment, the resident admits to having been abusing fentanyl for 6 months prior to the on-call incident. He identifies the stress of a recent divorce as a potential trigger for his descent into addiction. He claims he obtained fentanyl by administering less to his patients than he was charting and saving the excess. He would sometimes use β (beta) blockers to mask the physiologic signs of inadequate anesthesia. When inquiries were made into signs of abuse that might have been missed, other residents in the program were incredulous. They described this person as hyperconscientious and hardworking. They reported that he would often volunteer for extra call and decline relief for breaks.

### Analysis of Root Causes and Systems in Need of Improvement

The proximate cause of danger to patients under the care of this resident is clear. By injecting himself with a psychotropic medication while charged with supervising patient care, he was jeopardizing both his and patients' safety. All patients, but particularly patients in an intensive care setting, require vigilance and lucid decision-making. Had the junior resident been unable to respond appropriately, the consequences could have been catastrophic. Altered clinical decision-making capacity is a critical threat to patient safety.

The root causes, however, beyond this individual's breach of duty, lie in inadequate systems to prevent this from happening and being detected. The incident raises questions of how this resident was able to obtain narcotics and how his abuse of them continued in the workplace without raising the suspicion of his colleagues.

This particular resident admitted to obtaining fentanyl by charting its use, but administering less to his patients in the operating room. Subsequent review of his medication usage revealed a consistent pattern of using quantities of narcotics in excess of what would be typical for given procedures. He also admitted to several instances of withdrawing medications from Pyxis® machines remote in time and location from cases to which he was attributing them. Had a more rigorous system to track medication usage been in place, the department may have been alerted to these red flags.

Discussions with this resident's colleagues uniformly revealed shock and disbelief regarding their co-resident's addiction. In hindsight, he displayed behavior that could have been identified as subtle warning signs. Other residents described this individual as hardworking and hypervigilant. He would frequently volunteer for extra shifts and refuse relief for breaks. Some noted him to have become more withdrawn, but they surmised he was trying to deal with his divorce privately and did not wish to overstep the bounds of their professional relationship. If the residents had been more keenly aware of behaviors suggestive of substance abuse they might have been more inclined to intervene.

This program also did not employ routine drug screening for its residents. Faculty perceived this kind of action as intrusive and worried residents would balk at what might be considered an invasion of privacy.

## Discussion

Due to the ready availability of many medications with high potential for abuse, physician impairment has been identified as a possible hazard of anesthesia practice. The specialty tends to be overrepresented in substance abuse treatment programs compared to its contribution to the total pool of physicians. In 1987, Talbott et al. [8] examined data from the first thousand cases referred to the Medical Association of Georgia's Impaired Physician Program. They reported that anesthesia residents made up 33.7 % of those who presented for treatment while comprising only 4.6 % of residents in the state. While not as exaggeratedly, disproportionate rates of substance abuse appear to continue after residency. In 2009, Skipper et al. [9] analyzed data from 16 state physician health programs and excluded resident physicians from their analysis. Anesthesiologists represented 11.1 % of those enrolled in these programs, but accounted for only 4.1 % of physicians at that time. This study also showed anesthesiologists are much more likely to abuse intravenous narcotics than practitioners in other fields.

An impaired physician in the OR presents an obvious risk to patient safety. In spite of this risk, an analysis of closed claims in 1994 found substance abuse mentioned in only a small number of claims against anesthesiologists [10]. Still, these claims represent only instances when a patient has been demonstrably harmed due to substance abuse. Many cases where harm is less obvious or physician impairment has been overlooked likely go unreported. For example, scenarios involving inadequate analgesia or cardiovascular complications from patients not receiving narcotics due to anesthesiologists diverting drugs for personal use could be difficult to prove.

Not to be overlooked are the dangers to the anesthesiologist himself. Anesthesiologists have been found to have a relative risk of drug-related death of 2.79 (CI = 1.87–4.15, $P < 0.001$) when compared to general internists, with the highest risk of death occurring in the first 5 years of training [11].

Recognizing the issue of substance abuse in the anesthesia workplace, departments and institutions have developed ways to combat the problem. Efforts to prevent addiction have focused on drug control and education [12].

Easy access to narcotics and other potentially addictive drugs has logically been identified as a risk factor for substance abuse in anesthesiology [13]. Therefore, efforts have been made to restrict and monitor this access. The cornerstone of these efforts is detailed record keeping [14]. Records of medication usage can then be analyzed for patterns suggestive of drug diversion. Such patterns include high usage and wastage, transactions that occur at automated dispensers not located at the site of indicated use, and drugs obtained for completed, nearly completed, or canceled cases. Increasingly, automated systems are being developed to audit anesthesia records for these red flags [15]. Additionally, pharmacies now routinely screen returned wasted drugs to verify their contents [12].

Anesthesia departments have also instituted education programs aimed at highlighting the dangers of substance abuse and the importance of recognizing and reporting abuse in colleagues. Residency programs are now required by the Accreditation Council for Graduate Medical Education (ACGME) to have a substance abuse education program in place [16]. Residents are taught to identify behavior patterns that could easily be dismissed or thought unremarkable. More obvious signs of abuse include emotional lability, erratic behavior, and social withdrawal, but less glaring warnings are also highlighted. These include efforts on the part of the abuser to obtain and mask his addiction that are often interpreted as a strong work ethic. Substance abusers will often volunteer for extra call, decline relief breaks, or take frequent bathroom breaks [17]. Importantly, all members of the healthcare team must feel empowered to speak up about concerns, and lower ranking team members should not fear repercussions or reprisal for reporting suspected abuse [18].

Another potentially contentious method of identifying abuse is drug screening of those with access to narcotics. Use of random toxicology screening is not routinely employed due to reluctance to subject all personnel to what is perceived as an invasion of privacy. While many anesthesia departments have adopted drug screening, it is still more commonly used to confirm cases of suspected abuse.

While prevention is preferable to treatment of abuse that is ongoing, departments must be prepared to deal with abuse when it is discovered. The ACGME requires residency programs to have written policies in place to deal with cases of abuse [17]. Many states allow professional societies to divert impaired healthcare professionals into treatment and rehabilitation programs without the notification of licensing boards. Some degree of confidentiality is guaranteed contingent on successful completion of rehabilitation and compliance with all treatment requirements [19]. Unfortunately, the success rates of rehabilitation programs are low, and returning to the workplace often endangers patients and the returning physician. Relapse is all too often only discovered with the death of the anesthesiologist returning to practice [20]. Several authors have recommended redirection of anesthesiologists with substance abuse problems into other specialties with less access to narcotics [21, 22]. The decision to allow reentry should be made on a case-by-case basis, and when reentry is attempted, close monitoring with gradual reinstatement is advised [23]. The impaired physician highlights the duality of patient safety and its impact on the health system, its providers as well as its consumers.

## Case 2: Errors in Airway Management

### Timeline

*7:00 a.m.:* An anesthesiologist working in a freestanding ambulatory surgery center conducts his preoperative evaluation in the holding area for a 54-year-old male scheduled to undergo an elective inguinal hernia repair. The patient's only medical problem includes hypertension controlled with Enalapril. The anesthesiologist notes he is mildly obese (BMI 33) and has a short thyro-mental distance. Range of motion of the cervical spine and at the atlanto-occipital junction is fully intact. There are no issues with dentition. The Malampati score (a scaled score of 1–4 evaluating potential difficulty for intubation) is determined to be 3 [24]. The patient reports that his wife tells him he snores loudly, but has never seen a doctor for sleep apnea. He has a prior history of surgery for a broken humerus during a skiing accident. He remembers being kept overnight, but when asked if he was told about any complications from anesthesia he does not remember. He thinks his wife would remember better, but she is outside on the phone talking to their son. The anesthesiologist leaves before the patient's wife returns.

*7:35 a.m.:* The patient is brought to the room, monitors are placed, and preoxygenation is begun.

*7:40 a.m.:* Anesthesia is induced with midazolam, lidocaine, fentanyl, and propofol. Rocuronium is administered immediately following induction to ease intubation and provide paralysis for surgery.

*7:42 a.m.:* The anesthesiologist attempts to intubate with a size 7.0 endotracheal tube (ETT), but is unable to do so. He switches laryngoscope blades and makes another unsuccessful attempt to intubate the patient. He then asks the circulating nurse to place a shoulder roll under the patient and tries to intubate with a smaller tube, but now notices new-onset edema of the airway. He attempts to ventilate between intubation attempts, but the oxygen saturation drops to 70 %.

*7:46 a.m.:* After unsuccessfully attempting intubation with the smaller ETT the anesthesiologist now finds it increasingly difficult to ventilate the patient. The anesthesiologist asks the nurse to call for help and for the fiberoptic intubating endoscope.

*7:52 a.m.:* Help has not yet arrived, and the patient is now nearly impossible to mask ventilate. The patient's oxygen saturation levels have dropped into the teens, and he is bradycardic. After placing folded sheets to ramp up patient's head, the anesthesiologist makes a final unsuccessful attempt to intubate using a Miller laryngoscope and asks the surgeon to prepare for a surgical airway. The patient is hypotensive, bradycardic, and oxygen saturation is not accurately sensing.

*7:54 a.m.:* As the surgeon is to begin an invasive airway, another anesthesiologist arrives with a fiberoptic intubating endoscope. The scope is passed successfully through the patient's vocal cords and used to guide the placement of a size 6.5 ETT.

*7:59 a.m.: The patient's vital signs stabilize, and all believe that the crisis has been averted. The decision is made to proceed with the case.*

*9:20 a.m.: At the conclusion of the case efforts are made to arouse the patient, but even after no inhaled anesthetics are detectable in the patient's expired air he is unresponsive.*

This incident resulted in anoxic brain injury, and the patient remains in a persistent vegetative state. When the wife was informed she recalls being told after her husband's previous surgery that there was some difficulty with intubation. She had assumed this information would be in his chart or that her husband would have known to make his anesthesiologist aware of this.

## Analysis of Root Causes and Systems in Need of Improvement

This case exemplifies a scenario every anesthesiologist dreads. Securing the patient's airway during the induction of anesthesia is one of the anesthesiologist's most crucial responsibilities. Yet the overwhelming number of uneventful inductions may lead to lapses in vigilance and preparedness. This principle extends more broadly to the practice of anesthesia, where catastrophe must always be anticipated in spite of its infrequent occurrence.

The failures of the anesthesiologist in this case center on his lack of preparedness, beginning with not recognizing a potentially difficult airway. Several elements of this patient's preoperative history and physical exam should have alerted the anesthesiologist to this possibility. These include obesity, short thyro-mental distance, snoring, and most importantly, the patient's reference to previous complications. When asked to explain his decision not to clarify the patient's history with the patient's wife, the anesthesiologist reported being concerned about delaying the start of the case. In this way, financial concerns and perceived pressure from colleagues to proceed with cases can supersede proper regard for patient safety. This may be particularly true in a private practice setting.

As in nearly all cases of patient injury, the responsibility does not lie solely on the shoulders of the individual practitioner. Records of this patient's previous operative and postoperative course at an outside institution were not readily available to the anesthesiologist. Nor was there any system in place to alert subsequent providers to previous airway difficulties. Better interprovider information sharing may have prevented this incident, but benefits must be weighed against possible breaches in the security of protected health information.

Had this anesthesiologist anticipated a difficult airway, he might have altered his management plan. Standard induction protocols involve rendering the patient apneic before placement of the endotracheal tube. When faced with a high likelihood of difficulty securing a patient's airway, anesthesiologists will use modified protocols that avoid this situation. The anesthesiologist may have considered intubating this patient while he was still awake or using regional anesthesia, obviating the need for intubation altogether.

Beyond his failure to foresee difficulty, this anesthesiologist can also be faulted for ignoring the most current practice guidelines. These guidelines urge anesthesiologists to have a preformed plan for the possibility of a difficult intubation, including having rescue devices available should direct laryngoscopy fail. The anesthesiologist admitted to not feeling comfortable using some of the newer devices that are now available, having never been trained on them. In this way, he and his employer allowed suboptimal care to be delivered to their patients by not incorporating advancements in technology and techniques into their practice. It is all too easy to become complacent with one's level of training upon completing residency. Individual practitioners and provider organizations must develop ways to ensure that education and training continue throughout anesthesiologists' careers.

# Discussion

Respiratory system adverse events have historically been a major source of anesthesia malpractice claims. A 1990 analysis of closed claims found this type of injury to account for 34 % of claims, with 85 % of those resulting in brain damage and death. The authors noted that 17 % of respiratory events were rooted in difficult intubations [25], highlighting an area of concern. Recognizing the need to improve outcomes, the ASA developed practice guidelines in 1992 for managing difficult airways, which were updated in 2013 [26]. Other common sources of respiratory events identified in the 1990 study were inadequate ventilation and undiagnosed esophageal intubation, which were already being addressed with improved monitoring standards. Recent analyses of closed claims data show that with these improvements in place, the incidence of death and brain damage has declined significantly since the 1980s. Between 1990 and 2007, respiratory events were identified as the cause of 17 % of claims [6].

The ASA's guidelines recommend evaluation of the airway by history, physical examination, and, in certain cases, attempting to gather additional information. The single most important piece of information a patient can provide is a history of difficult intubation [27]. Unfortunately, patients are often unaware of a history of difficult intubation or the importance of conveying this information. Some institutions have developed policies to alert subsequent providers to a history of difficult intubation through a variety of means. Proposed methods for interinstitution communication of this information have included alert bracelets, registries [28], and wallet-sized identification cards [29].

Borrowing from the successful use of algorithms in the management of life-threatening cardiac events, the ASA has developed algorithms to illustrate key decisions points in the approach to a difficult airway (Fig. 18.1) [26]. The initial steps of the Difficult Airway Algorithm are designed to encourage practitioners to develop a preformed plan for each case. Inevitably, preparation will occasionally fail and patients will unexpectedly prove impossible to intubate and ventilate by face mask. Still, the anesthesiologist is not without recourse before resorting to an invasive

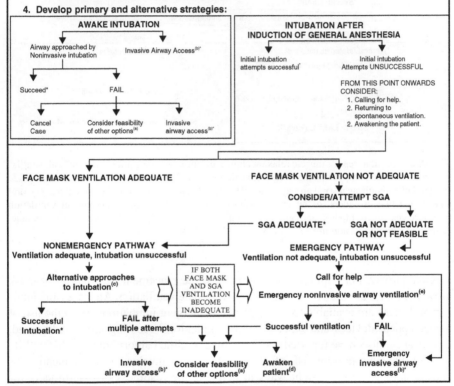

American Society of Anesthesiologists*

**DIFFICULT AIRWAY ALGORITHM**

1. **Assess the likelihood and clinical impact of basic management problems:**
   - Difficulty with patient cooperation or consent
   - Difficult mask ventilation
   - Difficult supraglottic airway placement
   - Difficult laryngoscopy
   - Difficult intubation
   - Difficult surgical airway access

2. **Actively pursue opportunities to deliver supplemental oxygen throughtout the process of difficult airway management.**

3. **Consider the relative merits and feasibility of basic management choices:**
   - Awake intubation *vs.* intubation after induction of general anesthesia
   - Non-invasive technique *vs.* invasive techniques for the initial approach to intubation
   - Video-assisted laryngoscopy as an initial approach to intubation
   - Preservation *vs.* ablation of spontaneous ventilation

4. **Develop primary and alternative strategies:**

*Confirm ventilation, tracheal intubation, or SGA placement with exhaled $CO_2$

a. Other options include (but are not limited to): surgery utilizing face mask or supraglottic airway (SGA) anesthesia (e.g., LMA, ILMA, laryngeal tube), local anesthesia infiltration or regional nerve blockade. Pursuit of these options usually implies that mask ventilation will not be problematic. Therefore, these options may be of limited value if this step in the algorithm has been reached via the Emergency Pathway.

b. Invasive airway access includes surgical or percutaneous airway, jet ventilation, and retrograde intubation.

c. Alternative difficult intubation approaches include (but are not limited to): video-assisted laryngoscopy, alternative laryngoscope blades, SGA (e.g., LMA or ILMA) as an intubation, conduit (with or without fiberoptic guidance), fiberoptic intubation, intubating stylet or tube changer, light wand, and blind oral or nasal intubation.

d. Consider re-preparation of the patient for awake intubation or canceling surgery.

e. Emergency non-invasive airway ventilation consists of a SGA.

**Fig. 18.1** American society of anesthesiologists difficult airway algorithm. With Permission from Anesthesiology. 2013; 118:251–70

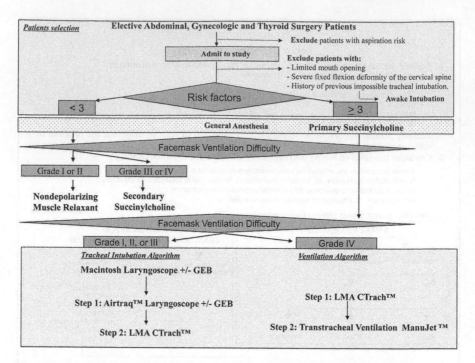

**Fig. 18.2** Decision tree for muscle relaxant choice and airway management. The difficult ventilation grading scale is the following: Grade I, ventilation without the need for an oral airway; grade II, ventilation requiring an oral airway; grade III, difficult and unstable ventilation requiring an oral airway and two providers, or an oral airway and one provider, using mechanical ventilation (pressure-controlled mode); and grade IV, impossible ventilation. *GEB* gum elastic bougie. Reprinted with permission from Anesthesiology. 2011; 114(1):25–33

surgical airway. For supraglottic obstructions, the placement of a laryngeal mask airway (LMA) can often be used to restore adequate ventilation. An increasing number of devices are available to assist with nonemergent intubations. With the use of newer optical devices and other tools for intubation becoming more widespread, European authors have reported success with a modified algorithm (Fig. 18.2) [30].

With the omnipresent risk of catastrophe, those involved in the education of anesthesiologists have sought ways to heighten the readiness of practitioners for uncommon, but critical events. Borrowing from other industries with similarly routine, but risky situations, simulation was introduced into the training of anesthesiologists beginning in the late 1980s [31]. Simulation training is now commonly used at all levels of education in anesthesiology, particularly in residency programs to develop a broad set of skills. These skills range from procedural and technical proficiency to team communication and reinforcement of protocols for rare events. While simulation seems intuitively well suited to training in these areas, its efficacy is difficult to prove. However, a growing body of evidence is supporting its use. The efficacy of simulation in the teaching of procedural skills is most easily measured and well supported [32]. It is more difficult to show improvement in performance in complex situations, such as team training [33]. However, anesthesiologists have reported feeling strongly influenced by simulator training when rare emergencies have been encountered subsequent to simulator preparation [34]. What is certain is that simulation training is gaining acceptance and its use will continue to grow.

## Conclusion

Anesthesiologists have served as pioneers in the medical profession embracing the principles of patient safety. Complications from anesthesia have declined dramatically over the last 50 years, and patient outcomes have improved. While perioperative deaths attributed to anesthesia were approximately 1 in 1,500 some 50 years ago, today that number has improved nearly tenfold; that is a dramatic increase in patient safety despite older and sicker patients being treated in operating rooms nationwide. At present, the chances of a healthy patient suffering an intraoperative death attributable to anesthesia is less than 1 in 200,000 when an anesthesiologist is involved in patient care. Therefore, vigilance and integrity coupled with medical knowledge and clinical skills are at the forefront of an anesthesiologist's goal in providing safe anesthesia care.

## Key Lessons Learned

### Case 1

- Physician wellness is an essential element of patient safety.
- While individuals are responsible for maintaining a state of physical and mental health that allows them to fulfill their professional obligations, colleagues and hospital systems can and should play an important role.
- Those in need of help may be identified before patients or practitioners are put at risk.
- Prevention is preferable to treatment, particularly when dealing with substance abuse.
- Prevention is best achieved through restriction of access to drugs of abuse.

### Case 2

- Better interinstitution information systems can help ease transmission of critical medical history.
- Practitioners must make a priority of staying current with the latest techniques, guidelines, and recommendations.
- Simulation training offers a way to develop procedural skills, team communication, and emergency preparedness in a safe environment.

# References

1. Beecher H, Todd D. A study of the deaths associated with anesthesia and surgery. Ann Surg. 1954;140(1):2–34.
2. Cooper JB, Newbower RS, Kitz RJ. An analysis of major errors and equipment failures in anesthesia management: considerations for prevention and detection. Anesthesiology. 1984;60(1):34–42.
3. Anesthesia Patient Safety Foundation. Foundation history. Available at http://www.apsf.org/about_history.php. Accessed 13 Jul 2013.
4. ASA Closed Claim Project. Overview. Available at http://depts.washington.edu/asaccp/welcome-anesthesia-closed-claims-project-its-registries. Accessed 13 May 2013.
5. Eichhorn JH. Prevention of intraoperative anesthesia accidents and related severe injury through safety monitoring. Anesthesiology. 1989;70(4):572–7.
6. Metzner J, Posner KL, Lam MS, Domino KB. Closed claims' analysis. Best Pract Res Clin Anaesthesiol. 2011;25(2):263–76. Review.
7. ASA standards, guidelines, statements, and other documents. Available at http://www.asahq.org/For-Healthcare-Professionals/Standards-Guidelines-and-Statements.aspx. Accessed 13 Jul 2013.
8. Talbott GD, Gallegos KV, Wilson PO, Porter TL. The Medical Association of Georgia's Impaired Physicians Program. Review of the first 1000 physicians: analysis of specialty. JAMA. 1987;257(21):2927–30.
9. Skipper GE, Campbell MD, Dupont RL. Anesthesiologists with substance use disorders: a 5-year outcome study from 16 state physician health programs. Anesth Analg. 2009;109(3): 891–6.
10. Sivarajan M, Posner KL, Caplan RA, Gild WM, Cheney FW. Substance abuse among anesthesiologists. Anesthesiology. 1994;80:704.
11. Alexander BH, Checkoway H, Nagahama SI, Domino KB. Cause-specific mortality risks of anesthesiologists. Anesthesiology. 2000;93(4):922–30.
12. Bryson EO, Silverstein JH. Addiction and substance abuse in anesthesiology. Anesthesiology. 2008;109(5):905–17.
13. Silverstein JH, Silva DA, Iberti TJ. Opioid addiction in anesthesiology. Anesthesiology. 1993;79(2):354–75.
14. Schmidt KA, Schlesinger MD. A reliable accounting system for controlled substances in the operating room. Anesthesiology. 1993;78(1):184–90.
15. Epstein RH, Gratch DM, Grunwald Z. Development of a scheduled drug diversion surveillance system based on an analysis of atypical drug transactions. Anesth Analg. 2007;105(4): 1053–60.
16. Accreditation Council for Graduate Medical Education. Institutional Requirements. Available at http://www.acgme.org/acgmeweb/tabid/158/ProgramandInstitutionalGuidelines/InstitutionalAccreditation/InstitutionalReview.aspx. Accessed 13 May 2013.
17. Bery AJ, Arnold WP. Chemical dependence in anesthesiologists: what you need to know when you need to know it. Park Ridge, Illinois: American Society of Anesthesiologists Task Force on Chemical Dependence of the Committee on Occupational Health of Operating Room Personnel; 1998.
18. Agency for Healthcare Research and Quality. Team Strategies and Tools to Enhance Performance and Patient Safety. Available at http://teamstepps.ahrq.gov/. Accessed 13 Jul 2013.
19. Walzer RS. Impaired physicians. An overview and update of the legal issues. J Leg Med. 1990;11(2):131–98.
20. Menk EJ, Baumgarten RK, Kingsley CP, Culling RD, Middaugh R. Success of reentry into anesthesiology training programs by residents with a history of substance abuse. JAMA. 1990;263(22):3060–2.
21. Collins GB, McAllister MS, Jensen M, Gooden TA. Chemical dependency treatment outcomes of residents in anesthesiology: results of a survey. Anesth Analg. 2005;101(5):1457–62.

22. Berge KH, Seppala MD, Lanier WL. The anesthesiology community's approach to opioid- and anesthetic-abusing personnel: time to change course. Anesthesiology. 2008;109(5):762–4.
23. Bryson EO, Levine A. One approach to the return to residency for anesthesia residents recovering from opioid addiction. J Clin Anesth. 2008;20(5):397–400.
24. Mallampati SR, Gatt SP, Gugino LD, Desai SP, Waraksa B, Freiberger D, et al. A clinical sign to predict difficult tracheal intubation: a prospective study. Can Anaesth Soc J. 1985;32(4):429–34.
25. Caplan RA, Posner KL, Ward RJ, Cheney FW. Adverse respiratory events in anesthesia: a closed claims analysis. Anesthesiology. 1990;72(5):828–33.
26. American Society of Anesthesiologists Task Force on Management of the Difficult Airway. Practice guidelines for management of the difficult airway: an updated report by the American Society of Anesthesiologists Task Force on Management of the Difficult Airway. Anesthesiology. 2013;118:251–70.
27. el-Ganzouri AR, McCarthy RJ, Tuman KJ, Tanck EN, Ivankovich AD. Preoperative airway assessment: predictive value of a multivariate risk index. Anesth Analg. 1996;82(6):1197–204.
28. Atkins R. Simple method of tracking patients with difficult or failed trachea intubation. Anesthesiology. 1995;83:1373–5.
29. Schaeuble JC, Caldwell JE. Effective communication of difficult airway management to subsequent anesthesia providers. Anesth Analg. 2009;109(2):684–6.
30. Amathieu R, Combes X, Abdi W, Housseini LE, Rezzoug A, Dinca A, et al. An algorithm for difficult airway management, modified for modern optical devices (Airtraq laryngoscope; LMA CTrach™): a 2-year prospective validation in patients for elective abdominal, gynecologic, and thyroid surgery. Anesthesiology. 2011;114(1):25–33.
31. Gaba DM, DeAnda A. A comprehensive anesthesia simulation environment: re-creating the operating room for research and training. Anesthesiology. 1988;69(3):387–94.
32. Nestel D, Groom J, Eikeland-Husebø S, O'Donnell JM. Simulation for learning and teaching procedural skills: the state of the science. Simul Healthc. 2011;6(Suppl):S10–3.
33. Eppich W, Howard V, Vozenilek J, Curran I. Simulation-based team training in healthcare. Simul Healthc. 2011;6(Suppl):S14–9.
34. Smith HM, Jacob AK, Segura LG, Dilger JA, Torsher LC. Simulation education in anesthesia training: a case report of successful resuscitation of bupivacaine-induced cardiac arrest linked to recent simulation training. Anesth Analg. 2008;106(5):1581–4.

# Chapter 19
# Patient Safety in Behavioral Health

**Renuka Ananthamoorthy and Robert J. Berding**

> *"Mistakes are a fact of life. It is the response to error that counts."*
>
> Nikki Giovanni

## Introduction

Behavioral health patients pose unique and complex safety challenges in the modern healthcare environment. They may enter the hospital setting with a psychiatric diagnosis in addition to medical comorbidities and/or co-occurring addictive disorders. Therefore, it is imperative that healthcare organizations have well-established policies and procedures to assess safety risks, provide targeted interventions, communicate across disciplines/departments, and include all necessary stakeholders in the process.

Overall, a culture of good teamwork should be fostered by the organization that places high value on respect, communication, role responsibility, and defined steps to escalate patient safety concerns. In addition, an organization should undertake a comprehensive risk analysis of potential safety pitfalls.

There are two basic analytic approaches that may be used to design safe systems for behavioral health patients. The first is a proactive approach involving

R. Ananthamoorthy, M.D.
Department of Behavioral Health, Kings County Hospital Center, 451, Clarkson Avenue,
Brooklyn, NY 11203, USA
e-mail: renuananth@aol.com

R.J. Berding, J.D., M.S. (✉)
Regulatory Compliance, Risk Management and Patient Safety, Kings County Hospital Center,
451 Clarkson Avenue, Brooklyn, NY 11203, USA
e-mail: rjberding@yahoo.com

A. Agrawal (ed.), *Patient Safety: A Case-Based Comprehensive Guide*,          295
DOI 10.1007/978-1-4614-7419-7_19, © Springer Science+Business Media New York 2014

multidisciplinary teamwork to examine the process of care from referral to discharge and then considering the possibilities for error at each step. The second is the "causal method," which involves learning from mistakes through a Root Cause Analysis (RCA) [1]. Of course, a cause is not something found but rather constructed from available evidence. Such causes of failure typically emerge from multiple sources [2]. These causes may range from direct to indirect, or from a true root cause to merely an opportunity for improvement. However, all causes should be appropriately addressed once identified through this process.

In this chapter, the causal method will be used by employing a "fishbone model" diagram in the following two cases to analyze systems breakdowns relating to (1) communication (2) staffing (3) education (4) medications (5) environment (6) patient (7) provider (8) treatment team (9) unit/hospital, and (10) Electronic Health Record (EHR).

## Case Studies

### Case Study 1: Aggressive Behavior Leading to Restraints and Patient/Staff Injury

#### Clinical Summary

*Albert is a 34-year-old male with a past psychiatric history of paranoid schizophrenia brought into the psychiatric Emergency Room (ER) in full body restraint (FBR) by Emergency Medical Transport accompanied by the police for menacing and aggressive behavior in a local park. The Registered Nurse (RN) knew Albert from prior admissions and simply triaged him as, "found agitated in park." Since he was not acting aggressive at the time, he was released from the FBR and police left the ER. At that time, Albert promised that he would sit quietly, so he was not given any medications and was left alone in a cubicle around the corner from the nursing station. After about 45 min, he again became agitated and began to spit at staff. He was seen by the physician-in-charge and was administered Haldol 5 mg and Ativan 2 mg, both intramuscular, while simultaneously being placed in four-point mechanical restraint for safety. During placement of the restraint, the patient kicked one Patient Care Technician (PCT) who fell to the floor, and bit another PCT. The PCTs left the ER to receive further evaluation. Albert continued to yell loudly at anyone passing by and violently attempted to remove the restraint. Thirty minutes later, he was given another dose of Haldol 5 mg and Ativan 2 mg by a second physician because staff were fearful of another violent outburst. He soon fell sleep and woke up 5 h later. The PCTs recorded all of these events on a "Q15 observation form." At that time, Albert asked to be released from the restraint. About an hour later, he complained to the RN of severe pain in the right wrist. An x-ray of the right wrist showed a fracture.*

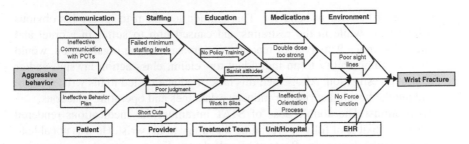

**Fig. 19.1** Case 1—Aggressive behavior leading to restraints and patient/staff injury

## Root Cause Analysis

The root cause analysis of the case revealed the following contributory factors that led to the adverse outcome of two staff members being injured and the patient sustaining an unintended injury to the wrist (Fig. 19.1).

1. Communication: The RN failed to verbally communicate Albert's prior aggressive behavior in the park to the PCTs. The PCTs had only occasionally observed him sitting quietly alone in the cubicle for a short time and did not consider him to be a threat to others. In general, it is beneficial for ER staff to share as much information as possible about newly arrived patients. This is because behavior may change from minute to minute depending on the patient's alternating moods. The shared knowledge of recent patient actions better prepares all staff to anticipate potential mood swings.

2. Staffing: The RN also failed to complete a comprehensive evaluation as required by policy. Instead, there was only the brief note describing the reason Albert was brought into the ER. The RN indicated that there was not sufficient time to fully complete the evaluation because the supervisor had not scheduled the minimum number of staff necessary for that shift. Due to the time constraint, the RN decided to skip Albert's full evaluation and spend time writing about other admissions. The full evaluation includes a standardized rating scale that would have resulted in a score that fell within the "high risk" range for aggression based on age, gender, diagnosis, involuntary admission status, and past psychiatric history including incidents of violence.

3. Education: The situation worsened when the restraints were misapplied by staff. The experienced PCTs in ER had recently been hastily replaced by two "agency" PCTs. Neither of the replacements had received behavioral health orientation training and did not know how to apply four-point mechanical restraints. Consequently, they applied the apparatus too tightly. Beyond that, the RN did not inform the PCTs about the 2-h limit and thus the restraints were not removed in a timely manner. Lastly, the PCTs failed to recognize the dangers associated with Albert's violent attempts to free himself. Based on subsequent staff interviews, it appeared they developed a "sanist" prejudice against Albert

because of his violent behavior. This attitude led them to ignore his obvious helplessness while in the restraints and caused him to suffer in a cruel and unusual manner. This type of reaction, if pervasive in the organization, would create a vulnerability to an individual tort claim, class action lawsuit, and/or federal investigation. Altogether, this lack of education led to his wrist fracture and put the hospital at risk for monetary damages and operating sanctions.

4. Medications: The double dose of STAT intramuscular medications rendered Albert incapable of fully appreciating the injury to his wrist. He was not able to recognize the damaging effects of his self-destructive movements in attempting to disengage the overly tight restraints or timely alert the staff to his injury.

5. Environment: Albert grew more agitated as time began to pass but the staff could not readily observe this behavior change because of poor sight lines into the cubicle. Consequently, no one on the team was in a position to anticipate his potentially dangerous behavior and prepare an adequate response strategy. As a result, the two PCTs sustained potentially serious occupational injuries and were no longer able to function as part of the ER treatment team.

6. Patient: Albert and the staff also missed out on opportunities to avert this adverse outcome through the "safe behavior plan." When he first arrived in the ER, there was no specific mention of how escalating behavior was to be identified and what countermeasures would be used by staff. Instead, there was only Albert's promise not to be violent if allowed to be freed from the FBR. There was no opportunity for Albert to describe how he might be calmed if the agitation began to manifest itself again. Perhaps he would have been more comfortable simply receiving the intramuscular medications. Furthermore, his active participation in the safe behavior plan might have provided some motivation for him to comply with the de-escalating efforts of the team.

7. Providers: Overall, the RN demonstrated poor judgment in the triage process. Instead of completing each task according to established procedures, shortcuts were taken to work around time constraints. The progress note was substituted for a comprehensive evaluation. Albert's promise "to behave" was substituted for a formal safe behavior plan. The handoff communication was omitted. Altogether, this attitude that shortcuts are permissible becomes a dangerous precedent in the workplace. Likewise, the second physician should have reassessed the patient instead of simply reordering medication based on escalating staff fears.

8. Treatment Team: There was an almost total lack of teamwork. Staff performed their own duties in silos. There was a lack of communication except during the emergency use of restraints. The team members did not offer important information or request it from one another. No one reviewed prior documentation. When staff compartmentalize their duties, it detracts from the team concept and increases the risk for adverse outcomes during transition periods.

9. Unit/Hospital: The hospital leadership was aware that PCTs routinely rotated to assignments without appropriate orientation but lacked an effective plan to ensure such targeted training. In general, there should be a system in place to assure all staff are appropriately oriented to the hospital and unit prior to assignment.

10. Medical Record: The RN was allowed to bypass the evaluation because there was no forcing function in the Electronic Health Record (EHR) that required its completion. In addition, the prior medical records were available in the EHR but there was no standard practice in place for staff review. In this case, it would have been helpful to have the EHR require the RN to complete a task rather than simply ignore it.

These root causes overlap to some extent and as such should not merely be approached in an isolated manner. It is beneficial to also consider any common threads that might exist among the identified factors. This concept will be explored later in the chapter under the heading of "Risk Reduction Strategies."

## Case Study 2: Multiple Factors Leading to a Psychotic Inpatient Committing Suicide

### Clinical Summary

*Beauregard is a 23-year-old male college graduate with a past psychiatric history of recurrent major depression with psychosis and no known history of substance abuse. He was last admitted to inpatient psychiatry a year ago for a suicide attempt in which his mother found him unconscious in the garage after inhaling exhaust fumes. On this occasion, he was brought in to the ER by EMS, after his mother called 911 for help. She reported that Beauregard called her at work to say that he was leaving New Jersey and going to Pennsylvania because the neighbors were tormenting him with fireworks. His mother begged EMS to take her son to the hospital because there was no one in Pennsylvania to care for him. Beauregard was evaluated and admitted to the inpatient psychiatry unit for increased paranoia, suspiciousness, anxiousness, restlessness, and depressed mood. His prior medical records were on paper and not available to inpatient physicians through their new EHR. An initial treatment plan was made by the team while Beauregard waited outside the conference room even though he had actively participated in the treatment planning during his prior stays. Due to his increased agitation, he was placed on routine observation and started only on antidepressant medication. The following day, Beauregard took his medications and participated in all assigned activities but was unable to see the social worker who was attending a mandated, all day in-service training program. He tried to contact his mother but was unable to do so. His mother called the unit to tell them that she had no transportation that evening but would visit Beauregard the next day. That message was taken by the unit clerk but no one informed the patient. She also asked to speak to the physician-in-charge who was too busy at the time and never returned her call. Shortly after visiting hours ended, another patient saw Beauregard hanging by his knotted bed sheets from the loopable door hinge (that was to be replaced but awaiting hospital funding). An emergency code was initiated but Beauregard could not be resuscitated and was pronounced dead.*

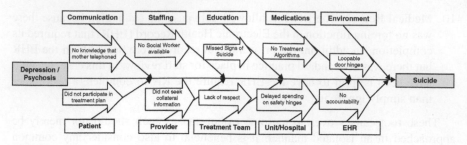

**Fig. 19.2** Case 2—Multiple factors leading to psychotic inpatient committing suicide

## Root Cause Analysis

The root cause analysis of the case revealed the following contributory factors (Fig. 19.2):

1. Communication: Despite his mother contacting the unit, Beauregard was never told of the telephone call. Perhaps this knowledge would have decreased his anxiety about her absence during visiting hours. In fact, there was no standard work in place to communicate outside information to patients. When creating communication protocols, it is necessary to include all stakeholders so that everyone has the information needed to support the treatment process.
2. Staffing: There was no back-up plan in place to fill the gap when the social worker was off the unit attending a training session. This could have been mitigated by rotating other staff onto the unit or planning the all-day training as two half-day sessions.
3. Education: When questioned about why the mother's telephone message was never shared with Beauregard, the clerk answered that she did not think it was as important as other duties. This demonstrated a lack of knowledge about the vital role that family members can play in the recovery effort. Also, staff's lack of understanding about the patient's agitation points to a gap in their clinical training. There is a need to provide ongoing education about the signs of impending suicide. If that type of training had been available, the staff may have made a better assessment about the potential for suicide in this case.
4. Medications: The patient was not started on anti-psychotics which would have reduced the potential of his command auditory hallucinations. It would have been helpful if appropriate treatment guidelines were used by the team.
5. Environment: In the Behavioral Health environment, it is imperative to minimize suicide risk by conducting an analysis of the potential environmental hazards. High on that list should be an assessment of door handles, hinges, and other loopable hardware. Likewise, close attention should be paid to sheets, blankets, towels, belts, and other items that may be fitted around the neck.
6. Patient: Beauregard was not invited to participate in the development of his treatment plan. He was aware of his role in the planning process but did not

proactively assert to have his voice heard by the team. While it is ultimately the team's responsibility to invite the patient into the process, the patient has the right to demand inclusion. This type of proactive participation is encouraged in the Wellness and Recovery literature [3].

7. Provider: The physician did not return the telephone call to seek out collateral information from Beauregard's mother. The information about his recent high risk behaviors would have fostered a better understanding of the seriousness of his condition.

8. Treatment Team: The treatment team should have included the patient in the planning process, especially because he was right outside the room at the time of discussion. This shows a lack of respect for the patient and his role as a team member.

9. Unit/Hospital: The administration was aware of the dangers associated with the current door hinge but decided to delay the purchase due to the costs. This type of purchase, especially identified through a proactive environmental risk analysis, should be prioritized or an alternate interim solution should be put in place.

10. Medical records: Although the staff were told to contact medical records for old paper charts, in practice no one ever called because there was no accountability built into the system. In such cases, it can be useful to add an attestation checkbox in the EHR that team members must check to affirm that they have received and reviewed the record.

## Discussion

The cases described above highlight some of the typical harm risks encountered in behavioral health settings. In a recently published handbook, the American Psychiatric Association (APA) Committee on Patient Safety identified and categorized six types of safety risks commonly associated with this population. These can be described using the SAFE MD mnemonic and include Suicide, Aggressive Behavior, Falls, Elopement, Medical Comorbidity and Drug Errors [1]. Suicide and any serious adverse outcome relating to the other safety risks rise to the level of a sentinel event which The Joint Commission (TJC) defines as "any unanticipated event in a healthcare setting resulting in death or serious physical or psychological injury to a patient or patients, not related to the natural course of the patient's illness." [4] TJC standard LD.04.04.05 requires each accredited organization to define sentinel event for its own purposes in establishing mechanisms to identify, report, and manage these events. At a minimum, an organization's definition must include any occurrence that meets any of the following criteria: (1) Any unanticipated death or major permanent loss of function, not related to the natural course of the individual's illness or underlying condition; (2) Suicide of any individual served receiving care, treatment, or services in a staffed around-the-clock setting or within 72 h of discharge; (3) Abduction of any individual served receiving care, treatment, or services, and (4) Rape.

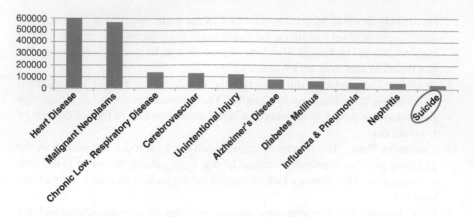

**Fig. 19.3** US top ten death causes (2009)

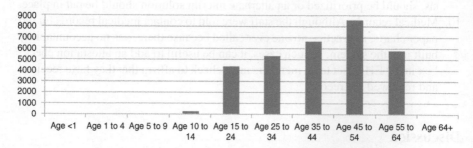

**Fig. 19.4** Number of US suicides by age category (2009)

## Suicide

According to the most recently published Centers for Disease Control and Prevention (CDC) reports, suicide ranks as the tenth leading cause of death in the USA and within the top four leading causes of death for persons from age 10 to 54 (Figs. 19.3 and 19.4) [5].

Among suicides, approximately six percent (6 %) are committed during an inpatient stay [6]. Inpatient suicide was the most common sentinel event reported to TJC over a 10-year period (1995–2005). Inpatient suicides are viewed as the most avoidable and preventable because they occur in close proximity to trained clinical staff. Early in the admission is a clear high-risk period, but risk declines more slowly for patients with schizophrenia. Other risk factors include absence of support and presence of family conflict. The greatest *clinical* root cause of inpatient suicide is a failure to perform a comprehensive and timely risk assessment [7]. In one study, risk was not adequately assessed in about 60 % of suicides, or else the risk level was not accorded appropriate precautions [8]. For all inpatients, the assessment should

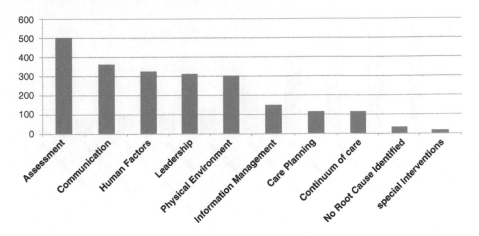

**Fig. 19.5** Top ten root causes relating to suicide

begin upon admission with the use of a standardized tool that ideally produces a rating of the suicide risk. This rating is often expressed in terms of a "score" that can be used in conjunction with an assessment of the patient's thoughts, plans, means, and ability to complete the suicidal act. For those at risk of suicide, the assessment should be repeated following any traumatic occurrence during the stay and upon discharge. The risk of suicide is higher during the period immediately following discharge from inpatient psychiatric care than at any other time in a service user's life [9]. TJC considers suicide as a sentinel event when occurring to an individual receiving care, treatment or services in a staffed around-the-clock setting or within 72 h of discharge. The root causes of suicides reported to TJC are displayed in Fig. 19.5 [10].

In the case of Beauragard, many of these factors existed. There was a poor assessment by the provider who did not recognize the presence of command auditory hallucinations. Concurrently, there was a clear breakdown in communication among team members and in failing to inform the patient about the contact from his mother. In addition, the physical environment risk could have been mitigated with proactive action by hospital leadership.

## Aggressive Behavior

Aggression in psychiatric settings is a complex workplace problem. Patient factors found to be related to violence include being a young male with a diagnosis of schizophrenia particularly with neurological impairment, having a history of violence, and being involuntarily admitted to the hospital [11]. Research examining staff factors found that the incidence of violence was higher on wards where staff

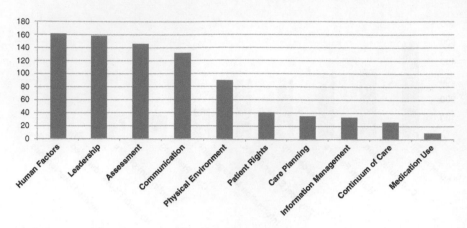

**Fig. 19.6** Top ten root causes relating to aggression

members were uncertain of their roles or where larger proportions of shifts were worked by substitute nursing staff [11]. Similar to assessing suicide risk, the treatment team should use a combination of standardized rating tools, observations, and interviews in order to identify the likelihood of aggression on the unit. TJC tracks the aggression events of assault, rape, and homicide under a category named Criminal Events. The root causes of aggression reported to TJC are displayed in Fig. 19.6 [10].

Beyond the obvious direct harms associated with aggression, there is also indirect risk of injury when attempting to manage this behavior, such as injuries resulting from attempts to subdue an aggressor. In addition, patients are at risk for self-injury if held in seclusion.

In the case of Albert, human factors played a major role in the injuries that occurred to staff and the patient. The RN should have completed the risk assessment instead of taking shortcuts. The PCTs should have completed the observation forms to better monitor Albert's condition. Also, the second physician should have checked the prior medication administration record before ordering a second dose.

## Falls

While the two cases above focused on suicide and aggression, there is a need to mitigate the other risks identified through SAFE MD. For example, falls may occur while patients are on behavioral health units or while experiencing altered mental status elsewhere in the hospital. There are many fall assessment tools available but the preferable ones will include the following risk factors: mental state impairment, gait and mobility, elimination problems, medications, and, fall history [12]. One study showed that behavioral health patients were more likely to fall if

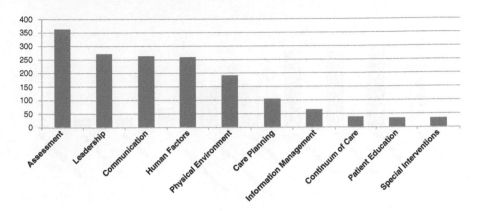

**Fig. 19.7** Top ten root causes relating to falls

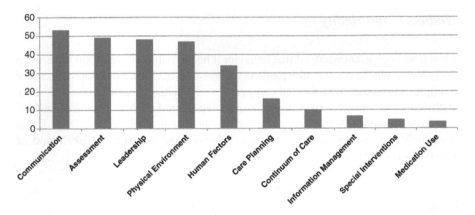

**Fig. 19.8** Top ten root causes relating to elopement

prescribed sedatives and/or hypnotics, experienced altered mental status, or elimination problems [13]. The root causes of falls reported to TJC are displayed in Fig. 19.7 [10].

## Elopement

Elopement is always a concern when persons are unwillingly detained through civil commitment and sometimes even when housed on a voluntary status. In order to minimize elopement risk, a healthcare organization should create an environment conducive to ongoing observation of potential elopers. In addition, there should be procedures in place for searching for successful elopers and returning them to the unit if found. The root causes of elopements reported to TJC are displayed in Fig. 19.8 [10].

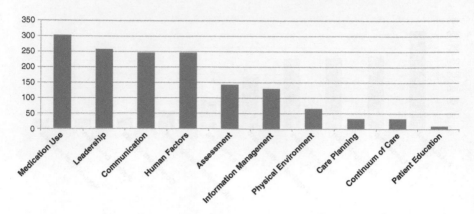

**Fig. 19.9** Top ten root causes relating to medication error

## Medical Comorbidity

It has long been acknowledged that behavioral health patients as a group were more likely than nonbehavioral health patients to have a co-occurring medical illness. For example, one recent study showed that persons with schizophrenia were more likely to have a greater number of conditions spanning several disease categories including cardiovascular, pulmonary, neurological, and endocrine disease [14]. These comorbidities pose greater prescribing challenges and increase the likelihood of adverse drug interactions.

## Drugs

The prevalence of unintended and untoward drug–drug interactions is increasing in concert with both the increasing number of pharmaceuticals available and the number of patients on multiple medications. The risk of poly-pharmacy is found to be greater for patients who are on psychiatric medications such as antidepressants [15]. Therefore, prescribers should consider how medications may interact on the basis of their pharmacodynamics and pharmacokinetics along with the intended therapeutic use. The root causes of drug errors reported to TJC are displayed in Fig. 19.9 [10]:

## Other Considerations

From a legal perspective, behavioral health patients may be admitted on a voluntary or involuntary basis, known as civil commitment. The general standard for involuntary civil commitment is whether or not the person poses a danger to self or others. An individual's "dangerousness" is clinically evaluated by one or more psychiatrists, but accurately predicting future harmful acts is far from an exact science [16]. It is the element of dangerousness that heightens the need for safety planning from prudent care management to legal obligation for this population. These legal standards have evolved through the power of the US Constitution, which provides 8th Amendment protection from Cruel and Unusual Punishment and gives Congress the 13th Amendment right to enact laws aimed to prevent harms stemming from discrimination. While not a specific protected class, behavioral health patients may be subjected to "sanism," which has been defined as, "the irrational prejudice that causes, and is reflected in, prevailing social attitudes toward persons with mental disabilities" [17]. These rights are generally protected by using "least restrictive alternatives" such as limiting the use of restraints and seclusion that might otherwise cause undue physical and/or psychological injury. This safety principle can be extended by the use of "safe behavior plans" in which patients contract to behave in a certain manner or else be subject to a consequence of a mutually agreed upon staff intervention. This approach can only be utilized if the patient exhibits the competence to complete a safe behavior plan. If the patient does not have such competence upon admission, then competence should be periodically reassessed throughout the stay.

## Risk Reduction Strategies

Once a root cause has been agreed upon, a corresponding corrective action plan should be put in place. This plan should reduce the risk of the occurrence repeating itself in the future. The following are some risk reduction strategies that may apply to a wide range of root causes.

### Establish Team Roles and Responsibilities

A well-delineated team structure assists all staff to work together. It is helpful to define the team membership, size, coordination of duties, and leadership lines. Often, it is just assumed that staff will perform their individual responsibilities and blend seamlessly together in the process. However, without clearly coordinated roles they are more likely to operate within the narrow silos of their clinical expertise. This lack of coordination could cause patients' needs to go unidentified or unattended thereby increasing safety risks.

### Establish Work Standards for Communicating Clinical Information

One method of sharing such information is through an interdisciplinary SBAR (Situation—Background—Assessment—Recommendation/Request) handoff among staff. This is a technique for communicating critical information that requires immediate attention and action concerning a patient's condition. SBAR provides a description of what is happening now, the clinical context, a general assessment of any problems, and an approach to correcting any problems. The SBAR is ideally given multiple times during the day in a short, huddle style. In addition to the SBAR technique, staff should be made aware of how to expeditiously escalate concerns when there is a change in patient behavior.

## Establish Clear Guidelines for Escalating Safety Concerns

Once the roles and work standards are in place, it is important for team members to have a mutually supportive method to escalate any perceived emerging safety issues. Sometimes staff are reluctant to challenge team leaders in fear of offending egos, overstepping professional boundaries, and/or retaliation. These fears must be put aside when they have an overriding safety concern. It becomes possible to allay such concerns if there is an organizational commitment to creating a culture whereby staff can respectfully advocate for the patient in a firm and assertive manner.

## Conduct Ongoing Environmental Risk Audits

Assemble a multidisciplinary team to periodically assess environmental risks. There are audit tools available such as the United States Department of Veteran Affairs National Center for Patient Safety's "Mental Health Environment of Care Checklist" [18]. This checklist was primarily designed to reduce the risk of suicide but is also useful for identifying objects that might be used in aggression toward others.

## Promote Culture of Respect and Sensitivity to Potential Sanist Attitudes

It is a fundamental principle that all persons deserve to be treated with dignity and respect. However, due to many largely unspoken myths about the underlying etiology of mental disability, staff may unwittingly dismiss important warning signs. For example, an increased volume of speech may be perceived as a sign of escalating aggression when in fact the patient is experiencing physical distress and simply lacks the cognition skills to identify and articulate the pain sensation. Beyond this, staff sometimes "blame" behavioral health patients for aggressive actions and feel justified in punishing them by using excessive force in return. This is not meant to minimize the importance of staff safety when it is necessary to resort to self defense. However, no force should be applied to satisfy angry motives or exceed the minimum amount of force required to maintain the safety of all persons in the behavioral health environment.

## *Utilize Safe Behavior Plans*

The use of safe behavior plans presumes that there is mutual respect between patient and staff to be able to honor their agreements. Furthermore, these plans reinforce that the behavioral health patient has choices and is willing to accept the agreed upon consequences if not adhering to the contract. Overall, it is a formidable tool for promoting self-determination, self-esteem, and status as an important decision-maker in treatment.

## Conclusion

The Behavioral Health patient poses unique safety risks as illustrated by the two case studies. The lessons learned from these cases include:

- Complete individualized risk assessments as a basis to formulate a clinical evaluation of potential for harm.
- Make sure all staff have received appropriate competency training.
- Use risk reduction strategies that balance safety concerns and individual liberty rights.
- Foster a culture that centers around respect, communication, and teamwork.
- Promote awareness of the insidious dangers of sanism.

## References

1. Jayaram G, Herzog A. SAFE MD: practical applications and approaches to safe psychiatric practice, in a resource document of the American Psychiatric Association's Committee on patient safety. Arlington, VA: American Psychiatric Association; 2008.
2. Dekker S. The field guide to understanding human error. Burlington, VT: Ashgate; 2011.
3. US DHHS. Recovery and wellness lifestyle: a self-help guide, SAMHSA, editor. Rockville, MD: Center for Mental Health Services; 2002.
4. Commission TJ. Sentinel Events, in CAMH; 2012.
5. CDC. 2009 Causes of death by age group; 2012.
6. Jabbarpour YM, Jayaram G. Suicide risk: navigating the failure modes. Focus. 2011;9:186–93.
7. Scott CL, Resnick PJ. Patient suicide and litigation. In: Simon R, editor. Textbook of suicide assessment and management. Arlington, VA: American Psychiatric Publishing; 2006.
8. Knowles J. Inpatient suicide: identifying vulnerability in the hospital Setting. Psychiatric Times. 2012.
9. Crawford M. Suicide following discharge from inpatient psychiatric care. Adv Psychiatr Treat. 2004;10:434–8.

10. The Joint Commission. Sentinel event data root causes by event type. 2012 [09 Dec 2012]. Available from http://www.jointcommission.org/Sentinel_Event_Statistics/.
11. Owen C, Tarantello C, Jones M, Tennant C. Violence and aggression in psychiatric units. Psychiatr Serv. 1998;49(11):1452–7.
12. Myers H. Hospital fall risk assessment tools: a critique of the literature. Int J Nurs Pract. 2003;9:223–5.
13. Estrin I, Goetz R, Hellerstein DJ, Bennett-Staub A, Seirmarco G. Predicting falls among psychiatric inpatients: a case-control study at a state psychiatric facility. Psychiatr Serv. 2009; 60(9):1245–50.
14. Carney C. Medical comorbidity in women and men with schizophrenia: a population-based controlled study. J Gen Intern Med. 2006;21:1133–7.
15. Preskorn S. Guide to psychiatric drug interactions. Prim Psychiatr. 2010;16(21):45–74.
16. Berding R. Involuntary civil commitment: protecting the public's right to safety or violating an individual's right to liberty? Healthc J Baton Rouge. 2008.
17. Perlin M. The hidden prejudice: mental disability on trial. Washington, DC: American Psychological Association; 2000.
18. VA National Center for Patient Safety. Mental Health Environment of Care Checklist [13 Jul 2013]. Available from http://www.patientsafety.va.gov/SafetyTopics.html#mheocc.

# Chapter 20
# Patient Safety in Outpatient Care

Urmimala Sarkar

*"Freedom is not worth having if it does not connote freedom to err."*

Mohandas K. Gandhi

## Introduction

### Defining Ambulatory Patient Safety

In conceptualizing patient safety in the outpatient setting, we employ the Institute of Medicine's (IOM) definition of patient safety: "the prevention of harm to patients." The IOM further specifies that both errors of commission, such as prescribing a contraindicated medication, and errors of omission, such as failure to perform recommended medication monitoring, can jeopardize patient safety. A unique aspect of outpatient settings is the central role of the patient and caregiver in ensuring safe delivery of care. While most definitions of patient safety do not directly address the patient as an active participant in care, in the outpatient setting, patients' self-management capacities and behaviors are critical for safety [1].

Ambulatory safety encompasses several distinct areas. First, safety risks exist for medication use, both for administration of medications in ambulatory care sites, and for patient/caregiver self-administration of medications at home [2]. Second, the prevalence of missed and delayed diagnosis in ambulatory settings constitutes a

U. Sarkar, M.D., M.P.H. (✉)
Department of Medicine, University of California, San Francisco, San Francisco General Hospital and Trauma Center, 1001 Potrero Avenue, Building 10, 3rd floor, San Francisco, CA 94110, USA
e-mail: usarkar@medsfgh.ucsf.edu

A. Agrawal (ed.), *Patient Safety: A Case-Based Comprehensive Guide*, 311
DOI 10.1007/978-1-4614-7419-7_20, © Springer Science+Business Media New York 2014

critical area of patient safety which is recently gaining attention and study [3]. Third, with increasing numbers of procedures performed in ambulatory settings, examining procedural errors has grown in importance [4]. Finally, because outpatients must actively recognize symptoms and seek care, as well as perform daily health-related activities, these patient roles are critical to safe outpatient care [1].

## Contrasting Acute Care and Ambulatory Settings

The patient safety movement emanated from adverse events in acute care [5], which differs substantially from the community and outpatient settings where the majority of healthcare takes place. In acute care, patients are under close observation and often passively receive care. In ambulatory care, patients must decide when to seek medical care, interact with outpatient health systems, and perform their daily health-related tasks. For those who have multiple chronic diseases, this includes following a disease-specific medication, diet, and exercise regimen. Some also adjust their medications based on various measurements, such as using glucose monitoring to adjust insulin dosing. When patients have difficulty with these self-management activities, they are at risk for adverse events. Moreover, ambulatory practices tend to lack specific organization structures to address quality and safety improvement. In addition, most outpatient practices are not subject to accreditation requirements such as strict staffing ratios and adherence to regulatory standards by organizations such as the Joint Commission [6].

## Epidemiology and Impact of Adverse Events in Ambulatory Care

Adverse events are frequent in ambulatory care. Nationally representative surveys suggest that approximately 4.5 million outpatient visits each year in the USA alone are related to adverse drug events [7]. Moreover, more than 701,547 emergency department visits are attributed to use of medications in home and community each year [8].

The types of errors that predominate in ambulatory care also differ from acute care. Treatment errors predominate in inpatients, whereas diagnostic errors do in outpatients [9]. In one study, about 10 % of preventable outpatient adverse events resulted in serious permanent injury or death [10].

Adverse events lead to significantly increased care utilization and associated healthcare costs. The burden associated with malpractice claims from ambulatory adverse events is also significant [11]. Finally, there are varying estimates of significant patient harm related to ambulatory adverse events. One study estimated that ambulatory care adverse events lead to ~400,000 hospitalizations per year [7] while another representative sample estimated that around 75, 000 hospitalizations per year are due to preventable adverse events that occur in the outpatient setting [10].

# Case Studies

## Case 1: Hyponatremia from Poor Outpatient Care Coordination

### Clinical Summary

*Mr. F. was a 66-year-old man with diabetes, hypertension, and heart failure. He presented to the primary care physician for a 3-month follow-up appointment, having seen his cardiologist and his endocrinologist in the interim since his last appointment. He reported increased fatigue for 1 month.*

*He reported that both the cardiologist and endocrinologist had made changes to his medication regimen, but he did not bring the medicines and could not report the changes. His primary care doctor did not have any documentation from the subspecialist visits.*

*The patient also had not had his electrolytes, BUN, and creatinine checked as ordered by his primary care physician at the prior visit, which were expected to be reviewed at today's visit. His daughter who cares for him stated that his endocrinologist had ordered laboratory tests the prior month, so she thought he did not need any more blood drawn. He reported feeling generally weak and unwell.*

*The primary care physician decided that the subspecialty visit information would be helpful and had his office call their offices to obtain it. The clinical documentation arrived by fax from the endocrinology office, and it was found that the endocrinologist increased the dose of metformin from 500 mg twice daily to 850 mg twice daily. The blood test the patient and daughter referred to was a hemoglobin A1c of 7.4 mg/dl obtained last month. The cardiology office had not yet faxed the last visit note.*

*Upon further history, the patient denied localizing symptoms. On physical examination, his vital signs were normal. In contrast with his usually elevated blood pressure, his blood pressure at this visit was 110/65 with a pulse of 70. A thorough physical examination was unrevealing, but the primary care physician elected to order a stat panel of electrolytes, BUN, and creatinine. While the patients' blood was being analyzed, the primary care office received the cardiology documentation from 2 months ago, which includes a dose escalation in furosemide from 20 mg daily to 40 mg daily. There was no mention of laboratory monitoring following this medication change. Mr. F and his daughter were able to corroborate the addition of a second "water pill."*

*A few hours later, the chemistry panel showed a serum sodium of 125 mg/dl, accounting for Mr. F's symptoms.*

### Root Cause Analysis: Why Did This happen?

Fundamentally, this adverse event stemmed from suboptimal self-management of chronic diseases [1] and, similar to most adverse events, can be attributed to multiple contributing factors [12]. The most important root causes and potential solutions are discussed below.

## Treatment Complexity

Mr F. has multiple comorbid conditions, and as such, multi-morbidity is known to be associated with adverse drug events and poor health outcomes [8]. Evidence suggests that adverse drug events are less related to any particular medication than to the overall number of medications prescribed [13], which implies that complex regimens, as well as "high-risk" medicines, should be considered a safety risk.

## Medication Understanding

Neither Mr. F nor his daughter is able to name his medications. This type of medication confusion is the norm rather than the exception due to the high cognitive demand in managing medications [14]. Literature shows that most patients cannot name all of their medications [15] or report medication changes accurately even immediately following an outpatient visit [16].

Mr. F's medication confusion could be due to limited health literacy that leads to a lack of medication understanding [17] or to visual impairment leading to difficulty reading medication labels. Individuals with limited health literacy and language barriers report greater problems across a range of communication domains, including informed consent, shared decision making, and elicitation of concerns. Mr. F could also have cognitive impairment, a common condition for which there is often a delay in diagnosis, further impairing medication understanding.

## Patient–Physician Communication

Mr. F's medication confusion also stems from inadequate patient–physician communication. The adequacy of communication between patients/caregivers and providers is crucial to patient safety. Suboptimal clinician–patient communication in chronic disease care is a consequence of multiple influences at the practice and system level, including medication labeling procedures and the communication practices of physicians and pharmacists. Most physicians fail to explain the four key aspects of a medication—name, dose, indication, and potential adverse effects—when initiating a new medication in the outpatient setting [18]. Time pressure in the outpatient visit is often cited as a reason for suboptimal medication communication [19].

## Aggressive Treatment Goals

Mr. F's various physicians were likely trying to achieve recommended blood pressure and glucose targets by intensifying his medications. Increased attention to stringent treatment goals may paradoxically lead to adverse events, as has been demonstrated for elders [20]. Aiming aggressively for lower blood glucose or blood pressure in hopes of reducing risk of future complications may increase adverse treatment effects such as symptomatic hypoglycemia or orthostasis, respectively, as reported by the ACCORD and ADVANCE clinical trials [21, 22].

## Symptom Recognition

Mr. F. experienced fatigue for 1 month after initiating a new medication, but did not report his symptoms either to the prescribing physician (cardiologist) or his primary care provider. Recognition of medication-related symptoms is part of self-management. Had Mr. F. reported his symptoms earlier, the medication could have been discontinued sooner without the resulting morbidity.

## Transitions Among Multiple Providers

### Communication

Mr F. sees multiple physicians who all adjust his medications and perform monitoring. It is well documented that transitions between care settings, and between primary care, specialty care, pharmacy, other providers, caregivers, and home care, pose a risk for adverse events [23]. In current outpatient practice, providers often rely on patients to report the outcome of subspecialty visits. However, many chronic disease patients like Mr. F cannot report the result of physician visits to the subsequent physician reliably and accurately. Therefore, Mr. F's primary care provider must rely on documentation from subspecialists, outside his practice, which he did not receive.

This case highlights the risk inherent in the transitions between ambulatory physicians (safety risks in care transition and handoff during inpatient settings is discussed in another chapter). Timely communication among providers is critical for comanagement of chronic conditions [24], and is known to be inadequate [25]. Although electronic health records (EHRs) are expanding rapidly, most ambulatory practices still use paper records and communicate with each other with faxed or mailed information. The lack of information about Mr. F's cardiology and endocrinology visits made it more difficult for his primary care provider to determine the cause of his fatigue, which turned out to be related to hyponatremia from an increased dose of his diuretic medication.

### Medication Monitoring

One would expect that changing the dose of a diuretic medication would require monitoring for symptoms and a blood test to ensure that electrolytes remain within normal limits. The cardiologist did not specifically document that she planned to monitor the patient following his medication change. This omission of medication monitoring is a frequent problem in the ambulatory setting [26]. A related concern is the patient and caregiver's lack of understanding about monitoring. Mr. F's daughter did not take him to have the blood tests ordered by his primary care physician because she assumed that the prior month's blood test ordered by the endocrinologist would be sufficient and would be communicated to the primary care physician. Had Mr. F undergone the blood test as scheduled prior to his primary care visit, he would have been diagnosed earlier.

Shared Physician Responsibility

When multiple providers are involved in a patient's care, it is often unclear which provider assumes responsibility for following up on a problem. It is possible that the cardiologist thought that the primary care provider would be checking the patients' electrolytes and thus decided not to order blood tests following the medication change. There is currently no clear standard about who should follow-up in an area of subspecialty-primary care overlap, and this lack of clarity leads to safety problems [27].

Table 20.1 summarizes the root causes and solutions/best practices applicable to Mr. F's case.

## Case 2: Delayed Diagnosis of Lung Cancer due to Poor Communication and Information Management

### Clinical Summary

*Mrs. J is a 71-year-old woman with hypertension, diabetes, and severe knee osteoarthritis. She lives in a rural area and is cared for by a primary care physician locally. Because her knee pain and immobility is affecting her functioning, she is referred to an orthopedic surgeon at the closest referral center, a teaching hospital about 2 h away from her home. She is deemed to be a candidate for knee replacement surgery and completes a preoperative evaluation at the referral hospital. As part of the evaluation, a chest X-ray is obtained which shows a suspicious mass, and the radiology report recommends a follow-up CT scan of the chest. In the radiology report, a telephone notification to Dr. X, the surgical intern rotating through the orthopedic surgery service, is documented. Mrs. J's surgery is cancelled because of the abnormal finding on chest X-ray. The intern rotates to another service, and the attending orthopedic surgeon is on vacation the following month. The radiology report is never sent to her rural primary care physician. Indeed, the primary care physician does not receive any documentation about the planned knee replacement or its cancellation. Mrs. J. follows up 3 months later with her primary care physician. She explains to her primary care physician that she had a "spot on her chest X-ray," which led to her surgery being cancelled. Her primary care physician contacts the radiologist, obtains the report and discusses the findings, and obtains a chest CT scan. The CT scan confirms the location and suspicious nature of the mass. On biopsy, Mrs. J is found to have a primary lung cancer, which is successfully resected with clear margins.*

**Table 20.1** Case 1—Inadequate medication monitoring: root causes and solutions/best practices

| Root cause | Recommendations |
| --- | --- |
| *Treatment complexity* | • Reconcile all medications at all ambulatory visits<br>• Consider simplifying medication regimen whenever possible |
| *Medication understanding* | • Use the Universal Medication Schedule, a validated template [28] with clear language. For example, use the instruction "take 1 pill in the morning and 1 pill at night," instead of "take one pill twice daily"<br>• Embed medication instructions with simple language as default choices into the electronic prescribing function of the EHR<br>• Provide medication counseling delivered by a pharmacy professional at the time of hospital discharge |
| *Patient–physician communication* | • When prescribing a new medication, ask the patient to "teach-back" to the prescriber the name, dosing, purpose, and potential adverse effects of the new medication |
| *Aggressive treatment goals*<br>*Symptom recognition* | • Tailor treatment targets, such as HbA1c in diabetes, to overall health status and patient preference<br>• Teach patients about potential adverse effects of treatments. For example, "if you feel sweaty, shaky, or lightheaded, your sugar may be too low. Please check it with your glucose meter" |
| *Transitions among multiple providers: communication* | • For subspecialist providers: promptly convey written medical records to the patients' medical home/primary care physician<br>• Use a single pharmacy for each patient so that potential drug interactions can also be assessed there<br>• Consider participating in a health information exchange program or implementing interoperable EHRs to facilitate seamless communication among ambulatory providers |
| *Transitions among multiple providers: medication monitoring* | • Prescribing provider should document the monitoring plan for all medications he/she prescribes |
| *Transitions among multiple providers: shared physician responsibility* | • A physician initiating a diagnostic or therapeutic intervention must assume responsibility for obtaining and acting on results unless another provider is made aware of the pending test and clearly agrees to take responsibility for follow up |

## Root Cause Analysis: Why Did This Happen?

Outpatient Health System Fragmentation

Mrs. J's delayed cancer diagnosis did not cause her harm, making this event a "near-miss," but only because the patient herself reported the test result to her physician. Many patients seek services at referral centers which they cannot obtain locally; however, a thorough documentation of clinical events at the referral center is infrequently sent to the referring primary care physician. Mrs. J's primary care physician does not have access to the sophisticated EHR used at the referral hospital. Although interoperability among EHRs is an explicit health policy goal, in current practice, geographically disparate providers and systems using different EHRs cannot easily share data.

## Patient Awareness of Abnormal Test Result

In this case, besides the patients' own awareness, there was no other mechanism for this critical abnormal test result to reach the primary care physician. Communication of abnormal results to patients is known to be suboptimal; clinically relevant abnormalities often are not conveyed to patients [29].

## Poor Information Availability

In ambulatory care, lack of real-time information is a common problem; it can lead to delays in diagnosis and treatment causing medical errors and poor health outcomes [30]. This is likely exacerbated in systems that are not integrated, although there is little comparative data.

## Gaps in Hospital Documentation

Mrs. J's primary care provider did not receive documentation from the orthopedic surgery service at the referral hospital. This lack of documentation is quite common following hospitalization; a meta-analysis of discharge summary availability revealed that discharge summaries were available to primary care physicians only 51–77 % of the time at 4 weeks following hospitalization [31]. It is important to note that the abnormal chest X-ray may not have come to light even if the documentations were provided as studies of hospital discharge summaries show that tests with either results pending or with clinically significant abnormal results requiring follow-up are often omitted [31]. Similarly, information important to primary care physicians, including medication regimen on discharge, planned outpatient follow-up, and main diagnosis, is often not included in discharge documentation [31]. This can contribute to delays in diagnosis and treatment in the outpatient setting.

## Notification of Abnormal Radiology Results

The radiologist who noted the mass on Mrs. J's chest X-ray performed a "warm handoff" by calling the ordering physician by telephone, and documented that he had done so. There is debate among radiologists about which abnormal findings warrant telephone notification; this disagreement leads to inconsistency about telephone notification among and even within institutions. However, it is generally

**Table 20.2**  Case 2—Inadequate follow-up on chest X-ray: root causes and solutions/best practices

| Root cause | Recommendations/best practices |
|---|---|
| *Outpatient health system fragmentation* | • Initiate standard information sharing such as implementation of regional health information exchange organizations (RHIO) or interoperable EHRs<br>• Use patient navigators to assist patients in working with multiple providers<br>• Use technological tools (such as an internet-based personal health record) to help patients manage their health information |
| *Patient awareness of abnormal test result* *Communication among providers* | • Provide written test result information to patients for all tests, for both abnormal and normal results. The "no news is good news" approach does not support patient safety<br>• For abnormal test results, inform patients of the next steps (such as attending their next visit), and of the responsible provider (either ordering provider or primary care provider)<br>• Primary care providers should counsel patients to request that all providers send records to their medical home<br>• Specialist providers should communicate major changes of plan (like a cancelled surgery) to the medical home in a timely fashion |
| *Gaps in hospital documentation* | • Abnormal test results or tests with pending results should be included in the hospital discharge summary |
| *Medical training and lack of experience* | • Medical trainees (residents and medical students) should have care transitions training and evaluation of the adequacy of their documentation |

agreed that urgent or unexpected findings warrant telephone notification. Another complication of delivering test results is, to whom? In this case, the ordering physician was the on-call orthopedics intern; he is clearly not the correct person to follow-up this abnormal finding. Many radiologists contend that it is the responsibility of the ordering physician to identify and inform the responsible physician, and in the absence of integrated or inter-operable EHRs, a clinician does have to take this responsibility.

## Medical Training and Lack of Experience

The intern who received the chest X-ray results relayed the results to the senior team members, who were sufficiently concerned about the lung mass to cancel the planned knee replacement surgery, but none on the orthopedics team conveyed the results to the primary care physician. The interns' lack of experience may have contributed to this error of omission on his part; however, the issue of shared responsibility may also play a role. The intern may view notification of the primary care physician as a senior resident or attending-level role; the attending physician may have assumed that a trainee sent the documentation to the primary care physician. In any case, the ambiguity about team responsibility increases potential for incomplete follow-up.

Table 20.2 summarizes the root causes and solutions/best practices applicable to Mrs. H's case.

# Discussion

## Chronic Diseases and Safety

Both cases above concern patients with chronic health conditions. Wagner's Chronic Care Model describes the factors needed to achieve optimal chronic disease health outcomes [32]. In Fig. 20.1, we apply this well-established Chronic Care Model to address patient safety issues in ambulatory care. This model addresses *underlying conditions,* which includes the community and health system; *individual context,* which includes communication between all participants in outpatient care, transitions in care, and patients' health status and disease burden; and *behaviors* (of patients and providers). These factors interact over time to affect safety among outpatients with chronic conditions. We believe that high-quality primary care is the cornerstone of patient safety in the outpatient setting, and recommendations below underscore the importance of those with chronic conditions having a longitudinal relationship with a primary care provider.

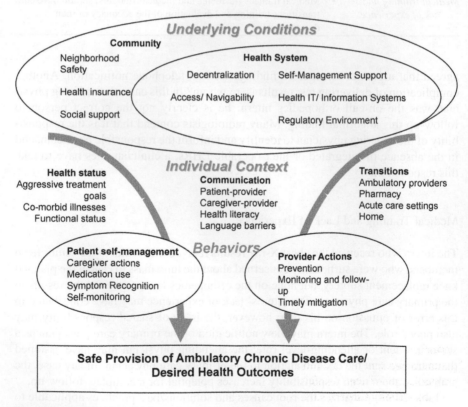

**Fig. 20.1** Ecological model for ambulatory patient safety in chronic disease

## Underlying Conditions: Health System and Community Factors

Although individual clinicians may not be able to address the health system and community factors associated with patient safety problems, an awareness of these issues can identify risky situations, prompt closer oversight, and inform processes of care. Both cases above reveal challenges inherent in the organization of outpatient healthcare systems. Because many ambulatory practices are small, patients often receive care at geographically and organizationally distinct locations: the primary care office, subspecialists, and ancillary services such as pharmacy care. Such complex systems of care can be confusing for patients and caregivers, and they make erroneous assumptions like Mr. F's daughter, about the flow of information among providers. Systems-oriented approaches such as patient navigators could address this complexity.

Lack of integration among outpatient providers and hospitals contributed to both cases, with lack of clinically relevant and timely information as a problem. Missing information contributes to diagnostic and treatment delays [11]. In prior studies, diagnostic delays [11] and lack of real-time information [33] have been shown to contribute to outpatient errors and resulting malpractice claims. Thus, best practices in clinical care include informing primary care providers of significant interventions, such as medications, and of abnormal test results. It is critical, moreover, to inform and educate patients about the need for monitoring and follow-up of abnormal results. The expectations for provision of results to patients vary widely; many patients never receive notification of normal test results. We recommend that all test results, regardless of results, are conveyed in written form to patients in a timely fashion.

Outpatient health systems often lack EHRs and are likely to lag behind acute-care settings even with recent legislation on "meaningful use" of health information technology [34]. Technologies such as computerized physician order entry and computer-based medication monitoring, which are integral to patient safety improvement, remain the exception rather than the rule in outpatient settings. Specific strategies to improve safety using health information technology include (1) requiring providers to acknowledge receipt of patient test results; (2) creating an "audit trail" for patient results; and (3) automating the provision of results to patients.

In the outpatient setting, in-depth investigation of adverse events seldom occurs. Accreditation is a driver for root cause analysis in inpatient settings, and most outpatient physician offices are not accredited by the Joint Commission [4]. In the absence of regulatory scrutiny, the actual prevalence and reporting of adverse events in the outpatient setting remains unclear. We recommend performing rigorous root cause analyses for adverse events in ambulatory care and using the results to implement systems changes.

In rural areas like Mrs. J's home, access to health care and lack of health system capacity remain important issues [35]. Similarly, community-level influences, such as insurance access, neighborhood safety, and social support, can constitute important barriers to provision of safe chronic disease management. Interventions directed at such community barriers, such as transportation assistance for follow-up appointments, may improve care for vulnerable chronic disease patients.

## Individual Context: Communication, Care Transitions, Health Status

In order for outpatient chronic disease care to be safely delivered, patients must be "activated and informed" and providers "prepared and coordinated" as the Chronic Care Model describes.

Patient–provider communication is essential to patient safety for outpatients with chronic diseases because patients and families are performing day-to-day self-management. Abundant evidence exists that patient–provider communication is suboptimal [36]. Many patients, like Mr. F, are unable to read and correctly interpret medication labels [37]. Clinicians often use jargon that is misinterpreted by patients, and there is a striking lack of agreement between patients and providers, even immediately after visits, about symptoms, medication changes, and barriers to self-management [18]. Best practices in communication, such as use of clear communication and techniques such as "teach-back," in which clinicians ask patients to repeat back information in order to confirm their understanding, should be routinely used. Similarly, medication instructions should be specified in plain language, using evidence-based wording such as Universal Medication Schedule [28].

Transitions between care settings, including primary care, specialty care, pharmacy, caregivers, and home care, carry an inherent risk for adverse events. At each point, patients must understand and carry out the plan of care, and providers must make clinical decisions within the limitations of available data. Communication among providers is critical for the provision of safe care in any setting, but in outpatient care, where brief visits are separated by months, such communication is all the more critical. Because most patients encounter disparate healthcare systems, clinicians must proactively communicate with each other, usually by sending clinical documentation via mail or fax. This requires clinicians to actively remember and act to share documentation; we know that, as in Mr. F's case, such documentation may not be sent. Moreover, even when it does occur, sharing of clinical documentation does not constitute a complete handoff between providers. Without the opportunity to ask and answer questions, quality of communication declines. Mechanisms to share and update clinical data among multiple clinicians, via inter-operable EHRs or a personal health record, could improve ambulatory safety by improving communication among clinicians.

Illness burden also plays into risk of adverse events for outpatients. Often patients with multiple chronic illnesses are at risk simply because of frailty, and aggressively treating one condition can worsen another, as when patients with heart failure experience worsening renal function with diuresis. Moreover, with each additional medication, the risk for adverse drug events increases [13]. This underscores the need for medication regimen simplification, whenever possible.

## Behaviors: Patient and Provider Actions

Both patient and provider behaviors, influenced by the context and interactions in care, directly affect patient safety. Ambulatory patients must perform a series of actions for appropriate medication use, including making decisions in an office encounter, obtaining a prescription, bringing the prescription to a pharmacy, receiving the medicines and instructions, taking the medication *correctly* at home on an ongoing basis, monitoring oneself for side effects, and following up with laboratory testing or provider visits. Problems at any of these junctures may lead to adverse drug events.

Mr. F's case illustrates that patient and caregiver errors can lead to harm, as Mr. F did not complete the requested blood tests, and he also did not recognize that his symptom of severe fatigue was related to a newly prescribed medication. Although it is not possible to avoid all adverse drug events, there are medications that are known to cause many adverse drug events, including insulin [13], warfarin [14], and others with known serious adverse effects, such as methotrexate and amiodarone. For these medications, symptom recognition is a crucial aspect of self-management, and appropriate communication must be the standard of care. In addition, medication management is only one aspect of patient self-management, which also includes appropriate diet and exercise, appointment adherence, and recognition of symptoms. Because appropriate patient behaviors are needed to ensure outpatient safety, we recommend provision of self-management support to foster safety, particularly for chronic disease populations.

## Conclusion and Key Lessons Learned

- Patients and caregivers are critical patient safety champions in the outpatient setting.
- Promoting effective patient–provider communication is critical to improving outpatient safety.
- "Warm" handoffs (interactive communication) among outpatient care providers can prevent adverse events.
- Management of abnormal test results constitutes an important aspect of patient safety.
- The implementation of interoperable EHRs that enable seamless sharing of information among providers and personal health records (PHRs) that enable information sharing between providers and patients present important opportunities to improve safety through technology.

## References

1. Sarkar U, Wachter RM, Schroeder SA, et al. Refocusing the lens: patient safety in ambulatory chronic disease care. Jt Comm J Qual Patient Saf. 2009;35(7):377–83. 341.
2. Lorincz C, Drazen E, Sokol P, et al. Research in ambulatory patient safety 2000–2010: a 10-year review. Chicago, IL: American Medical Association; 2011.
3. Schiff GD, Klass D, Peterson J, et al. Linking laboratory and pharmacy: opportunities for reducing errors and improving care. Arch Intern Med. 2003;163(8):893–900.
4. The Joint Commission. http://www.jointcommission.org/. Accessed 13 Jul 2013.

5. Institute of Medicine. To err is human: building a safer health system. Washington, DC: National Academy Press, Institute of Medicine Committee on Quality of Health Care in America; 2000.
6. Wachter RM. Is ambulatory patient safety just like hospital safety, only without the "stat"? Ann Intern Med. 2006;145(7):547–9.
7. Sarkar U, Lopez A, Maselli JH, et al. Adverse drug events in U.S. adult ambulatory medical care. Health Serv Res. 2011;46(5):1517–33.
8. Budnitz DS, Pollock DA, Weidenbach KN, et al. National surveillance of emergency department visits for outpatient adverse drug events. J Am Med Assoc. 2006;296(15):1858–66.
9. Gandhi TK, Lee TH. Patient safety beyond the hospital. N Engl J Med. 2010;363(11):1001–3.
10. Woods DM, Thomas EJ, Holl JL, et al. Ambulatory care adverse events and preventable adverse events leading to a hospital admission. Qual Saf Health Care. 2007;16(2):127–31.
11. Gandhi TK, Kachalia A, Thomas EJ, et al. Missed and delayed diagnoses in the ambulatory setting: a study of closed malpractice claims. Ann Intern Med. 2006;145(7):488–96.
12. Sarkar U, Handley MA, Gupta R, et al. What happens between visits? Adverse and potential adverse events among a low-income, urban, ambulatory population with diabetes. Qual Saf Health Care. 2010;19(3):223–8.
13. Budnitz DS, Shehab N, Kegler SR, et al. Medication use leading to emergency department visits for adverse drug events in older adults. Ann Intern Med. 2007;147(11):755–65.
14. Schillinger D, Wang F, Rodriguez M, et al. The importance of establishing regimen concordance in preventing medication errors in anticoagulant care. J Health Commun. 2006;11(6):555–67.
15. Schillinger D, Piette J, Grumbach K, et al. Closing the loop: physician communication with diabetic patients who have low health literacy. Arch Intern Med. 2003;163(1):83–90.
16. Sarkar U, Schillinger D, Bibbins-Domingo K, et al. Patient-physicians' information exchange in outpatient cardiac care: time for a heart to heart? Patient Educ Couns. 2010;85(2):173–9.
17. Schillinger D, Barton LR, Karter AJ, et al. Does literacy mediate the relationship between education and health outcomes? A study of a low-income population with diabetes. Public Health Rep. 2006;121(3):245–54.
18. Sarkar U, Piette JD, Gonzales R, et al. Preferences for self-management support: findings from a survey of diabetes patients in safety-net health systems. Patient Educ Couns. 2008;70(1):102–10.
19. Peyton L, Ramser K, Hamann G, et al. Evaluation of medication reconciliation in an ambulatory setting before and after pharmacist intervention. J Am Pharm Assoc. 2010;50(4):490–5.
20. Laiteerapong N, Karter AJ, Liu JY, et al. Correlates of quality-of-life in older adults with diabetes: the diabetes & aging study. Diabetes Care. 2011;34(8):1749–53.
21. Action to Control Cardiovascular Risk in Diabetes Study Group, Gerstein HC, Miller ME, et al. Effects of intensive glucose lowering in type 2 diabetes. N Engl J Med. 2008;358(24):2545–59.
22. Patel A, MacMahon S, Chalmers J, et al. Intensive blood glucose control and vascular outcomes in patients with type 2 diabetes. N Engl J Med. 2008;358(24):2560–72.
23. Forster AJ, Clark HD, Menard A, et al. Adverse events among medical patients after discharge from hospital. CMAJ. 2004;170(3):345–9.
24. American College of Physicians. The Patient-Centered Medical Home Neighbor: The Interface of the Patient-Centered Medical Home with Specialty/Subspecialty Practices. Philadelphia: American College of Physicians; 2010: Policy Paper.
25. Singh H, Esquivel A, Sittig DF, et al. Follow-up actions on electronic referral communication in a multispecialty outpatient setting. J Gen Intern Med. 2011;26(1):64–9.
26. Selby JV, Ray GT, Zhang D, et al. Excess costs of medical care for patients with diabetes in a managed care population. Diabetes Care. 1997;20(9):1396–402.
27. Kripalani S, Robertson R, Love-Ghaffari MH, et al. Development of an illustrated medication schedule as a low-literacy patient education tool. Patient Educ Couns. 2007;66(3):368–77.
28. Wolf MS, Davis TC, Curtis LM, et al. Effect of standardized, patient-centered label instructions to improve comprehension of prescription drug use. Med Care. 2011;49(1):96–100.

29. Casalino LP, Dunham D, Chin MH, et al. Frequency of failure to inform patients of clinically significant outpatient test results. Arch Intern Med. 2009;169(12):1123–9.
30. Gandhi TK, Sittig DF, Franklin M, et al. Communication breakdown in the outpatient referral process. J Gen Intern Med. 2000;15(9):626–31.
31. Kripalani S, LeFevre F, Phillips CO, et al. Deficits in communication and information transfer between hospital-based and primary care physicians: implications for patient safety and continuity of care. JAMA. 2007;297(8):831–41.
32. Wagner EH "Chronic disease management: what will it take to improve care for chronic illness?" Eff Clin Pract. 1998;1(1):2–4.
33. Poon EG, Gandhi TK, Sequist TD, et al. "I wish I had seen this test result earlier!": dissatisfaction with test result management systems in primary care. Arch Intern Med. 2004;164(20):2223–8.
34. Blumenthal D, Tavenner M. The "meaningful use" regulation for electronic health records. N Engl J Med. 2010;363(6):501–4.
35. Arcury TA, Preisser JS, Gesler WM, et al. Access to transportation and health care utilization in a rural region. J Rural Health. 2005;21(1):31–8.
36. Schillinger D, Machtinger EL, Wang F, et al. Language, literacy, and communication regarding medication in an anticoagulation clinic: a comparison of verbal vs. visual assessment. J Health Commun. 2006;11(7):651–64.
37. Davis TC, Wolf MS, Bass PF, et al. Literacy and misunderstanding prescription drug labels. Ann Intern Med. 2006;145(12):887–94.

# Part IV
# Organizational Issues

# Chapter 21
# Error Disclosure

Bryan A. Liang and Kimberly M. Lovett

> *"For, confession of error is like a broom which sweeps away*
> *dirt and leaves the surface brighter and cleaner than before,*
> *I feel stronger for confession."*
>
> Mohandas K. Gandhi

## Case Studies

### Case 1: Patient Death Due to Medication Error

*Patient SR is undergoing herniated disk repair under general anesthesia. A new anesthesiology resident is assigned the case without orientation training. During surgery, SR receives phenylephrine, which induces severe hypertension, a known side effect of the drug. While responding to the event, the resident makes a syringe-swap error due to similarly colored vials and accidentally gives labetalol, inducing an intraoperative cardiac event. SR decompensates and dies.*

B.A. Liang, M.D., Ph.D., J.D. (✉)
Institute of Health Law Studies, California Western School of Law, San Diego, CA, USA

Department of Anesthesiology, San Diego Center for Patient Safety, University of California, San Diego School of Medicine, 350 Cedar Street, San Diego, CA 92101, USA
e-mail: baliang@alum.mit.edu

K.M. Lovett, M.D.
Department of Family Medicine, Southern California Permanente Medical Group,
1630 East Main Street, El Cajon, CA 92012, USA
e-mail: klovett@ucsd.edu

A. Agrawal (ed.), *Patient Safety: A Case-Based Comprehensive Guide*, 329
DOI 10.1007/978-1-4614-7419-7_21, © Springer Science+Business Media New York 2014

## Case 2: Patient Harm Due to Misdiagnosis

*AR, a 62-year-old female florist, experiences conjunctival injection, pain, and itching in right eye on night of Christmas Eve. AR goes to the Emergency Department (ED) for care. During questioning with the intern, AR's job as a florist is neither disclosed nor elucidated. The intern diagnoses an unspecified bacterial disorder, provides topical antibiotic, and sends AR home. AR uses the prescribed antibiotic for 24 h, subsequently becomes febrile, and is later found unconscious by her family. The family rushes AR to ED, where she is found to be septic, and it is determined that her eye must be removed.*

## Introduction

Following the release of the Institute of Medicine report, *To Err Is Human* [1], the systems nature of healthcare delivery has become broadly recognized. Important efforts have emerged to address clinical systems of care through assessment of process and outcomes measures, as well as analytics relying on tools from aviation, engineering, and systems sciences [2, 3]. This process of quality measurement and error investigation facilitates quality and safety improvement, root cause analysis for medical errors, as well as prospective efforts in designing systems that are more resistant to the inevitable occurrence of human error. Recent healthcare reform efforts have also supported quality and safety endeavors, although there have been some frustrations in the pace and effectiveness of these safety and quality improvement efforts [4].

Despite these important efforts, attention to disclosure of medical errors has been lacking in systemic strategies to improve patient safety. Although there have been significant clinical efforts, research, and interventions involving patient injury avoidance and an acceptance of ethical mandates to disclose [5–12], relatively less attention has been focused upon system-based disclosure and approaches to addressing patient needs when error occurs [13–17]. Further, the potential for using error disclosure as a learning and culture tool has also been limited. In this chapter, we provide a systems-based approach to medical error disclosure (in contrast to traditional ad hoc, legally oriented means) and illustrate how the flawed legacy of error disclosure may be accounted for in system activities to teach lessons and facilitate community and facility cultural change.

## Error Disclosure Systems

### Traditional Medical Error Disclosure

When an error results from system failure, the system is accountable to those people impacted by the failure. Consistent with a philosophy of mutual respect, trust, responsibility, and partnership between patients and providers, system errors must be disclosed to those adversely affected by them, particularly the patient.

One central tenet of medicine is avoidance of harm to patients; we can achieve this goal through blocking inevitable error and avoiding preventable error. However, error *does* occur throughout the medical system, and patients do get injured. Hence, any disclosure process should address both patient and system needs as well as inform future system activities. Using the disclosure process not just to ethically inform patients but also to enhance medical system safety is our duty. We are bound to integrate learning from systems errors and responsible actions toward patients—key members of the healthcare team who represent the last barrier to harm [18].

This strategy is in stark contrast to the traditional "deny and defend, shut up and fight" model of handling medical error. Traditionally and lingering throughout contemporary medical culture, when patient harm results from medical error, both the administration and providers tend to avoid communication with the patient or family in anticipation of litigation; the error and injury are treated as a "risk management" concern. Under this traditional model, "risk management" refers to managing risk of loss to the medical facility/providers rather than managing risk of injury to patients within the medical facility [19]. This is clearly in conflict with ethical and fiduciary duties of medical providers/facilities to patients.

Further, the traditional model of dealing with medical error tends to focus responsibility for error on individual providers, usually the last provider to touch the patient. This provider-based focus limits the opportunity for system-based improvements or change [20]. And, although a focus on individuals is sometimes appropriate, an approach of individual "shame and blame" is never appropriate. Focusing on the roles and perspectives of individuals contributing to medical error should instead be approached as an opportunity for system improvement in an environment of team cooperation [18]. Indeed, mandating the last person who touched the patient to assume a shame and blame persona of humiliation to disclose only supports traditional, ineffective reactions antithetical to promoting system improvement and patient safety.

An alternative, more progressive model is an open system of medical error disclosure. In an open model, medical error is disclosed using a systematic process to promote communication with the patient and/or family as well as a transparent culture [18]. An open model focuses on the needs of the patient and family and gathers information to promote system learning with the goal of improving operations and outcomes. This is an important approach, since perceptions of adequate disclosure by clinicians are much more circumscribed and limited than that desired by patients [21–23]. In addition, there is limited empirical guidance on these processes [24]. Although there is competing evidence as to the potential reduction or increase in litigation costs [5, 25, 26], open models of disclosure can benefit system processes and safety. We describe one open model variant below.

## An Open System of Medical Error Disclosure

At baseline under this open system, all healthcare entities should have *error disclosure teams*. The error disclosure team would be charged with disclosing errors, addressing patient-focused needs, and identifying patient-centered systems lessons for improvement of future care. Creation of a standing error disclosure team should represent a central aspect of the quality and safety strategy for a healthcare entity as outlined in its policies and procedures. In addition, consistent with an open system of medical error disclosure, all healthcare entities should form a parallel *error investigation team* charged with performing root cause analysis for system improvement by tracing pathways of care. This two-pronged strategy permits a shift of ethical and cultural assessment from individuals to a systems-based focus that includes error disclosure as well as error investigation [11].

Additionally, both the error investigation team and the error disclosure team must be trained. It is important to understand that contemporary medical training does not automatically amount to being well versed in effective medical error disclosure. Members of error disclosure teams must have legitimate and recurrent training in delivering bad news [27–33]. Further, error disclosure practice should be a mandated part of medical training curriculum, as error disclosure represents one of the most difficult forms of communication in a provider's career. Error disclosure curricula should employ both low and high fidelity simulations. These training programs should include computer simulations, video observation of error disclosure, and participation in live patient-actor simulations [34, 35]. Indeed, the first time a provider discloses a medical error should not be the first time he or she is involved in a patient safety incident. Some principles on disclosure discussions are noted in Table 21.1.

Further, a disclosure record should be created and maintained by the error disclosure team for each and every instance of medical error disclosure. The error disclosure record should include: an objective description of the error (when, where, who, and

**Table 21.1** Principles of error disclosure

Before meeting with patient/family
- Know all facts up to the point of disclosure
- Assemble error disclosure team (including an identified person that has a trust relationship with the patient/family, if available)
- Identify information and materials that will be presented to patient/family and who will communicate each concern
- Identify who will answer specific questions
- Identify support information to be provided to patients/families (e.g., hotels in area; telephone/internet/communication needs)
- Ensure a private setting is identified and secured for discussion with no interruption

Timing of disclosure
- Serious errors should be disclosed as soon as possible
- Errors without full explanations/analysis at the time of disclosure should be provided with the message that updates will be provided at a later time

Content of disclosure
- Disclosure should focus upon the patient's needs, including issues such as the patient's clinical condition, concerns, and treatment plan (both for the underlying condition and, if applicable, remediation from the error)
- Provide information on specific steps being taken to trace the pathway of care and available data on the "what, how, and why" as available
- Ensure that the "Three C's" are always communicated (Concern, Commitment, and Compassion)

what), all actions preceding and resulting from the error, a full description of all communications regarding the error between facility employees, and a full description of all subsequent contacts between members of the error disclosure team and the family/patient. Personal observations may be included, but objective and descriptive language should be the standard without any accusations or attributions of fault or blame therein. Importantly, to control use and access to this information, states may consider these records protected from discovery under peer review/quality assurance privilege. Further, under the federal Patient Safety and Quality Improvement Act, the information would likely be considered part of a patient safety evaluation system. If the entity works with a Patient Safety Organization, then the information itself is deemed protected from use in lawsuits [36].

## Team and Technique

The error disclosure team members should include prominent members of the medical care delivery team and administration. This would include relevant specialty physicians (e.g., if the injury involved anesthesia errors, an anesthesiologist should be included), senior facility administrators, and a patient/family liaison. We believe that the provider(s) most directly involved in the error and injury should *not* be a part of the initial error disclosure; the emotional turmoil associated with firsthand proximity to a medical error can potentially impair one's effectiveness in error disclosure [18, 36–40]. Indeed, other systems, such as the Veterans Affairs hospital in Lexington, Kentucky also do not have the involved provider(s) at the initial disclosure meeting [41]. However, the errant provider(s) should be involved in the error investigation, both for system informational reasons as well as to address the personal impact of errors on the provider(s).

The error disclosure team should request an early intervention mediation meeting with the family and/or patient as expediently as possible once an error is recognized. Early intervention mediation should be the standard approach when disclosing an error to the patient and/or family. In this process, a neutral third party proactively assists each stakeholder in assessing and creating potential resolutions for conflicts [42].

Throughout the mediation process, all provided information and explanations should be objective, descriptive, and devoid of finger-pointing or blame. We believe that the senior healthcare provider or medical staff leader and facility administrator should lead the disclosure effort in most circumstances; these individuals are collectively aware of both clinical and administrative ramifications of the error as well as the administrative resources to address it. The healthcare provider is necessarily focused on clinical details related to the error while both the provider and administrator can provide information and explanation of steps that are being taken to address the issue, including medical error analysis, system assessments, and root cause analysis.

Further, the "Three C's" should be always remembered and employed throughout mediation and all error-related communications. The "Three C's" are: Concern, Commitment, and Compassion [43]. Since communication style in error disclosure is critical in determining how it will be received, the need for appropriate communication training using the Three C's is paramount. Promoting effective communication involves active and empathetic listening, avoidance of defensive reactions, assuming a culturally and gender competent approach, and keeping the Three C's at the forefront.

Mediation provides many advantages to the medical system and personnel over the traditional litigation course. First, mediation allows for more open, stakeholder-driven (rather than lawyer-driven) dynamic discussions focused upon identifying interests and goals to be reached rather than the alternative threats and posturing [44, 45]. Second, the transparent and effective communication involved throughout the mediation process can prevent the well-known reaction of patients and families of turning to litigation as a response to poor provider communication. Third, mediation can mitigate the "shame and blame" approach that litigation encourages and potentially even lead to greater healthcare system quality and safety [43].

The mediation process also provides several advantages to patients over the tradition course of litigation. First, it allows the patient and family to vent and express emotion, acknowledges their suffering, allows them to tell their story, and provides a patient-centered explanation of the event. Second, mediation provides the ability for the patient/family to participate in the safety effort, which provides catharsis to patients/families while also providing error investigation and disclosure teams insights as to system safety weaknesses. Indeed, the patient and family witness virtually the entire spectrum of care, whereas each healthcare provider generally is only narrowly focused on respective clinical responsibilities. Finally, both patients and providers have reported satisfaction with mediation processes [46].

The final essential component of this error disclosure system is the patient/family liaison—the link between the patient/family and the healthcare system. This liaison is critical and serves as the primary contact and "face" of the healthcare entity during the error disclosure effort. The liaison should report to the patient and family regularly regarding the progress of the error investigation team (e.g., every 72 h). The liaison contact should be consistent, on schedule, and reliable—even if to only report that the team is still working on the assessment.

## Use of Apology

Early intervention mediation provides the opportunity for system representatives to apologize to the patient and family for the event. Apology is not always synonymous with admission of wrongdoing. In general, sincerely expressed team-based apologies for a family or patient loss are appropriate, such as "We are so sorry you are going through this traumatic event," which is in contrast to the incorrect individually based apologetic admission such as "I'm so sorry I made the mistake that injured you." This contrasts with traditional clinical perceptions that apology is and should be conflated with an acknowledgment of responsibility as "an offender," focused on the individual and generally ignoring the system [47, 48].

From a legal standpoint, there is a tradition of concern regarding the use of apology [49]. Although some states exclude apologies and "expressions of regret" from use in court, it should be noted that some courts may consider these statements an admission of liability [50]. Consequently, state statutes may and can permit apology to be used in some settings, but specific jurisdictions will have nuances that may require crafting apologies so as to avoid unintended legal consequences [51].

Hence, it is important for error disclosure teams and individual providers to consult with legal counsel before offering an apology. However, regardless of whether or not apology is employed, the Three C's should always be used in communications with patients.

## Application 1: Traditional Disclosure

In Case 1, the hospital employed the "deny and defend, shut up and fight" model against the patient. After the syringe-swap error, the hospital's general counsel lawyer alone went to speak with the patient's family. He indicated that there was a "problem" with the surgery and that the patient died. In parallel, the hospital risk management representative, guided by the general counsel, told the surgeon, anesthesiologist, resident, and all operating room personnel not to speak or communicate with the family. Further, the general counsel indicated to these providers that all statements regarding the case must be cleared through him.

The patient's family reacted by requesting information on the death. The hospital representative and providers avoided the family and provided little information, although they believe they were "polite" when doing so.

In response, the family hired an attorney, and told lawyer that they want him to "find out what happened." Hence, the lawyer filed suit against the hospital, surgeon, anesthesiologist, and resident.

The result for each stakeholder was the following:

- The anesthesiologist settled the lawsuit with the family independently, in part by saying he would "make the resident pay for her error by making sure she loses her license," and would "testify against the hospital for its incompetence."
- The anesthesiology resident, distraught, quit her residency.
- The hospital settled with the family for an undisclosed sum.
- The surgeon did not settle, litigated, and won—after spending roughly ~$300,000 in legal fees.
- The family endured 4 years of painful and emotional litigation, dislikes the providers, and appeared on local media to tell their story.

Beyond individual costs, the "deny and defend, shut up and fight" model also has similarly suboptimal outcomes for the delivery system. There are no assessments of system weaknesses; there are no changes in new resident orientation (or other staff) to prevent a similar outcome; and there are no changes to ad hoc disclosure policies. Indeed, there is no discussion, identification, or improvement in the storage of syringes or education on syringe swap errors to prevent similar outcomes. Further, the patient and family—key members of the healthcare team—never receive a full explanation of the event and are hence unable to achieve full closure or understanding. Finally, from a societal standpoint, loss of a resident from the training program is costly both financially—a societal investment of hundreds of thousands of dollars for training a physician—and systematically—due to increased burdens upon remaining staff leading to increased systematic vulnerabilities and threats to patient safety.

Hence, the traditional disclosure approach creates tremendous downside risk to all stakeholders: patients, providers, and the delivery system. Under these circumstances, nothing evolves to improve safety or quality, and useful learning is negated.

## Application 2: Open System of Disclosure

In Case 2, the use of a system of medical error disclosure is employed. After the emergency surgical intervention of removing the patient's eye is completed, a hospital disclosure team of the general counsel, internal medicine physician, ophthalmologist, and patient/family liaison call family members for a mediation session regarding the event. Each member of the team is trained in medical error disclosure, difficult news delivery, and use of the Three C's.

The team discloses to the family that the patient was septic, and as a result lost her eye. They also indicate that a medical error was potentially associated with the event. The error disclosure team communicates to the family that the error investigation team is evaluating the event as they speak and will not stop until they have an explanation.

The team indicates that the patient/family liaison will contact them every 72 h for updates on the event investigation. The team also communicates that questions may be asked at any time, and the patient/family care liaison will have a pager and will be available 24/7.

Additionally, the team asks if there is anything that can be done to further address family/patient needs, including access to phones, places to stay, etc. (any related costs should be covered by the healthcare entity). The team also indicates that the family may wish to consult with a lawyer for protection of their legal rights.

During the mediation, the team encourages the family to vent, express emotion, and empathizes with the family. The team apologizes for the patient's and family's losses (this particular state prohibits apology from being used in court as a liability admission). The team also asks for the family's (and later, the patient's) help for determining systems concerns and weaknesses in care delivery. The family and patient then meet with the team two more times to discuss Emergency Department conditions, residents on call, and questions/methods that might have helped the patient give better information for clinical purposes.

The result for each stakeholder was the following:

- The family settled the conflict in 8 months with the hospital assuming costs of care.
- The family was allowed to participate in hospital policy system reform procedures and therefore felt a commitment from the hospital to create new systems to improve safety.

- The errant intern was able to proactively learn from the error, change practice, and inform colleagues to permit learning from the error and analysis.
- The hospital created a new call policy and new communications training, informally naming the policy after the patient.
- The hospital publicly thanked the family and patient for their help in improving patient safety.
- The patient and family became advocates for the facility.

The system benefits are also significant, including:

- The provider–patient/family relationship is maintained.
- The system of error disclosure is responsive to the needs and provides catharsis for the patient and family.
- Systems issues of on call scheduling and communications methods are discovered, discussed, and now able to be addressed.
- A disclosure record is established for institutional memory.
- Compensation and remedy are relatively fast and allow all parties to move forward.
- Facility costs are likely lower than traditional settlement or full litigation.

## Community and Culture Integration

Medical error disclosure can evolve to become an open, positive, therapeutic systematic process or remain an ad hoc, poorly orchestrated event that serves few, if any, stakeholders. Using a systematic, proactive, transparent approach provides the ability to serve the patient and family's needs, and also provides significant opportunity for system learning that may be integrated into an entity's culture and community characteristics.

As a general matter, creating a system of medical error disclosure would allow healthcare entities to publicize the systems nature of medical care and patient safety focus within their communities. This community involvement would facilitate harnessing of patients and families as partners in patient safety as well as branding its importance. Indeed, entities can use the opportunity to showcase their own transparency and safety systems as competitive characteristics. The result is a broader discussion and consciousness of patient safety systems with an entity's community.

Further, this process can be strategically coordinated with individual provider support and activities. For example, during error disclosure publicizing efforts, providers may begin a process of renewing open communications with patients, emphasizing a culture of transparency, and encouraging mutual provider–patient engagement in the patient safety process. This, in combination with entity efforts to provide information on medical errors and disclosure, will allow a mutually reinforcing patient safety environment to emerge by coordinated efforts of patient-centered activities.

Additionally, establishing a culture of error transparency and system-based patient safety would negate the often perpetuated, destructive misconception that provider perfection is the only method to avoid medical error. It is important to accept that a system run by humans is a system inevitably bound for error; consequently, it is more productive to create a system to catch, mitigate, and prevent errors rather than unrealistically hope that errors will not occur and then respond with the ad hoc "shame and blame" reaction when they do.

In the shorter term, the medical error disclosure system can provide cultural benefits for the entity. By maintaining a disclosure record for review by current and new employees, the entity can emphasize a policy strategy focused on transparency and growth from error disclosure. Indeed, as part of annual reviews and/or retreats, staff can lead discussions on error disclosure events in which they have participated, as well as assist new employees in simulated disclosure events as coaches or simulated patients. Broader dissemination through publication of redacted summaries and lessons learned should also be considered and supported by entities. This would further reinforce a culture engaged in systems improvement in clinical activities as well as in error disclosure.

Finally, it bears noting that engaging in patient-centered systems approaches may result in greater patient satisfaction and experience. This increased patient satisfaction may have significant financial value and ramifications for healthcare facilities: under healthcare reform, up to 30 % of incentive payments for facilities may be based on patient satisfaction scores [4]. Indeed, effective disclosure is associated with higher quality ratings by patients [52].

# Conclusion

Overall, it behooves all stakeholders to shift the patient safety focus toward systematic assessment of medical errors and disclosure of medical errors. Unfortunately, however, systems for medical error disclosure have yet been relatively ignored. By creating and implementing a system of medical error disclosure, benefits to the system, its stakeholders, and its beneficiaries can continually evolve and finally result in an optimal, patient-centered health delivery system.

# References

1. Kohn LT, Corrigan JM, Donaldson MS, editors. To err is human: building a safer health system. Washington, DC: Institute of Medicine, National Academy Press; 1999.
2. Liang BA, Hamman W, Riley W, Beaubien J, Rutherford W. In situ simulation: using aviation principles to identify relevant teamwork and systems issues to promote patient safety. Int J Saf High Consequences Ind. 2011;1(1):53–64.
3. Carayon P, Wood KE. Patient safety – the role of human factors and systems engineering. Stud Health Technol Inform. 2010;153:23–46.
4. Liang BA, Mackey T. Quality and safety in medical care: what does the future hold? Arch Pathol Lab Med. 2011;135:1425–31.

5. Kachalia A, Kaufman SR, Boothman R, Anderson S, Welch K, Saint S, et al. Liability claims and costs before and after implementation of a medical error disclosure program. Ann Intern Med. 2010;153:213–21.
6. Gallagher TH, Studdert D, Levinson W. Disclosing harmful medical errors to patients. N Engl J Med. 2007;356:2713–9.
7. Levinson W, Gallagher TH. Disclosing medical errors to patients: a status report in 2007. CMAJ. 2007;177:265–7.
8. Clinton HR, Obama B. Making patient safety the centerpiece of medical liability reform. N Engl J Med. 2006;354:2205–8.
9. Snyder L, Leffler C, Ethics and Human Rights Committee, American College of Physicians. Ethics manual: fifth edition. Ann Intern Med. 2005;142:560–82.
10. Mazor KM, Simon SR, Yood RA, Martinson BC, Gunter MJ, Reed GW, et al. Health plan members' views on forgiving medical errors. Am J Manag Care. 2005;11:49–52.
11. Liang BA. A policy of system safety: shifting the medical and legal paradigms to effectively address error in medicine. Harvard Health Policy Rev. 2004;5(1):6–13.
12. Blendon RJ, DesRoches CM, Brodie M, Benson JM, Rosen AB, Schneider E, et al. Views of practicing physicians and the public on medical errors. N Engl J Med. 2002;347:1933–40.
13. Loren DJ, Klein EJ, Garbutt J, Krauss MJ, Fraser V, Dunagan WC, et al. Medical error disclosure among pediatricians: choosing carefully what we might say to parents. Arch Pediatr Adolesc Med. 2008;162:922–7.
14. Gallagher TH, Waterman AD, Garbutt JM, Kapp JM, Chan DK, Dunagan WC, et al. US and Canadian physicians' attitudes and experiences regarding disclosing errors to patients. Arch Intern Med. 2006;166:1605–11.
15. Weissman JS, Annas CL, Epstein AM, Schneider EC, Clarridge B, Kirle L, et al. Error reporting and disclosure systems: views from hospital leaders. JAMA. 2005;293:1359–66.
16. Lamb RM, Studdert DM, Bohmer RM, Berwick DM, Brennan TA. Hospital disclosure practices: results of a national survey. Health Aff (Millwood). 2003;22:73–83.
17. Gallagher TH, Waterman AD, Ebers AG, Fraser VJ, Levinson W. Patients' and physicians' attitudes regarding the disclosure of medical errors. JAMA. 2003;289:1001–7.
18. Liang BA. A system of medical error disclosure. Qual Saf Health Care. 2002;11:64–8.
19. Liang BA. Dr. Arthur W. Grayson distinguished lecture in law & medicine: promoting patient safety through reducing medical error: a paradigm of cooperation between patient, physician, and attorney. South Ill Univ Law J. 2000;24:541–68.
20. Liang BA. The adverse event of unaddressed medical error. J Law Med Ethics. 2001;29:346–68.
21. Gallagher TH, Garbutt JM, Waterman AD, et al. Choosing your words carefully: how physicians would disclose harmful medical errors to patients. Arch Intern Med. 2006;166:1585–93.
22. Fein SP, Hilborne LH, Spiritus EM, et al. The many faces of error disclosure: a common set of elements and a definition. J Gen Intern Med. 2007;22:755–61.
23. Espin S, Levinson W, Regehr G, Baker GR, Lingard L. Error or "Act of God"? A study of patients' and operating room team members' perception of error definition, reporting, and disclosure. Surgery. 2006;139(1):6–14.
24. Mazor KM, Simon SR, Gurwitz JH. Communicating with patients about medical errors. Arch Intern Med. 2004;164:1690–7.
25. Pelt JL, Faldmo LP. Physician error and disclosure. Clin Obstet Gynecol. 2008;51(4):700–8.
26. Studdert DM, Mello MM, Gawande AA, Brennan TA, Wang YC. Disclosure of medical injury to patients: an improbable risk management strategy. Health Aff (Milwood). 2007;26(1):215–26.
27. Little P, Everitt H, Williamson I, et al. Observational study of effect of patient centredness and positive approach on outcomes of general practice consultations. BMJ. 2001;323:908–11.
28. Rogers AE, Addington-Hall JM, Abery AJ, et al. Knowledge and communication difficulties for patients with chronic heart failure: qualitative study. BMJ. 2000;321:605–7.
29. Lefevre FV, Waters TM, Budetti PP. A survey of physician training programs in risk management and communication skills for malpractice prevention. J Law Med Ethics. 2000;28:258–65.

30. Allen J, Brock S. Health care communication using personality type: patients are different! London: Routledge; 2000.
31. Edwards A, Elwyn G, Gwyn R. General practice registrar responses to use of different risk communication tools in simulated consultations: a focus group study. BMJ. 1999;319:749–52.
32. Coiera E, Tombs V. Communication behaviours in a hospital setting: an observational study. BMJ. 1998;316:673–6.
33. Buckman R. How to break bad news: a guide for healthcare professionals. Baltimore, MD: Johns Hopkins Press; 1992.
34. Wayman KI, Yaeger KA, Sharek PJ, Trotter S, Wise L, Flora JA, et al. Simulation-based medical error disclosure training for pediatric healthcare professionals. J Healthc Qual. 2007;29(4):12–9.
35. Ziv A, Wolpe PR, Small SD, Glick S. Simulation-based medical education: an ethical imperative. Acad Med. 2003;78(8):783–8.
36. The Patient Safety and Quality Improvement Act of 2005. Pub. L. 109–41; 2005.
37. Wu AW. A major medical error. Am Fam Physician. 2001;63:985–8.
38. Brazeau C. Disclosing the truth about a medical error. Am Fam Physician. 2000;62:315.
39. Christensen JF, Levinson W, Dunn PM. The heart of darkness: the impact of perceived mistakes on physicians. J Gen Intern Med. 1992;7:424–31.
40. Leape LL, Swankin DS, Yessian MR. A conversation on medical injury. Public Health Rep. 1999;114:302–17.
41. Kraman SS, Hamm G. Risk management: extreme honesty may be the best policy. Ann Intern Med. 1999;131:963–7.
42. Liang BA. Alternative dispute resolution. In: Liang BA, editor. Health law & policy. Boston, MA: Butterworth-Heinemann; 2000. p. 257–70.
43. Liang BA. Themes for a system of medical error disclosure: promoting patient safety using a partnership of provider and patient. In: Davies HTO, Tavakoli M, editors. Health care policy, performance, and finance: strategic issues in healthcare management. Aldershot: Ashgate; 2004. p. 92–105.
44. Liang BA. The perils of law and medicine: avoiding litigation to promote patient safety. Prev Law Rep. 2001;19:10–2.
45. Goldberg SB, Sanders FA, Rodgers NH, Cole SR. Dispute resolution: negotiation, mediation, and other processes. New York, NY: Aspen Press; 2007.
46. Dauer EA, Becker DW. Conflict management in managed care. In: Dauer EA, Kovach KK, Liang BA, et al., editors. Health care dispute resolution manual: techniques for avoiding litigation. Gaitherburg, MD: Apsen Publishers; 2000. p. 1:1–1:68.
47. Lazare A. Apology in medical practice. JAMA. 2006;296(11):1401–4.
48. Leape LL. Full disclosure and apology—an idea whose time has come. Physician Exec. 2006;32:16–8.
49. Campaigne C, Costantino J, Guarino G, et al. Evidence; particular types of evidence; admissions and declarations; person making or affected by statement; agents and employees. In: American jurisprudence. Evidence. 2nd ed. Rochester, NY: Lawyers Co-operative Publishing Co; 2000.
50. Liang BA. Error disclosure for quality improvement: authenticating a team of patients and providers to promote patient safety. In: Sharpe VA, editor. Accountability: patient safety and policy reform. Washington, DC: Georgetown University Press; 2004. p. 59–82.
51. O'Rourke PT, Hershey KM. The power of "Sorry": know how state statutes work before apologizing for an error. Hospitalist. 2007;17.
52. Lopez L, Weissman JS, Schneider EC, Weingart SN, Cohen AP, Epsein AM. Disclosure of hospital adverse events and its association with patients' rating of the quality of care. Arch Intern Med. 2009;169(20):1888–94.

# Chapter 22
# The Culture of Safety

Alberta T. Pedroja

> *"Knowledge and error flow from the same mental sources, only success can tell the one from the other."*
>
> Ernst Mach

## Case Studies

### Case 1

*The patient was a 92-year-old male with a previous history of peptic ulcer disease requiring multiple surgeries for internal bleeding. At 8 p.m. on a Friday evening before the start of the Labor Day weekend, the surgeon was performing a procedure to insert a Jackson Pratt drain to remove excess fluids from the body. As per protocol, the staff performed the first of three sponge and instrument counts at the start of the surgery. The second count performed before the closure of the wound indicated a sponge may be missing. The staff looked in the operating room (OR) but did not find the sponge. The surgeon gingerly checked inside the patient but was unable to feel the sponge, so he called for a radiology technologist to take an X-ray. The radiologist saw a foreign object, and since the procedure called for a drain, he erroneously concluded that he was looking at a Penrose Drain when he was actually looking at the missing sponge. He wrote a brief note on the film stating, "No foreign object other than the drain." Though the team was very reluctant to close, keeping the patient under anesthesia any longer was the greater risk, so the surgeon closed the*

A.T. Pedroja, Ph.D. (✉)
ATP Healthcare Services, LLC, 8624 Louise Avenue, Northridge, CA 91325, USA
e-mail: apedroja@atphs.com

A. Agrawal (ed.), *Patient Safety: A Case-Based Comprehensive Guide*,     341
DOI 10.1007/978-1-4614-7419-7_22, © Springer Science+Business Media New York 2014

patient, and completed the surgery, Staff performed the final of three sponge and instrument counts and documented the missing sponge. Next day, the Chief of Surgery reviewed the X-ray and located the sponge that was the source of the confusion. A CT confirmed the location of the sponge. The patient was taken back to OR and the sponge was removed.

## Case 2

A 50-year-old patient arrived in the Emergency Department (ED) via ambulance with a diagnosis of pulmonary embolism. The ED physician ordered intravenous unfractionated heparin that requires weight-based dosing. Since the patient was not ambulatory and unable to step on the scale, the ED nurse estimated the weight to be 80 kg and ordered the heparin dose accordingly. She initiated the heparin per protocol based on the estimated weight of the patient and the patient was transferred to ICU. The patient's actual weight taken in the ICU was 60 kg; however, no one made a correction in the heparin dose being administered to the patient. The lab reported the PTT result, taken 6 h after the loading dose, to be 113.3, well above the normal therapeutic range. The ICU staff recognized the error and adjusted the dose based on the actual weight of the patient. The error was classified as a Class E medication error, i.e., the error reached the patient and required treatment but did not cause permanent harm.

## Introduction

The landmark report from the Institute of Medicine, *To Err is Human* [1], states that evidence-based practices are critical, but the contextual framework in which care is delivered also contributes to patient safety. By 2004, articles describing the Culture of Safety [2–4] concluded that preventing adverse incidents depends as much on cultural changes as on structural changes in healthcare organizations. Evidence-based medicine provides the rules, often in the form of policies and procedures. The culture determines how we behave when the rulebook is gone, a situation that occurs on a regular basis given the exigencies of patient care.

According to James Reason [5], much of the work performed in health care can be categorized into three types: skill-based, rule-based, and knowledge-based. Skill-based work is performed automatically and takes little conscious thought. Taking vital signs is skill-based work. Activities performed infrequently are rule-based, as are complicated processes that need a series of reminders to be sure that every step is performed as expected. On a regular basis, staff follows the rules enumerated via guidelines, protocols, and hospital policies. The protocol for dosing unfractionated heparin in the ED case study above offers an example of rule-based work and the potentially serious consequences that can ensue if the rules are not followed. Knowledge-based work is required in circumstances where the situation is unique and rules do not apply. Professionals draw on previous experience, similar situations, other team members, or the literature in the field to devise a course of action. Case Study 1 is an example of knowledge-based work. The policy for sponge and instrument count did not anticipate a situation in which the sponge count was off and the staff could not account for it in the OR or in the patient via the radiologic image since it mistook the sponge for a drain. The policy could have dictated a response but it was assumed that if the sponge was not in the OR, it was in the patient, and that the radiologic image would be conclusive. An organization committed to patient safety offers skills training to support skills-based work; ready access to the steps in the process to support rule-based work; and a Culture of Safety to encourage staff to make good decisions when the rules no longer apply and they are required to use critical thinking skills to perform knowledge-based work.

The Culture of Safety is defined as "the product of individual and group values, attitudes, perceptions, competencies, and patterns of behavior that determine the commitment to, and the style and proficiency of, an organization's health and safety management [6]." The aviation industry has contributed important ideas to the Culture of Safety in health care because flying a plane is also considered a high-risk, complex endeavor, dependent on human factors and reliable systems. Their investigations established the significance of leadership, teamwork, situational awareness, and safety by design [7, 8]. In health care, as in the aviation, integral to systems designed for safety is the understanding that human error is inevitable, and only through systems that support safe practices will the risk of human error reaching the patient and causing harm be reduced.

Teamwork is the lynchpin of the Culture of Safety. Effective team performance requires team members to cooperate in a shared vision, i.e., patient safety and demands that there is good communication free of the authority gradient [9]. The "Time Out" process before operative and invasive procedures where all members of the team must acknowledge a common understanding of the procedure about to be performed is an example of a teamwork technique borrowed from aviation. In the case study above, the surgical team, deeply affected by the failure of their system to protect the patient from a retained foreign body, instituted a "Count Pause." Now, surgery is halted while the surgical technician performs the instrument count to minimize the risk of error. More importantly, direct physician-to-physician communication is the key. The attending surgeon must directly communicate with the radiologist to make sure they share an understanding of the indication and interpretation of the radiological image findings.

Crew Resource Management (CRM) has been the approach to teamwork in aviation. The Agency for Healthcare Research and Quality (AHRQ) used the CRM principles to develop a program called TeamSTEPPs© that focuses on the knowledge, skills, and attitudes needed for teamwork in health care. The AHRQ Web site offers a review of the literature, a patient safety culture survey, and a variety of other resources to adapt the principles of teamwork into the challenges of clinical practice (http://teamstepps.ahrq.gov/).

Other characteristics of the Culture of Safety have been identified by studying high-reliability organizations (HROs) such as nuclear power generation plants, firefighters, and hostage negotiating teams. Weick and Sutcliffe [10] found that HROs track small failures, resist oversimplification, remain sensitive to operations, maintain capabilities for resilience, and take advantage of shifting locations of expertise. Small failures are treated as symptomatic of larger and potentially more serious problems in the system and hence a timely resolution of small failures can avert adverse safety events. In both case studies above, the patient sustained no permanent harm. However, staff treated each incident as a sentinel event because it was clear that the systems were not fail-safe. Resilience speaks to the ability to change focus and adapt to changing realities. Given the number of specialties involved and the frequently unexpected turns in the patients' conditions, the locus of expertise also often changes from one situation to the next.

James Reason [11] attributes additional characteristics to the Culture of Safety, including a "Reporting Culture" that fosters a nonpunitive environment encouraging incident reporting; a "Just Culture" that assures staff that mistakes will be handled fairly; a "Learning Culture" that encourages everyone to learn from their mistakes and adverse events; and a "Flexible Culture," where staff quickly adapt to changing circumstances.

In the Just Culture model proposed by David Marx [12], individuals have three fundamental duties: the duty to avoid causing unjustified risk or harm, the duty to produce an outcome, and the duty to follow a procedural rule. Against this background, a mistake can be classified into three categories. The first is the human error—inadvertently doing what should not have been done, also referred to as slips and lapses. The second is the at-risk behavior where risk is not recognized or mistakenly believed to be justified. The third is reckless behavior, a choice to consciously disregard a substantial and justifiable risk. The model proposes the following actions: console for human error, coach for at-risk behavior, and punish for recklessness.

## Safeguards: Prevention, Mitigation, Recovery

Strategies to promote the Culture of Safety can be categorized into three phases: prevention, mitigation, and recovery. The prevention phase focuses on proactively anticipating potential risks in the system and correcting them. Mitigation occurs when there are known risks. Finally, when patient harm does occur, recovery includes a series of steps which often result in strategies that prevent or mitigate these risks in the future. Taken together these strategies support the Culture of Safety.

### Prevention

Reliability is the "probability of a product performing a specified function without failure under given conditions for a specified period of time [13]." Reliability is usually reported as a defect rate, e.g., $10^{-1}$, $10^{-2}$, $10^{-3}$, and so forth. $10^{-1}$ is one error in ten tries; $10^{-2}$ is one error in 100 tries; $10^{-3}$ is one error in 1,000 tries, and so on.

**Table 22.1** If 99.9 % were good enough[a]

| IRS lost documents | Two million per year |
|---|---|
| Major plane crashes | Three per day |
| Lost items in the mail | 16,000 per hour |
| ATM errors | 37,000 per hour |
| Pacemakers incorrectly installed | 291 per year |
| Babies given to the wrong parents | 12 per day |
| Erroneous medical procedures | 107 per day |

[a]With permission from the Massachusetts QIO

Table 22.1 provides examples of what would occur if we were content with a $10^{-3}$ defect rate, i.e., 99.9 % accuracy. With so much at stake, healthcare professionals hold themselves to an even higher standard; consequently, "six sigma" or $10^{-6}$ is the goal in many healthcare organizations.

It is estimated that unconstrained human performance guided by discretion only is generally at a reliability level between $10^{-1}$ and $10^{-2}$. Constrained human performance with limits on discretion such as alerts built into the system or forcing functions can reach levels between $10^{-2}$ and $10^{-3}$. Strategies likely to bring clinical practice to a level of $10^{-1}$ reliability include training and awareness, checklists, information/feedback mechanisms on compliance, and standardization of equipment and supplies. $10^{-2}$ strategies necessitate more sophisticated failure prevention such as decision aids and reminders built into the system, defaults to the desired actions, multiple layers of redundancy, habituated patterns, standardization of processes, opt-out vs. opt-in choices and forcing functions [14].

Every time another check or another signature is required, such as with the use of checklists, we are reducing the probability of human error using forced redundancy [15–17]. The use of automation improves those odds further. For example, computer-based physician ordering systems (CPOE) have built-in forcing functions to freeze the order entry screen until medication allergy information is entered and to provide warning alerts and reminders in the case of drug–drug interactions. Forcing functions essentially stop the process from moving forward to prevent a step from occurring thus improving the likelihood that evidence-based practices known to improve outcomes and reduce patient harm will be utilized.

## Mitigation

In the mitigation phase, the Culture of Safety is characterized by teamwork and communication using patient safety as the organizing principle. Well-functioning teams demonstrate a common purpose of safe patient care. The roles of various team members are clear but not overly rigid so that members can easily adapt when needed. Power is decentralized and the autonomy eschewed to prevent error. The importance of teamwork is particularly acute when circumstances deviate from the norm, when the rules are absent and the team must rely on an educated guess. This is also known as "critical thinking." Each member of the team must be free to act or contribute, because sometimes the hierarchy is unwieldy or worse, an impediment.

Good teamwork relies on good communication in order to achieve desired outcomes. Regulatory and accreditation agencies such as The Joint Commission require standardization of communication between providers to ensure that it is comprehensive and complete. SBAR, a commonly used process to standardize communication in health care is an example of a risk mitigation strategy [18].

Another initiative to mitigate risk is the team huddle where staff regularly convenes, typically at the start of the shift, to review risks associated with patient care such as wound care, surgical procedures, restraints, etc. This alerts the staff to watch for problems that may arise over the course of the shift and increases situational awareness. In the OR, a "Time Out" is required by regulation before the start of a procedure to achieve the same effect.

## Recovery

In a Culture of Safety, the recovery phase after a near miss or an actual adverse event is focused on learning from the event. A full investigation that includes individual interviews with staff and a rigorous analysis of the processes associated with the failure is required for all sentinel events, but if the organization is a "fanatic for failure" [10], process analysis is used more widely for near misses as well. The Root Cause Analysis (RCA) is employed for sentinel events and an Intensive Analysis, a streamlined process investigation, is used for any case that did not go as planned even though there was no harm to the patients. An RCA is a systematic review of every structure and process associated with patient care including staffing, communication, leadership, training, information, and the environment to name a few. An intensive analysis will review some, though not all, of the issues specified in an RCA. Intensive analyses vary, but one such process relies on staff preparation of the case including a timeline and a description of the incident including time, date, and patient condition. Then staff reviews selected processes that need a drill-down, such as, the equipment, staffing, education, communication, information, environment of care, or leadership. This information is taken to a weekly risk meeting where the cases are discussed and recommendations made. These may go out to the entire organization if it is seen as a weakness. The RCA or intensive analysis process during the recovery phase provides an opportunity to learn from the potential system vulnerabilities and develop policy and protocols to effectively transition the knowledge-based work into rule-based work.

During the RCA of case study 2, one manager recommended counseling action against the employees who ordered and administered the heparin without getting a weight on the patient. Hearing that there are no gurneys in the ED that have built-in scales, the team recommended the purchase of new equipment. Upon further consideration, this solution, too, was rejected as impractical. There is no guarantee that this gurney would be available when a patient needing to be weighed arrived in the ED which may be why this is not the community standard. The solution devised by the team, therefore, was that heparin dosed in the ED with an estimated weight will include an alert in the system for the unit staff indicating that heparin was dosed with estimated weights. The patient then needs to be weighed immediately upon arrival in the unit and heparin dose must be adjusted accordingly. The policy now includes a procedure in which the pharmacist will adjust the dosing if the estimated weight is more than ten pounds off in either direction.

Thus, the recovery phase often leads to additional steps for prevention and mitigation of risks, completing the cycle. None of this would be possible without a staff willing to report the error in an environment promoting transparency. To foster transparency, institutional leadership must ensure that those reporting adverse events are safe from unfair retribution, that the process for reporting is easy and well understood, and that the process analysis is just. In addition, staff must be confident that the purpose of the discussion is to learn from the experience and not to unjustifiably prosecute those that were involved [11].

Storytelling is also becoming an important part of patient safety armamentarium in the recovery phase. Dennis Quaid [19], Sorrel King [20], Linda Kenney [21], and others have had a national impact telling their stories to large audiences of healthcare workers. At the local level, hospitals across the country are using stories to facilitate the implementation of new patient safety policies and procedures; sometimes patients are also included in the discussions so that they can provide staff with firsthand accounts. The quality reports to the Board that include "Lessons Learned" or "Stories from the Field" provides Board members with a deeper understanding of the complexities associated with delivering safe patient care. One of the six recommendations from the Institute for Healthcare Improvement (IHI) on engaging boards in improving quality and safety includes storytelling [22]. Specifically, they recommend, "Select and review progress toward safer care as the first agenda item at every board meeting, grounded in transparency, and putting a 'human face' on harm data."

## Measuring the Culture of Safety

The truism, "you manage what you measure," prompted AHRQ to sponsor the development of a Culture of Safety survey; the Joint Commission and other regulatory agencies also require that the organizations administer such a survey on a regular basis. The dimensions on the AHRQ survey that can be found on its Web site [23] include leadership, the learning environment, willingness to report, teamwork, and communication to enumerate a few critical ones. The purpose of the survey is to raise staff awareness, assess the current situation of the organization, and support the improvement efforts.

## Barriers to the Culture of Safety

Competing priorities, fragmentation of work among different disciplines, and hierarchical structures are a few of the long-standing challenges for organizations that are striving to create a Culture of Safety. Steep authority gradient is still common in hospital operations that must give way to shared responsibility needed for patient safety. The constant production pressures may lead to greater efficiency but can also create obstacles to the checks and double-checks on high-risk operations by an increasingly busier staff. Finally, the redesign of processes is a costly endeavor and is often undertaken after an adverse event rather than proactively in an effort to design safe systems.

In a Culture of Safety, autonomy and trust in an individual professional is not enough; it must be supplemented by fail-safe processes designed to prevent errors. A double-check when transfusing blood products or administering high-risk medications is not an inefficiency but a precaution that serves to protect patients from harm due to healthcare error.

Other traditional viewpoints have had to change as we have become more sensitized to patient safety. When guidelines and protocols were introduced, they were disparagingly called "cookbook medicine" and were seen as a threat to the autonomy of the clinicians. Now we understand them as important tools to facilitate the implementation of best practices. The acknowledgement of human fallibility still remains problematic in health care. Transparency has had an uphill battle for acceptance. Physicians and staff are well aware of the threat of litigation, and it may seem that to admit wrongdoing is to put themselves and the hospital in financial jeopardy if the patient sues. And, the courts continue to search for someone to blame. It seems counterintuitive to many that disclosure may actually reduce the overall risk of patient dissatisfaction and litigious behavior.

The greatest dilemma facing the Culture of Safety has been the need to balance accountability while promoting a nonpunitive environment that encourages reporting and transparency [24]. Hospital administrators have sought to strike a balance using James Reason's types of work (skills-, rule-, and knowledge-based) in conjunction with Just Culture algorithms to determine appropriateness and type of staff counseling and disciplinary action. Table 22.2 displays one method to determine accountability for human error is by first determining the type of work performed and asking relevant questions. If all questions can be answered in the positive, then the staff is believed to have acted in a responsible manner. If any are answered in the negative, then it is reasonable to hold the staff accountable and offer solutions such as counseling, coaching, or other disciplinary actions.

## Building and Improving the Culture of Safety

Despite these challenges, changes have occurred, some voluntarily and others under duress. The Leapfrog Group [25], IHI's "100,000 Lives Campaign" [26], and the "5 million Lives Campaign" [27] are voluntary initiatives that have affected sweeping

**Table 22.2** Determining accountability for medical error. To determine whether staff should be counseled, review the criteria for each type of work. If all can be answered in the affirmative, staff is not held accountable. If any of the questions is negative, staff is accountable for the error

**Type of work: Skill-based**

**Questions for skill-based work**

  1. Did staff assigned to the task have the appropriate skill?

  2. Was the skill something that could be expected for this job category?

  3. Did the hospital adequately train staff to ensure competencies are present?

  4. Was the activity known to carry risk?

  5. Were safeguards performed properly?

**Example: Staff held accountable**

*The Case*: A nurse was dosing insulin for a diabetic patient. Hospital policy requires a second signature because insulin is considered a high-risk medication. However, the unit was very busy and the nurse was a seasoned professional so she handed the chart to the second nurse who cosigned without checking.

*Analysis*:

  1. Did staff assigned to the task have the appropriate skill? Yes.

  2. Was the skill something that could be expected for this job category? Yes.

  3. Did the hospital adequately train staff to ensure competencies are present? Yes.

  4. Was the activity known to carry risk? Yes.

  5. Were safeguards performed properly? No.

*Result*: Both nurses were counseled.

*Discussion*: "Busy" cannot be an excuse for unsafe care.

**Example: Staff not held accountable**

*The Case*: A patient with blood type AB needed fresh frozen plasma (FFP) at 2 a.m., but the Blood Bank did not have the AB type. The blood bank technician (BBT) removed the informational chart from the wall and erroneously noted that Type A FFP was a clinically appropriate substitution. After discussing with the supervisor, he released the FFP to the clinical area where an astute nurse caught the error and prevented patient harm. The analysis revealed that the BBT had mistakenly read the informational chart for packed cells where Type A is an appropriate substitution.

*Analysis—BBT*:

  1. Did staff assigned to the task have the appropriate skill? Yes.

  2. Was the skill something that could be expected for this job category? Yes.

  3. Did the hospital adequately train staff to ensure competencies are present? Yes.

  4. Was the activity known to carry risk? Yes.

  5. Were safeguards performed properly? Yes.

*Result*: The technical was appraised of the mistake but was not counseled. However, the supervisor was counseled as he failed to double-check the work of the technician.

*Discussion*: The technician committed a slip, but slips are a part of the human condition. Hospital processes include double, triple, and quadruple checks to accommodate this reality.

**Type of work: Rule-based**

**Questions for rule-based work**

  1. Did staff know the rules?

  2. Should staff have known them?

  3. Were the rules available for review if needed?

  4. Was it reasonable to make an exception in this circumstance?

(continued)

**Table 22.2** (continued)

| Type of work: Rule-based |
| --- |

**Example: Staff held accountable**

*The Case*: The surgical checklist includes verifying the presence of a valid history and physical (H&P) performed within 30 days. The nurses were responsible for assuring the completeness of the surgical checklist. The H & P on the chart was 35 days old and the physician had little tolerance for rules he thought were foolish; so, the nurse let the patient go through.

*Analysis*:

1. Did staff know the rules? Yes.
2. Should staff have known them? Yes.
3. Were the rules available for review if needed? Yes.
4. Was it reasonable to make an exception in this circumstance? No.

*Result*: The nurse was counseled and this was included in the physician's Ongoing Professional Practice Evaluation (OPPE).

*Discussion*: Staff knew the rules and the extenuating circumstances were not sufficient for ignoring them. A current H & P is a patient safety concern. If the nurse was uncomfortable, she should have spoken to her supervisor.

**Example: Staff not held accountable**

*Case*: Nurses were asked to provide gentle reminders to physicians to sign their telephone orders within 48 h. One physician did not take kindly to these and let the nurses know it, but the Joint Commission had recently cited the hospital for this offense. When the physician came on the floor, the staff nurse looked for her supervisor but she was not available. So she let him go through.

*Analysis*:

1. Did staff know the rules? Yes.
2. Should staff have known them? Yes.
3. Were the rules available for review if needed? Yes.
4. Was it reasonable to make an exception in this circumstance? Yes.

*Result*: The nurse was not counseled.

*Discussion*: The hospital took the position that it has a responsibility to protect its staff from disruptive physicians. She discussed the situation with the nurse, and the supervisor approached the physician in an alternative venue.

| Type of work: Knowledge-based |
| --- |

**Questions for knowledge-based work**

Given the choices this person made, did s/he show good judgment?

**Example: Staff held accountable**

*The Case*: An ICU nurse floating to the ED had an order for intravenous methylprednisolone. Methylprednisolone was in the ICU smart pump library, but not in the ED library. Hence, she delivered the medication free-flow. The error was discovered when the patient received an overdose.

*Analysis*:

Given the choices this person made, did s/he show good judgment? No.

*Result*: The nurse was counseled.

*Discussion*: Given the risks of the medication, the nurse did not show good judgment protecting the patient from harm because no attempt was made to contact a physician or the supervisor. She was floating from another unit and could be expected to encounter slightly different circumstances which she had the responsibility to check.

**Example: Staff not held accountable**

*The Case*: See Case Study 1. The sponge was inside the patient but they closed anyway.

*Analysis*:

Given the choices this person made, did s/he show good judgment? Yes.

*Result*: Staff was not counseled.

*Discussion*: Staff followed the policy and acted in the best interest of the patient under the circumstances. The risk of prolonged anesthesia was greater than the risk of the sponge. A CT performed the next day provided the location of the retained sponge.

changes. For example, The Leapfrog Group was among the first to recommend the implementation of computerized physician order entry (CPOE) to reducing medication errors. Federal funding is now available for CPOE implementation through incentive payments for the use of certified electronic health records. Rapid Response Teams (RRTs), a voluntary initiative in the 100,000 Lives Campaign, was considered so valuable that it is now incorporated in the Joint Commission's regulations requiring that hospitals recognize and respond to a patient's change in condition using RRTs (Hospital Accreditation Standards, PC.02.01.19).

Regulation has played an important part in promoting a Culture of Safety. The Joint Commission requires a staff climate survey that includes questions on willingness to report errors and other dimensions associated with the Culture of Safety and the leadership standards for accreditation require hospital administration to provide the resources needed for a patient safety program. A number of states have laws that require hospitals to report their serious adverse events and publish their findings on the Web. In 2005, the federal government authorized the creation of Patient Safety Organizations (PSOs) to encourage reporting of adverse events by hospitals without the fear of reprisals. The goal of the PSOs is to improve quality and safety through the collection and analysis of data on adverse events [28].

Leadership engagement has taken a number of forms; one example includes the implementation of executive walkabouts where members of the executive team walk around the units to directly hear patient safety concerns from the staff [29]. Many have embraced transparency and a balanced view of the responsibility of the organization and the individual.

## Conclusion and Lessons Learned

The following are key considerations in building and sustaining an organizational culture that promotes safety:

*Patient Safety as an Organizing Principle*: Given that there are inherent risks in patient care are the processes designed to keep patients free from harm due to medical mistakes? Does staff hold patient safety as an inviolable principle?

*Leadership*: Does the organization commit the resources need to address safety concerns? Do the leaders encourage transparency?

*Teamwork and Communication*: When faced with a problem, does everyone within and between departments step forward to help regardless of the roles and hierarchy? Is everyone free to speak to alert the team about threats to patient safety?

*Transparency*: Is your team willing to report errors without fear of reprisals?

*A Learning Environment*: When an error occurs, does the team come together to understand what happened and how this can be prevented in the future? Can the organization adapt to the changes needed when a risk to patient safety is uncovered?

If your organization has a Culture of Safety, you are likely to find a team willing to work together, to see good communication within and between departments, and to have a robust process for analyzing process; in short, you will have patient safety as an organizing principle pervasive throughout the organization.

# References

1. Kohn LT, Corrigan JM, Donaldson MS. To err is human: building a safer health system, in committee on quality and healthcare in America. Washington, DC: National Academy Press; 2000.
2. Sammer C, Lykens K, Singh K, Mains DA, Lackan NA. What is a patient safety culture? A review of the literature. J Nurs Scholarsh. 2010;42(2):156–65.
3. McCarthy D, Blumenthal D. Stories from the sharp end: case studies in safety improvement. Milbank Q. 2006;84(1):165–200.
4. Walshe K, Shortell SM. When things go wrong: how health care organizations deal with major failures. Health Aff. 2004;23(3):103–11.
5. Reason J. Human error. Cambridge, MA: University Press; 1992.
6. Sorra JS, Nieva VF. Hospital survey on patient safety culture. Rockville, MD: Agency for Healthcare Research and Quality; 2004. Prepared by Westat under Contract No. 290-96-0004. AHRQ Publication No. 04-0041.
7. Denham CR, Sullenberger CB, Quaid DW, Nance JJ. An NTSB for healthcare: learning from innovation: debate and innovate or capitulate. J Patient Saf. 2012;8(1):3–14.
8. Lewis GH, Vaithianathan R, Hockey PM, Hirst G, Bagian JP. Counterheroism, common knowledge, and ergonomics: concepts from aviation that could improve patient safety. Milbank Q. 2011;89(1):4–38.
9. Baker DP, Gustafson S, Beaubien J, et al. Medical teamwork and patient safety: the evidence-based relation. [Internet] Rockville, MD: AHRQ Publication No. 05-0053, April 2005. Available from http://www.ahrq.gov/qual/medteam/. Last accessed 28 Oct 2012.
10. Weick KE, Sutcliffe KM. Managing the unexpected. Resilient performance in an age of uncertainty. San Francisco, CA: Wiley; 2007.
11. Reason J. Managing the risk of organizational accidents. Burlington, VT: Ashgate; 1997.
12. Marx D. Patient safety and the "Just Culture": a primer for health care executives. New York, NY: Columbia University; 2001.
13. Reliability of military electronic equipment, report by the Advisory Group on Reliability of Electronic Equipment (AGREE). Washington, DC: US Government Printing Office; 1957.
14. Amalberti R, Hourlier S. Human error reduction strategies in health care. In: Carayon P, editor. Handbook of human factors and ergonomics in health care and patient safety. Boca Raton, FL: CRC; 2011.
15. Pinsky HM, Taichman RS, Sarment DP. Adaptation of airline crew resource management principles to dentistry. J Am Dent Assoc. 2010;141(8):1010–8.
16. Weiser TG, Haynes AB, Lashoher A, et al. Perspectives in quality: designing the WHO surgical safety checklist. J Qual Health Care. 2010;22(5):365–70.
17. Gawande AA. The checklist manifesto: how to get things done right. New York, NY: Picador; 2009.
18. Heinrichs WM, Bauman E, Dev P. SBAR "flattens the hierarchy" among caregivers. Stud Health Technol Inform. 2012;173:175–82.
19. Quaid D, Thao J, Denham CR. Story power: the secret weapon. J Patient Saf. 2010;6(1): 5–14.
20. What happened. 2007. The Josie King Foundation. Available from http://www.josieking.org/page.cfm?pageID=10. Last accessed 24 Apr 2012.

21. Kenney LK, van Pelk RA. To err is human; the need for trauma support is, too. A story of the power of patient/physician partnership after a sentinel event. Patient Safety and Quality Healthcare: Marietta, GA; 2005. Available from http://www.psqh.com/janfeb05/consumers.html. Last accessed 28 Oct 2012.
22. Conway J. Getting boards on board: engaging governing boards on quality and safety. Jt Comm J Qual Patient Saf. 2008;34(4):214–20.
23. Surveys on patient safety culture. Rockville, MD: Agency for Healthcare Research and Quality; May 2012. Available from http://www.ahrq.gov/qual/patientsafetyculture/index.html. Last accessed 28 Oct 2012.
24. Wachter RM, Pronovost PJ. Balancing "no blame" with accountability in patient safety. N Engl J Med. 2009;361(14):1401–6.
25. The leapfrog group [Internet]. Washington, DC; 2012. Available from http://www.leapfroggroup.org. Last accessed 28 Oct 2012.
26. Berwick DM, Calkins DR, McCannon CJ, Hackbarth AD. The 100,000 lives campaign: setting a goal and a deadline for improving health care quality. JAMA. 2006;295(3):324–7.
27. Protecting 5 million lives from harm [Internet]. Cambridge, MA: Institute for Healthcare Improvement; 2012. Available from http://www.ihi.org/offerings/Initiatives/PastStrategicIniti atives/5MillionLivesCampaign/Pages/default.aspx. Last accessed 28 Oct 2012.
28. Patient safety organization information [Internet]. Rockville, MD: Agency for Healthcare Research and Quality. Available from http://www.pso.ahrq.gov/psos/overview.htm. Last accessed 28 Oct 2012.
29. Thomas EJ, Sexton JB, Neilands TB, Frankel A, Helmreich RL. The effect of executive walk rounds on nurse safety climate attitudes: a randomized trial of clinical units. BMC Health Serv Res. 2005;5(1):4.

# Chapter 23
# Second Victim

Susan D. Scott and Kristin Hahn-Cover

> *"We are all steeped in weaknesses and errors; let us forgive one another's follies, it is the first law of nature."*
>
> Voltaire

## Introduction

The Institute of Medicine report, *To Err is Human*, projected that as many as 44,000–98,000 individuals die in the USA each year as a result of adverse events in hospitals [1]. Serious adverse events not only cause harm to patients and their families but also put well-intending care providers at risk for significant emotional duress. There is now growing recognition that when a serious unanticipated adverse event occurs, while the patient as the recipient of the harm is clearly the "first victim," clinicians often also experience a harsh emotional response in the aftermath and may be described as "second victims" [2]. The second victim phenomenon, described as an emotional aftershock or stress reaction, may take an immense professional and personal toll on involved healthcare providers. At times, this may be a life-altering experience leaving a permanent scar on the clinician's professional identity and psyche [3]. Even in the absence of a mistake in care, clinicians may be affected by their patients' outcomes because of their relationship with a particular

S.D. Scott, R.N., M.S.N. (✉)
Sinclair School of Nursing, Office of Clinical Effectiveness – Patient Safety, University of Missouri Health System, One Hospital Drive – Room 1E29C, Columbia, MO 65203, USA
e-mail: scotts@health.missouri.edu

K. Hahn-Cover, M.D., F.A.C.P.
Department of Internal Medicine, Office of Clinical Effectiveness, University of Missouri Health System, One Hospital Drive, 1W24, Columbia, MO 65212, USA
e-mail: hahncoverk@health.missouri.edu

A. Agrawal (ed.), *Patient Safety: A Case-Based Comprehensive Guide*,
DOI 10.1007/978-1-4614-7419-7_23, © Springer Science+Business Media New York 2014

patient, past clinical experiences, or the similarity of a patient to a member of the clinician's own family [4].

Despite an increasing awareness and understanding of the second victim phenomenon, many healthcare organizations and clinicians are not familiar with the concept. Consequently, most second victims do not receive emotional and social support from their organizations after a momentous unanticipated clinical event [5]. Without appropriate support and guidance, the distress experienced by excellent healthcare providers may lead to long-term consequences such as leaving their chosen fields prematurely or experiencing prolonged professional/personal suffering.

The impact of medical errors on healthcare providers originally appeared in the medical literature as depictions of the personal hardships encountered by individual clinicians [6–9]. Information garnered from personal accounts provided keen insights into the agonizing nature and complexity of clinician experiences in the aftermath of adverse clinical events [10–15]. Clinicians drawn to health care by their aspiration to help others may be traumatized when they are involved in situations that bring harm rather than healing to the patients for whom they care. Adverse patient experiences may trouble even the most self-assured providers [16]. Suffering healthcare providers often hold themselves personally responsible for unexpected outcomes, feeling as if they have failed their patients. Clinicians frequently replay the clinical events in their minds, puzzling whether they could have done something differently to avoid the outcome. The mismatch between their values and motivation to enter their field, and an adverse outcome in which they played a role, leads clinicians to second-guess their career choices. Ultimately, the clinician may question his or her clinical skills, knowledge base, and professional identity [17]. Such suffering is not limited to the time immediately following the event; second victim clinicians may experience this emotional trauma for months, or even years, thereafter [18].

Second victims respond to an adverse clinical event in a variety of ways—emotionally, behaviorally/physically, and cognitively [19, 20]. Although healthcare providers may experience second victim responses even in instances of "near misses" that do not actually harm patients, significant reactions are more likely when responders are involved in a serious error [21]. Second victims often describe feelings of shock, helplessness, worry, depression, guilt, fear, shame, inadequacy, and anger [4, 22]. Somatic symptoms such as fatigue, headache, rapid heart rate, increased blood pressure, muscle tension, and rapid breathing are not uncommon [3, 4]. Second victims also describe difficulty concentrating and may develop sleep disturbances such as insomnia. Reactions can also be intensified in instances where a prior belief in clinician infallibility—a perception that good clinicians will never make mistakes—exists, and when the error is tied more directly to an individual clinician's actions rather than to system failures [23]. Many clinicians describe a common emotion of fear after unanticipated clinical events, particularly fear of lawsuits and loss of reputation due to judgment by professional colleagues [24]. In a nursing survey regarding medication errors, respondents often voiced apprehension over the possibility of punishment, censure, and even job loss [25]. Every second victim study to date reveals that these unanticipated clinical events may have a profound and lasting impact on clinicians.

The prevalence of second victims has been estimated to vary between 10.4 and 43.3% of clinicians [17, 26, 27]. During the past decade, numerous research initiatives have deepened our understanding of the second victim phenomenon and have yielded valuable insights into the definition, identification, experience, and risk factors for a second victim response [25]. In general, responses of clinicians across professions are more similar than different. Differences in responses relate more to individual differences rather than professional differences in coping with the second victim experience.

The second victim experience is not limited to practicing clinicians. Numerous studies focusing on student learners have identified that this group can also be deeply affected by adverse events [19, 28–31]. Perceived personal responsibility coupled with poor patient outcomes is associated with more intense responses and greater personal distress among resident physicians [28]. Schools of health-related professions must be aware of the second victim phenomenon and have action plans to address the unique needs of their learners, to complement the response and support provided by healthcare facilities for second victims.

## Case Studies

### Case Study 1: ED Resident Involved in a Missed Diagnosis of Acute Myocardial Infarction

#### Clinical Summary

*Gary Boyd, a 58-year-old man, presented to the emergency room (ER) with chest discomfort. The resident physician, KP, was initially concerned about angina and ordered an EKG and cardiac enzymes. The cardiac enzymes and EKG were interpreted as normal. Dr. KP discussed the case with the attending physician and discharged Mr. Boyd to home with instructions to return if symptoms worsened. The next day, her chief resident informed her that Mr. Boyd was brought to the hospital after his wife activated EMS when he passed out at home. He was found to have an acute myocardial infarction and was sent directly to the cardiac catheterization lab for intervention in an unstable condition. The chief resident asks why Dr. KP discharged him home with an abnormal EKG. Dr. KP reviews the EKG in Mr. Boyd's electronic record and realizes she had never seen it. She reviewed a normal EKG when she saw him yesterday, but now realizes she had mistakenly reviewed the EKG from a different patient. Dr. KP is devastated.*

#### Discussion

Dr. KP is shocked to realize that she discharged a patient with an evolving acute coronary syndrome. She is sure her chief resident has lost confidence in her and that her fellow residents and the Emergency Department faculty and staff will no longer trust her abilities. She wonders if she will be fired and worries she will be sued. She feels she let Mr. Boyd down and blames herself for his serious decline. She suffers in silence but desperately needs support and guidance. She would benefit from social support from a colleague who understands her experience. The one-on-one support of a colleague in the aftermath of an adverse event is a powerful healing tool and is a valued and desired intervention by many second victims. In addition, she would benefit from comprehensive support from her institution.

## Case Study 2: Unsuccessful Resuscitation of a Colleague's Daughter

### Clinical Summary

*Katie Donnell, an active 8-year old, was involved in a motor vehicle accident and was being transported by ambulance with probable bilateral femur fractures. She was awake and alert at the scene but in severe pain. She asked the paramedics to call her mother, who was at home. After learning of the accident, the mother raced to the level two Trauma Center where she had worked for the past ten years as ER's evening supervisor. The mother arrived prior to the ambulance and was comforted to see a senior orthopedic surgeon awaiting her daughter's arrival. As the ambulance pulled up to the ER entryway, Katie suffered a sudden cardiac arrest. Paramedics initiated chest compressions and continued them as they wheeled Katie into the ER. Katie's mother was grief stricken the moment she saw Erin, a senior paramedic and a good friend, rhythmically performing chest compressions on her precious child. The entire team felt an intense pressure to save this child as her mother—and their colleague— wept uncontrollably in the corner of the trauma bay. Despite exhaustive resuscitative efforts that lasted more than an hour, Katie was pronounced dead. The trauma team had lost their battle and their young patient. But this case hurt more than most because this patient was also the daughter of "one of their own." They had let their colleague and friend down when she needed them the most.*

### Discussion

Although each clinician involved in a patient's care responds uniquely to the event, there are times when entire teams of clinicians are impacted by the case. The death of a pediatric patient almost always elicits greater distress even among the most seasoned clinicians. Intensifying this particular experience was the fact that the team "connected" with Katie as "one of their own" as the daughter of their coworker. In this type of scenario, a group debriefing may be beneficial for the entire team. Group debriefings, facilitated by trained individuals knowledgeable and skillful in group crisis intervention, provide involved clinicians with an opportunity to share their inner most feelings about the case and how the event has impacted them. When focused on the clinicians' emotional response to the event, these debriefings can greatly facilitate second victim recovery.

# Discussion

## *The Six Stages of the Second Victim Phenomenon*

The following description of the six stages is based on a systematic study of the second victim phenomenon at the University of Missouri Health Care System in Columbia, Missouri. Since 2006, we have led the multidisciplinary research team at the University of Missouri to gain insights into the second victim phenomenon [17, 32]. The research team interviewed healthcare workers who had experienced second victim responses as a consequence of adverse patient events. The following types of clinical scenarios were found to be associated with a higher risk for precipitating a second victim response: events which "connect" the patient to the clinician's family, medical error leading to *preventable* harm to a patient, failure to identify patient deterioration, serious outcomes involving a pediatric patient, first experience with the death of a patient, and clinical areas that experience numerous patients with poor outcomes within a short period of time [33].

The research participants' emotionally charged descriptions of their personal suffering revealed a predictable recovery trajectory consisting of six distinct stages (Table 23.1) (1) chaos and accident response, (2) intrusive reflections, (3) restoring personal integrity, (4) enduring the inquisition, (5) obtaining emotional first aid, and (6) moving on [3].

**Table 23.1** Six stages of the second victim recovery trajectory

| | | Stage | Characteristics |
|---|---|---|---|
| **Impact realization** (Can occur individually or concurrently) | 1 | Chaos and accident response | Event realization<br>Patient stabilization<br>A "wave" of emotion |
| | 2 | Intrusive reflections | Haunted re-enactments<br>Self-isolation<br>Internal inadequacy |
| | 3 | Restoring personal integrity | Fear is prevalent<br>Work/Social structure angst |
| | 4 | Enduring the inquisition | Reiterates case scenario<br>Responds to multiple "why's" from numerous employees<br>Sensemaking begins |
| | 5 | Obtaining emotional first aid | "Hinting/Hoping" for support<br>Attempts to seek guidance<br>Receives social support |
| **Moving on** (One of Three chosen) | 6A | Dropping out | Transfers to another unit, department or hospital<br>Considers leaving profession |
| | 6B | Surviving | Coping but doesn't return to pre-event "baseline" of performance |
| | 6C | Thriving | Gains insight/perspective/wisdom from event<br>Learns from event to help others<br>Advocates for patient safety |

Stage 1: Chaos and Accident Response

Clinicians describe initial chaotic and often confusing scenes with intense external and internal turmoil in the first moments after the adverse event or clinical outcome is identified. The clinician begins to grasp the severity of the event that has transpired under his/her watch. This realization is quite harsh. Simultaneously, the patient may be unstable and require intensive monitoring and potentially a higher level of care. Frequently, additional clinicians and resources are called upon to provide support with procedures and/or testing. The clinician may find himself/herself unable to cognitively or physically support the healthcare team's resuscitation or other treatment efforts at his/her usual level of performance because of the extreme internal turmoil.

Stage 2: Intrusive Reflections

The initial chaos is followed by periods of social isolation where the clinician attempts to more fully understand the clinical event. During this stage, the clinician re-evaluates the situation repeatedly with "what if" questions, describing periods of haunted re-enactments, often with feelings of personal and professional inadequacy accompanied by a desire to self-isolate from colleagues and fellow team members. The clinician is frequently distracted and immersed in self-reflection, while also trying to manage a patient in crisis. Some clinicians may benefit from relief of patient care duties for a brief period to help them collect their thoughts prior to resuming patient care.

Stage 3: Restoring Personal Integrity

The third stage consists of seeking insights from an individual with whom the clinician has a trusting relationship such as a colleague, supervisor, personal friend, or family member. Many struggle to identify the person they should turn to because of the belief that no one can relate to their experience or understand the impact the event had on them professionally and personally. A consuming doubt regarding their future professional career plagues many professionals. Some second victims describe an inability to move forward when the event is followed by nonsupportive or negative departmental "grapevine gossip" which triggers additional memories and intensifies the self-doubt, lack of clinical confidence, and worry about being perceived as a "not to be trusted" or "incompetent" clinician. One of the clinician's numerous fears is that he/she will be perceived as a 'weak link' among the healthcare team members. Progressing beyond this stage requires the clinician to decide he/she has the clinical skills and the personal strength to return to their professional duties.

Stage 4: Enduring the Inquisition

After an initial focus on stabilizing the patient and personal reflections, there is a growing awareness that the institution will be reacting to the event in uncertain ways. Specifically, the clinician starts to wonder about repercussions affecting employment, licensure, state board notification, and potential future litigation. There is an intense fear of the unknown in the investigational process. The clinician will frequently meet with unfamiliar institutional leaders who are reviewing the case to better understand what transpired. Not understanding exactly what to expect in the process coupled with interacting with unfamiliar investigators is described as "extremely painful" by the clinicians who are extraordinarily sensitive to the case events.

Stage 5: Obtaining Emotional First Aid

In the fifth stage of recovery, second victims crave emotional support beyond that necessary during the immediate aftermath of the event. However, the busy and hectic clinical environment coupled with a perceived stigma among healthcare clinicians to reach out for assistance usually results in the clinician secretly hoping for someone to approach them about the adverse event. The vast majority of second victims quietly wait and hope that someone will offer support. Many express concerns about not knowing who is a "safe" person in whom to confide. Approximately one-third of the second victims seek support from loved ones but indicate they are cautious when doing so because of privacy and legal considerations. Others note that their loved ones would not be able to comprehend their professional life and should be protected from the unbelievable hurt [3]. A few of our second victim respondents actively sought support on their own accord from coworkers, supervisors, or department chairs. However, the amount and duration of support provided may be insufficient as negative feelings may linger for days, weeks, and even longer.

Supervisory personnel can play a vital role in provision of emotional first aid for the clinician during this stage. However, many clinicians report that they did not receive the support they craved. The second victims describe specific interventions they desired from their supervisors, which included connecting with the involved clinician as soon as possible after the event, determining ways to provide time away from the clinical area (if necessary) so the clinician could compose themselves prior to resuming patient care, voicing the message that they are still a trusted and valued member of the care team, informing the second victim of next steps in the institution's investigations of the case, and continuing to periodically check on them after the event.

Stage 6: Moving on—Dropping Out, Surviving, or Thriving

Although numerous clinicians describe the event as affecting their work practices, some feel the event will stay with them throughout their careers. There is a push internally (from the second victim) and externally (from coworkers, colleagues, supervisors) to "move on" and sequester the event to the past. However, many clinicians find it difficult, if not impossible, to completely put the event behind them.

This is a unique stage for recovery as it has three potential paths: dropping out, surviving, and thriving. The **dropping out** path involves changing the professional role, leaving the profession, or moving to a different practice location. Among those clinicians who are not able to reconcile their involvement in the unanticipated clinical event despite ongoing social support from supervisors and peers, the event may take an extreme toll. Other researchers have identified that involvement in healthcare errors can lead to measureable decreases in quality of life scores and empathy and increased risks for burnout and depression [21, 34].

In the second potential path, **surviving**, the individual performs at expected levels and is "doing okay," but continues to be plagued by the event. These individuals do not reach pre-event baseline performance levels and can be described as "just not the same since their event," or as "just existing."

Fig. 23.1 Elements of an ideal second victim support—What a second victim desires

The last potential path of recovery is **thriving**. In this path, recovering clinicians who found something good from the adverse clinical experience are identified as thriving. Second victims who thrive describe gaining incredible insights and perspectives from the event which improve their current and future practice. Individuals experiencing a thriving outcome frequently report strong social support from numerous individuals and even their institutions.

## Responsibility of Healthcare Institutions Toward Second Victims

Tragically, most clinicians suffering as second victims do not receive adequate support from their colleagues or institutions. Many articles recommend that healthcare organizations assume accountability for social support for clinicians, but there is little published guidance on the design of specific supportive interventions [17, 20, 33, 35–39]. The Institute for Healthcare Improvement's white paper titled *Respectful Management of Serious Clinical Adverse Events* provides one helpful roadmap [38]. Additionally, developing an institutional infrastructure requires insights into what the second victim needs and desires. In our study, research participants were asked to describe 'ideal' support for someone experiencing the second victim phenomenon. The "word cloud" in Fig. 23.1 depicts a summary of their responses. The larger the word, the more often it appeared within the responses. Healthcare facilities designing a support infrastructure would benefit from understanding these keen insights as they design support to address the needs of their clinicians.

We recommend that institutions design a structured response plan that ensures ongoing surveillance for the identification of potential second victims as well as actions to mitigate emotional suffering immediately upon second victim identification. They should develop processes to actively assess staff involved in high risk clinical events in an attempt to intervene quickly and not allow clinicians to suffer in silence. Assistance should be made available to address identified second victim responses of varying severity, ranging from simple one-on-one peer support encounters to prolonged professional counseling support for more severe responses. We believe that most of today's healthcare institutions have internal resources that can be organized in a manner to provide this spectrum of surveillance to support.

## Conclusion

Medical professionals frequently experience a significant emotional toll after unanticipated clinical events, especially those involving medical errors. Feelings of isolation, shame, guilt, anger, loss of empathy, lack of confidence, and depression are all potential responses [22]. Physical responses, such as rapid heart rate, elevated blood pressure, muscle tension, and difficulty in sleeping occur as well. Immediate social support in the aftermath of an unanticipated clinical event is crucial [35–37]. Interventions to help the second victim cope must be offered by colleagues and supervisors immediately after the event as well as periodically long after the event [17].

As the patient safety movement continues to evolve, addressing the professional and personal impact of unanticipated clinical events must be considered as part of an all-inclusive patient safety program. If clinicians feel personally and professionally vulnerable because of a medical error or other adverse event, they may be less likely to report the events which will ultimately undermine any patient safety program.

Institutions can help prevent second victims as well as guide them toward the optimal recovery stage of "thriving," by recognizing the potential impact that negative, unanticipated clinical events can have on clinicians. An understanding and appreciation of the serious implications of the second victim phenomenon provides an opportunity for healthcare institutions to respond in a constructive way, enabling second victims to return to their respective professional roles and not only to survive but also to thrive in the aftermath of their experiences. As new programs are designed, programmatic research evaluations and critiques can guide future team development and ultimately define standards of compassionate support for healthcare workers dealing with the emotional aftermath of unanticipated adverse patient events [40].

## *Key Lessons Learned*

There are numerous challenges in providing clinician support:

- There is a perceived stigma related to a clinician reaching out for help. Many clinicians prefer that someone reach out to them before they ask for help.
- In high acuity clinical areas, there is little protected time for clinicians to comprehend what has transpired under their watch and collect themselves before they move on to the next clinical task/assignment.

- There is an intense fear of the unknown. All clinicians yearn for information about the institution's response to the clinical event and what to expect from the investigation process; medical staff also tend to worry about the litigation process and wonder when the feared subpoena will arrive.
- Clinicians frequently fear damage to professional relationships with colleagues and peers. They worry that they will no longer be a trusted and valued member of the healthcare team.

Second victims each respond in unique ways.

We have found that no two clinicians will respond in the same way, including individuals who are involved in the same clinical event.

The first intervention at any healthcare facility should be an educational campaign regarding the second victim phenomenon.

It is not surprising that most clinicians have not heard of the term "second victim." However, when they hear the description of the second victim phenomenon, most can readily relate to it and recall specific events experienced by themselves or by colleagues. Awareness of the second victim phenomenon helps "normalize" the pain and suffering that is experienced by the clinician and can help move recovery forward.

Healthcare facilities should proactively design a plan of action to provide care for the patient/family as well as the clinician involved in the unanticipated clinical outcome.

There are often many resources available to provide social support/guidance for clinicians suffering as second victims; however, those resources—peers, chaplains, social workers, employee assistance programs, psychologists—may be unknown to the clinical staff. Designing a specific plan with contact information for key supporters is important before a high-risk adverse clinical event occurs.

# References

1. Corrigan JM, Donaldson MS, Kohn LT, et al. To err is human: building a safer health system. Washington, DC: National Academy Press; 2000.
2. Wu A. Medical error: the second victim. BMJ. 2000;320:726–7.
3. Scott SD, Hirschinger LE, Cox KR. Sharing the load: rescuing the healer after trauma. RN. 2008;71:38–43.
4. Scott SD, Hirschinger LE, Cox KR, et al. The natural history of recovery for the health care provider 'second victim' after adverse patient events. Qual Saf Health Care. 2009;18:325–30.
5. Wu AW, Steckelberg RC. Medical error, incident investigation and the second victim: doing better but feeling worse? BMJ Qual Saf. 2012;21(4):267–70.
6. Anonymous. The mistake I will never forget. Nursing. 1990;20:50–1
7. Levinson W, Dunn PM. A piece of my mind: coping with fallibility. JAMA. 1989;261:2252.
8. Hilfiker D. Healing the wounds: a physician looks at his work. New York, NY: Pantheon; 1985.

9. Hilfiker D. Facing our mistakes. New Engl J Med. 1984;310:118–22.
10. Kalra J. Medical errors: overcoming the challenges. Clin Biochem. 2004;37:1063–71.
11. Serembus JF, Wolf ZR, Youngblood N. Consequences of fatal medication errors for health care providers: a secondary analysis study. Medsurg Nurs. 2001;10(4):193–201.
12. Newman MC. The emotional impact of mistakes on family physicians. Arch Fam Med. 1996;5:71–5.
13. Wu AW, Folkman S, McPhee SJ, et al. How house officers cope with their mistakes. West J Med. 1993;159:565–9.
14. Christensen JF, Levinson W, Dunn PM. The heart of darkness: the impact of perceived mistakes on physicians. J Gen Int Med. 1992;7:424–31.
15. Wu AW, Folkman S, McPhee SJ, et al. Do house officers learn from their mistakes? JAMA. 1991;265(16):2089–94.
16. Vander Zyl SK, Hohneke L. The 'battlefield of caring'. Nurs Matters. 2006;17:7–8.
17. Scott SD, Hirschinger LE, Cox KR, et al. Caring for our own: deploying a systemwide second victim rapid response team. Jt Comm J Qual Saf. 2010;36(5):233–40.
18. Treiber LA, Jones JH. Devastatingly human: an analysis of registered nurses' medication error accounts. Qual Health Res. 2010;20(10):1327–42.
19. Fischer MA, Mazor KM, Barilo J. Learning from mistakes. Factors that influence how students and residents learn from medical errors. J Gen Int Med. 2006;21(5):419–23.
20. Wolf ZR. Stress management in response to practice errors: critical events in professional practice. PA-PSRS Patient Saf Advis. 2005;2(4):1–2.
21. Schwappach DB, Bouarte TA. The emotional impact of medical error involvement on physicians: a call for leadership and organizational accountability. Swiss Med Wkly. 2009; 139(1–2):9–15.
22. Weiss PM. Medical errors and the second victim. Female Pat. 2011;36:29–32.
23. Manser T. Managing the aftermath of critical incidents: meeting the needs of health-care providers and patients. Best Pract Res Clin Anaesthesiol. 2011;25:169–79.
24. Gazoni FM, Amato PE, Malik ZM, et al. The impact of perioperative catastrophes on anesthesiologists: results of a national study. Anesth Analg. 2012;114(3):596–603.
25. Sirriyeh R, Lawton R, Gardner P, et al. Coping with medical error: a systematic review of papers to assess the effects of involvement in medical errors on healthcare professionals' psychological well-being. Qual Saf Health Care. 2010;19:1–8.
26. Lander LI, Connor JA, Shah RK, et al. Otolaryngologists' responses to errors and adverse events. Laryngoscope. 2006;363:1114–20.
27. Wolf ZR, Serembus JF, Smetzer J, et al. Responses and concerns of healthcare providers to medication errors. Clin Nurs Spec. 2000;14:278–87.
28. Engel KG, Rosenthal M, Sutcliffe KM. Residents' responses to medical error: coping, learning, and change. Acad Med. 2006;81(1):86–93.
29. Martinez W, Lo B. Medical students' experiences with medical errors: an analysis of medical student essays. Med Educ. 2008;42:733–41.
30. Muller D, Ornstein K. Perceptions of and attitudes towards medical errors among medical trainees. Med Educ. 2007;41:645–52.
31. West CP, Huschka MM, Novotny PJ, et al. Association of perceived medical errors with resident distress and empathy: a prospective longitudinal study. JAMA. 2006;302:1294–300.
32. University of Missouri Health Care forYOU Team. Available at http://www.muhealth.org/secondvictim. Accessed 15 Aug 2012.
33. Scott SD, Hirschinger LE, McCoig M. The second victim. Textbook of rapid response systems. New York, NY: Springer; 2011.
34. Shanafelt TD, Balch CM, Bechamps G, et al. Burnout and medical errors among American surgeons. Ann Surg. 2010;251(6):995–1000.
35. Paparella S. Caring for the caregiver: moving beyond the finger pointing after an adverse event. J Emerg Nurs. 2011;37(3):263–5.
36. VanPelt F. Peer support: healthcare professionals supporting each other after adverse medical events. Qual Saf Health Care. 2008;17:249–52.

37. Institute for Safe Medication Practices. Too many abandon the 'second victims' of medical errors. ISMP medication safety alert; 2011.
38. Conway J, Federico F, Stewart F, et al. Respectful management of serious clinical adverse events. IHI Innovation Series White Paper. Institute for Healthcare Improvement, Cambridge, MA; 2010.
39. Denham C. TRUST: the 5 rights of the second victim. J Patient Saf. 2007;3:107–19.
40. White AA, Waterman A, McCotter P, Boyle D, Gallagher T. Supporting health care workers after medical error: considerations for health care leaders. J Clin Outcomes. 2008;15(5): 240–7.

# Appendix

## Preface

- **Joint Commission National Patient Safety Goals.** Available at http://www.joint commission.org/standards_information/npsgs.aspx. Last accessed 5 Jan 2013.
- **National Quality Forum: Serious Reportable Events in healthcare.** Available at http://www.qualityforum.org/projects/hacs_and_sres.aspx. Last accessed 5 Jan 2013.
- **Center for Medicare and Medicaid Services: Hospital-acquired conditions.** Available at http://www.cms.gov/Medicare/Medicare-Fee-for-Service-Payment/HospitalAcqCond/downloads/hacfactsheet.pdf. Last accessed 5 Jan 2013.
- **AHRQ patient safety indicators.** Available on http://www.qualityindicators.ahrq.gov/modules/psi_overview.aspx. Last accessed 13 Jan 2013.
- **National Patient Safety Foundation (NPSF).** The website offers a wealth of educational resources and tools. Available at http://www.npsf.org/. Last accessed 13 Jan 2013.
- **Root cause analysis.** US Department of Veterans Affairs. Available on http://www.patientsafety.gov/rca.html. Last accessed 13 Jan 2013.
- **Framework for conducting a root cause analysis and action plan.** The Joint Commission. Available on http://www.jointcommission.org/Framework_for_Conducting_a_Root_Cause_Analysis_and_Action_Plan/. Last accessed 13 Jan 2013.
- **Adverse events in hospitals: National incidence among Medicare beneficiaries.** Office of Inspector General. Department of Health and Human Services. 2010. Available on https://oig.hhs.gov/oei/reports/oei-06-09-00090.pdf. Last accessed 13 Jan 2013.
- **Hospital incident reporting systems do not capture most patient harm.** Office of Inspector General. Department of Health and Human Services. 2012. Available on https://oig.hhs.gov/oei/reports/oei-06-09-00091.pdf. Last accessed 13 Jan 2013.

A. Agrawal (ed.), *Patient Safety: A Case-Based Comprehensive Guide*,
DOI 10.1007/978-1-4614-7419-7, © Springer Science+Business Media New York 2014

- **What happened to Josie?** Available on http://www.ihi.org/offerings/ IHIOpenSchool/resources/Pages/AudioandVideo/WhatHappenedtoJosieKing. aspx. Last accessed 5 Jan 2013.
- **Leadership guide to patient safety.** Institute for Healthcare Improvement. Available on http://www.ihi.org/knowledge/Pages/IHIWhitePapers/Leadership GuidetoPatientSafetyWhitePaper.aspx. Last accessed 13 Jan 2013.

## Chapter 1: Patient Identification

- **Five rules of causation.** VA National Center for Patient Safety. Available on http://www.patientsafety.gov/CogAids/Triage/index.html#page-9. Last accessed 13 Jan 2013.
- Hall LW. **Mistaken identity.** AHRQ Web M&M. October 2008. Available on http://www.webmm.ahrq.gov/case.aspx?caseID=187. Last accessed 13 Jan 2013.
- Dougherty D. **Mother's milk, but whose mother?** AHRQ Web M&M. November 2010. Available on http://www.webmm.ahrq.gov/case.aspx?caseID=228. Last accessed 13 Jan 2013.
- **Sample Policy: Correct patient and procedure identification for inpatient and outpatient radiology services.** Pennsylvania Patient Safety Advisory. Available on http://patientsafetyauthority.org/EducationalTools/PatientSafety Tools/upradiology/Pages/policy.aspx. Last accessed 13 Jan 2013.

## Chapter 2: Teamwork and Communication

- **Can teamwork enhance patient safety?** Risk Management Foundation. Available at http://www.rmf.harvard.edu/~/media/Files/_Global/KC/PDFs/ Forum_V23N3_teamworksafety.pdf. Last accessed 5 Jan 2013.
- **Patient safety is enhanced by teamwork**. Pennsylvania Patient Safety Advisory report. 2010. Available at http://patientsafetyauthority.org/ADVISORIES/ AdvisoryLibrary/2010/jun16_7(suppl2)/Pages/14.aspx. Last accessed 5 Jan 2013.
- Weinger MB, Blike GT. **Intubation mishap**. AHRQ Web M&M. September 2003. Available on http://www.webmm.ahrq.gov/case.aspx?caseID=29. Last accessed 13 Jan 2013.
- **Medical teamwork and patient safety: The Evidence-based relation.** AHRQ report. Available at http://www.ahrq.gov/qual/medteam/. Last accessed 5 Jan 2013
- **SBAR technique for communication: A situational briefing model.** Institute for Healthcare Improvement. Available on http://www.ihi.org/knowledge/Pages/ Tools/SBARTechniqueforCommunicationASituationalBriefingModel.aspx. Last accessed 13 Jan 2013.
- **MedTeams.** Available on http://teams.drc.com/Medteams/Home/Home.htm. Last accessed 13 Jan 2013.

# Chapter 3: Handoff and Care Transitions

- **Communication during patient hand-overs.** WHO Collaborating Center for Patient Safety Solutions. Available on http://www.ccforpatientsafety.org/common/pdfs/fpdf/presskit/PS-Solution3.pdf . Last accessed 6 Jan 2013.
- **Handoffs and Transition Learning Network. Strategies to improve handoffs.** Available on http://www.marylandpatientsafety.org/html/learning_network/hts/materials/resources/handoffs/HandoffsStrategiesChart.pdf. Last accessed 13 Jan 2013.
- **Handoffs and transitions in care: An inpatient perspective.** Available on http://jdc.jefferson.edu/cgi/viewcontent.cgi?article=1087&context=pehc. Last accessed 13 Jan 2013.

# Chapter 4: Graduate Medical Education and Patient Safety

- **Gaining ground: Quality improvement and U.S. medical residency.** Institute for Healthcare Improvement. Available on http://www.ihi.org/knowledge/Pages/AudioandVideo/WIHIGainingGroundQIandUSMedicalResidency.aspx. Last accessed 5 Jan 2013.
- **ACGME duty hours.** Available on http://www.acgme.org/acgmeweb/GraduateMedicalEducation/DutyHours.aspx. Last accessed 5 Jan 2013.
- **Strategies to improve trainee supervision.** Slides from the MERGE (Medical Education Research Growth and Experience) Lab. Available on http://www.slideshare.net/MergeLab/final-arora-farnansupervision. Last accessed 5 Jan 2013.
- Kashner TM, Byrne JM, Gilman S. **Resident supervision index: Assessing feasibility and validity**. Department of Veterans Affairs. 2008. Available on http://www.va.gov/oaa/archive/rsi-report.pdf. Last accessed 13 Jan 2013.

# Chapter 5: Electronic Health Record and Patient Safety

- **Health information technology and patient safety: Building safer systems for better care.** Institute of Medicine. Available on http://www.iom.edu/Reports/2011/Health-IT-and-Patient-Safety-Building-Safer-Systems-for-Better-Care.aspx. Last accessed 13 Jan 2013.
- **Safely implementing health information and converging technologies.** The Joint Commission Sentinel Events Alert #42. 2008. Available on http://www.jointcommission.org/assets/1/18/SEA_42.PDF. Last accessed 13 Jan 2013.

- **Medicare and Medicaid EHR incentive program.** Available on http://www. cms.gov/Regulations-and-Guidance/Legislation/EHRIncentivePrograms/ Getting_Started.html. Last accessed 13 Jan 2013.
- **Certified Health IT Products List.** Available on http://oncchpl.force.com/ ehrcert?q=CHPL. Last accessed 13 Jan 2013.
- Electronic health record (EHR) usability

  - **Selecting an EHR for your practice: Evaluating usability.** Health Information and Management Systems Society (HIMSS). Available on http:// www.himss.org/content/files/HIMSS%20Guide%20to%20Usability_ Selecting%20an%20EMR.pdf. Last accessed 5 Jan 2013.
  - **EHR usability 101.** Health Information and Management Systems Society (HIMSS). Available on http://www.himss.org/asp/topics_FocusDynamic. asp?faid=559. Last accessed 5 Jan 2013.
  - **Health IT usability collaboration site.** National Institute of Standard and Technology. Available on http://collaborate.nist.gov/twiki-hit/bin/view/ HealthIT/WebHome . Last accessed 13 Jan 2013.

- **Copied and pasted and misdiagnosed (or cloned notes and blind alleys).** MedScape. 2012. Available on http://www.medscape.org/viewarticle/763617. Last accessed 13 Jan 2013.

# Chapter 6: Clinical Ethics and Patient Safety

- **Sample POLST form from Center for Ethics in Healthcare**. Oregon Health and Sciences University. Available on http://www.ohsu.edu/polst/programs/doc uments/POLST.JUNE.2009sample.pdf. Last accessed 5 Jan 2013.
- **American Medical Association's Code of Medical Ethics.** Available on http:// www.ama-assn.org/ama/pub/physician-resources/medical-ethics/code-medical- ethics.page. Last accessed 5 Jan 2013.
- **Crackdown on physician disruptive behavior.** MedScape. 2010. Available on http://www.medscape.com/viewarticle/731384. Last accessed 13 Jan 2013.
- **The Center for Ethics and Professionalism.** American College of Physicians. Available on http://www.acponline.org/running_practice/ethics/. Last accessed 13 Jan 2013.

# Chapter 7: Medication Error

- **Index for categorizing medication errors.** NCC MERP (National Coordinating Council for Medication Error Reporting and Prevention). Available on http:// www.nccmerp.org/pdf/indexColor2001-06-12.pdf. Last accessed 13 Jan 2013.
- **NCC MERP (National Coordinating Council for Medication Error Reporting and Prevention) taxonomy of medication errors.** Available on http://www.nccmerp.org/pdf/taxo2001-07-31.pdf. Last accessed 13 Jan 2013.

- **Institute for Safe Medication Practices (ISMP)** has a wealth of resources. Available on http://www.ismp.org/. Last accessed 13 Jan 2013.
- **A list of error-prone abbreviations, symbols, and dose designations.** ISMP. Available on http://www.ismp.org/tools/errorproneabbreviations.pdf. Last accessed 5 Jan 2013.
- **Massachusetts Coalition for the Prevention of Medication Errors** has a wealth of resources. Available on http://www.macoalition.org/. Last accessed 13 Jan 2013.
- **Preventing Medication Errors: Quality Chasm Series.** The National Academies Press. Available on http://www.nap.edu/catalog.php?record_id=11623. Last accessed 13 Jan 2013.
- **Medication Errors.** The U.S. Food and Drug Administration. Available on http://www.fda.gov/drugs/drugsafety/medicationerrors/default.htm.          Last accessed 13 Jan 2013.
- **Medication safety program.** Center for Disease Control and Prevention. Available on http://www.cdc.gov/medicationsafety/. Last accessed 13 Jan 2013.
- **IHI trigger tool for measuring adverse drug events.** Available on http://www.ihi.org/search/pages/results.aspx?k=medication%20error. Last accessed 13 Jan 2013.

# Chapter 8: Medication Reconciliation Error

- **Medications at Transitions and Clinical Handoffs (MATCH) Toolkit for Medication Reconciliation.** AHRQ. Available on http://www.ahrq.gov/qual/match/match.pdf. Last accessed 6 Jan 2013.
- **IHI has a wealth of resources on medication reconciliation.** Available on http://www.ihi.org/explore/adesmedicationreconciliation/Pages/default.aspx. Last accessed 13 Jan 2013.
- **The physician's role in medication reconciliation.** American Medical Association (AMA) monograph. Available on http://www.ama-assn.org/resources/doc/cqi/med-rec-monograph.pdf. Last accessed 13 Jan 2013.
- **Reconciling medications: safe practice recommendations.** Massachusetts Coalition for the Prevention of Medication Errors. Available on http://www.macoalition.org/Initiatives/RecMeds/SafePractices.pdf. Last accessed 13 Jan 2013.

# Chapter 9: Retained Surgical Items

- **NoThing Left Behind** has a wealth of resources. Available on http://www.nothingleftbehind.org/. Last accessed 5 Jan 2013.
- **The prevention of retained foreign bodies after surgery.** Bulletin of the American College of Surgeons. Available on http://www.facs.org/fellows_info/bulletin/2005/gibbs1005.pdf. Last accessed 5 Jan 2013.

- **Update on the prevention of retained surgical items**. Pennsylvania Patient Safety Advisory. 2012. Available on http://patientsafetyauthority.org/ADVISORIES/ AdvisoryLibrary/2012/Sep;9(3)/Pages/106.aspx. Last accessed 5 Jan 2013.

# Chapter 10: Wrong-Site Surgery

- **Preventing wrong-site surgery.** Pennsylvania Patient Safety Advisory. Available on http://patientsafetyauthority.org/EDUCATIONALTOOLS/ PATIENTSAFETYTOOLS/PWSS/Pages/home.aspx. Last accessed 5 Jan 2013.
- **The Joint Commission. Universal protocol.** Available at http://www.jointcommission.org/standards_information/up.aspx. Last accessed 5 Jan 2013.
- **WHO surgical safety checklist and implementation manual.** Available at http://www.who.int/patientsafety/safesurgery/ss_checklist/en/index.html. Last accessed 5 Jan 2013.
- Gawande A. **The Checklist Manifesto**. 2009. Metropolitan Books, New York, NY. Available on http://gawande.com/the-checklist-manifesto. Last accessed 13 Jan 2013.

# Chapter 11: Transfusion-Related Hazards

- Astion M. **Right patient, wrong sample.** AHRQ Web M&M. December 2006. Available on http://www.webmm.ahrq.gov/case.aspx?caseID=142. Last accessed 13 Jan 2013.
- **Improving the safety of the blood transfusion process.** Pennsylvania Patient Safety Advisory. June 2010. Available on http://patientsafetyauthority.org/ ADVISORIES/AdvisoryLibrary/2010/Jun7(2)/Pages/33.aspx. Last accessed 13 Jan 2013.
- LaRocco M, Brient K. **An interdisciplinary approach to safer blood transfusion.** Patient Safety and Quality in Health Care. 2008. Available on http://www. psqh.com/marapr08/transfusion.html. Last accessed 13 Jan 2013.

# Chapter 12: Hospital-Acquired Infections

- **Hand hygiene in healthcare settings.** Centers for Disease Control and Prevention. The website offers a wealth of resources. Available on http://www. cdc.gov/handhygiene/. Last accessed 6 Jan 2013.
- **Infection control.** World Health Organization. The website offers a wealth of resources. Available on http://www.who.int/topics/infection_control/en/. Last accessed 6 Jan 2013.

- **Preventing central line-associated blood stream infection**. IHI's How-to Guide. Available on http://www.ihi.org/knowledge/Pages/Tools/HowtoGuide PreventCentralLineAssociatedBloodstreamInfection.aspx. Last accessed 13 Jan 2013.
- **Using care bundles to improve healthcare quality.** IHI. Available on http://www.ihi.org/knowledge/Pages/IHIWhitePapers/UsingCareBundles.aspx. Last accessed 13 Jan 2013.
- **Overview of the patient safety component, device-associated module (CLABSI, VAP, CAUTI).** National Healthcare Safety Network. Available on http://www.cdc.gov/nhsn/wcOverviewNHSN.html. Last accessed 6 Jan 2013.
- **Demonstrating return on investment for infection prevention and control.** Pennsylvania Patient Safety Advisory. 2010. Available on http://patientsafetyauthority.org/ADVISORIES/AdvisoryLibrary/2010/Sep7(3)/Pages/102.aspx. Last accessed 6 Jan 2013.

# Chapter 13: Hospital Falls

- **Reducing patient injuries from falls.** IHI's How-to Guide. Available on http://www.safetyandquality.health.wa.gov.au/docs/squire/IHI%20Guide_Reducing_Patient_Injuries_from_Falls.pdf. Last accessed 13 Jan 2013.
- Healey, F. **Implementing a fall prevention program.** AHRQ Web M&M; 2011. Available on http://webmm.ahrq.gov/perspective.aspx?perspectiveID=114. Last accessed 6 Jan 2013.

# Chapter 14: Pressure Ulcers

- **Prevent pressure ulcers.** Institute for Healthcare Improvement. The website offers a prevention guide, measures, and tools for prevention. Available on http://www.ihi.org/explore/pressureulcers/pages/default.aspx. Last accessed 13 Jan 2013.
- Barbour S. **The forgotten turn.** AHRQ Web M&M. December 2010. Available on http://webmm.ahrq.gov/case.aspx?caseID=230. Last accessed 13 Jan 2013.
- **Pressure ulcers: New staging, reporting, and risk reduction strategies.** Pennsylvania Patient Safety Advisory. December 2008. Available on http://patientsafetyauthority.org/ADVISORIES/AdvisoryLibrary/2008/Dec5(4)/Pages/118.aspx. Last accessed 13 Jan 2013.

## Chapter 15: Diagnostic Error

- Groopman J. **How doctors think**. Houghton Mifflin Harcourt, New York, NY. 2007.
- Kahneman D. **Thinking, fast and slow.** Farrar, Stratus, and Giroux. New York, NY. 2011.
- MacDonald OW. **Physician perspectives on preventing diagnostic errors.** QuantiaMD. September 2011. Available on http://www.quantiamd.com/q-qcp/QuantiaMD_PreventingDiagnosticErrors_Whitepaper_1.pdf. Last accessed 13 Jan 2013.
- **Diagnostic errors.** AHRQ Patient Safety Network. Available on http://psnet.ahrq.gov/primer.aspx?primerID=12. Last accessed 13 Jan 2013.
- **Diagnostic error in acute care.** Pennsylvania Patient Safety Advisory. 2010. Available on http://www.patientsafetyauthority.org/ADVISORIES/Advisory Library/2010/Sep7(3)/Pages/76.aspx. Last accessed 13 Jan 2013.
- **Why diagnostic errors don't get any respect… and what can be done about it.** Wachter's World. Available on http://community.the-hospitalist.org/2008/06/02/why-diagnostic-errors-don-t-get-any-respect-and-what-can-be-done-about-it/. Last accessed 13 Jan 2013.

## Chapter 16: Patient Safety in Pediatrics

- **Policy statement - Principles of pediatric patient safety: Reducing harm due to medical care.** American Academy of Pediatrics. Pediatrics. 2011;127(6):1199-210. Available on http://pediatrics.aappublications.org/content/early/2011/05/25/peds.2011-0967. Last accessed 13 Jan 2013.
- **Patient safety in pediatric emergency care.** Illinois Emergency Medical Services for Children. Available on http://www.luhs.org/depts/emsc/PatientSafety_APPENDICES_FINAL.pdf. Last accessed 13 Jan 2013.
- **What pediatricians should know about medical radiation safety.** American Academy of Pediatrics. Available on http://www2.aap.org/sections/radiology/RadiologyPediatricianPage.pdf. Last accessed 13 Jan 2013.
- **AHRQ health care innovations exchange: Pediatric care.** The website offers links to numerous innovations and tools to improve pediatric safety. Available on http://www.innovations.ahrq.gov/innovations_qualitytools.aspx?categoryID=54805&taxonomyID=54814. Last accessed 13 Jan 2013.

## Chapter 17: Patient Safety in Radiology

- **Image Wisely®**. Radiation safety in adult medical imaging. Available on http://www.imagewisely.org/. Last accessed 6 Jan 2013.
- **Image Gently®.** The Alliance for Radiation Safety in Pediatric Imaging. Available on http://www.pedrad.org/associations/5364/ig/. Last accessed 6 Jan 2013.

- **ALARA (as low as reasonably achievable)**. United States Nuclear Regulatory Commission. Available on http://www.nrc.gov/reading-rm/basic-ref/glossary/alara.html. Last accessed 6 Jan 2013.
- **Radiation risks of diagnostic imaging.** The Joint Commission Sentinel Events Alert #47. 2011. Available on http://www.jointcommission.org/assets/1/18/sea_471.pdf. Last accessed 6 Jan 2013.
- **Applying the Universal Protocol to improve patient safety in radiology services.** Pennsylvania Patient Safety Advisory, 2011. Available on http://patient-safetyauthority.org/ADVISORIES/AdvisoryLibrary/2011/jun8(2)/Pages/63.aspx. Last accessed 6 Jan 2013.

## Chapter 18: Patient Safety in Anesthesia

- **Anesthesia Patient Safety Foundation** – the website has a wealth of resources including a medication safety video and clinical safety tools. Available on http://www.apsf.org/. Last accessed 13 Jan 2013.
- **Reporting impaired, incompetent, or unethical colleagues.** American Medical Association Code of Ethics. Available on http://www.ama-assn.org/ama/pub/physician-resources/medical-ethics/code-medical-ethics/opinion9031.page. Last accessed 13 Jan 2013.
- **Policy on physician impairment (2011).** Federation of State Medical Boards. Available on http://www.fsmb.org/pdf/grpol_policy-on-physician-impairment.pdf. Last accessed 13 Jan 2013.
- **Federation of State Physician Health Programs.** Includes a directory of physician health programs in various US states. Available on http://www.fsphp.org/. Last accessed 13 Jan 2013.
- Sahr JS, Hosseini P. **Secured but not always safe.** AHRQ Web M&M. 2006. Available on http://www.webmm.ahrq.gov/case.aspx?caseID=139. Last accessed 13 Jan 2013.
- Lee C. **Who nose where the airway is?** AHRQ Web M&M. 2009. Available on http://www.webmm.ahrq.gov/case.aspx?caseID=208. Last accessed 13 Jan 2013.

## Chapter 19: Patient Safety in Behavioral Health

- **Behavioral health patient room: Common hazards: An interactive graphic.** Pennsylvania Patient Safety Advisory. Available on http://patientsafetyauthority.org/EducationalTools/PatientSafetyTools/behavioral_health/Documents/pt_room.swf. Last accessed 6 Jan 2013.

- Geetha J, Herzog A. SAFE MD: **Practical applications and approaches to safe psychiatric practice. Committee on patient safety**: American Psychiatric Association. June 2008.
- **Screening tools.** The SAMHSA-HRSA Center for Integrated Health Solutions. Available at http://www.integration.samhsa.gov/clinical-practice/screening-tools. Last accessed 6 Jan 2013. The website provides screening tools for general population, depression, drug and alcohol use, bipolar disorder, suicide risk, and anxiety disorders.

## Chapter 20: Patient Safety in Outpatient Care

- **A patient safety handbook for ambulatory care providers**. The Joint Commission Resources. 2009.
- **Meeting the challenge of patient safety in the ambulatory care setting**. Medical Group Management Association Patient Safety and Quality Advisory Committee White Paper. Available on http://mail.ny.acog.org/website/PtSafety/WhitePaper.pdf. Last accessed 13 Jan 2013.
- **The Chronic Care Model.** Improving Chronic Illness Care. Available on http://www.improvingchroniccare.org/index.php?p=The_Chronic_Care_Model&s=2. Last accessed 6 Jan 2013.

## Chapter 21: Error Disclosure

- Conway J, Federico F, Stewart K, Campbell MJ. **Respectful Management of Serious Clinical Adverse Events** (Second Edition). IHI Innovation Series white paper. Cambridge, Massachusetts: Institute for Healthcare Improvement; 2011. (Available on http://www.IHI.org). Last accessed 6 Jan 2013. This white paper provides

  - A Disclosure Culture Assessment tool
  - Words to use in apology
  - Respectful management of adverse event checklist for organization

- **A national survey of medical error reporting laws**. Yale Journal of Health Policy, Law and Ethics; 2008(9):201-286.
- **Teaching module: Talking about harmful medical errors with patients**. Available on http://depts.washington.edu/toolbox/errors.html. Last accessed 6 Jan 2013.

## Chapter 22: The Culture of Safety

- **Pursuing perfection: Report from McLeod Regional Medical Center on leadership patient rounds.** Institute for Healthcare Improvement. Available on http://www.ihi.org/knowledge/Pages/ImprovementStories/PursuingPerfection ReportfromMcLeodRegionalMedicalCenteronLeadershipPatientRounds.aspx. Last accessed 6 Jan 2013.
- **Adverse drug event (ADE) drill scenario.** Institute for Healthcare Improvement. Available on http://www.ihi.org/knowledge/Pages/Tools/AdverseDrugEventDrill Scenario.aspx. Last accessed 6 Jan 2013.
- Conway J. **Getting boards on board**: Engaging governing boards in quality and safety. Joint Commission Journal on Quality and Patient Safety. 2008 Apr;34(4):214-220. Available on http://www.ihi.org/knowledge/pages/publica tions/gettingboardsonboard.aspx. Last accessed 6 Jan 2013.
- **Case study: Sustaining a culture of safety in the U.S. Department of Veterans Affairs Health Care System.** April / May 2010; The Commonwealth Fund. Available on http://www.commonwealthfund.org/Newsletters/Quality-Matters/2010/April-May-2010/Case-Study.aspx. Last accessed 6 Jan 2013.
- **AHRQ: Surveys on patient safety culture.** Available on http://www.ahrq.gov/ qual/patientsafetyculture/. Last accessed 6 Jan 2013.
- **The Joint Commission Sentinel Events Alert #43. Leadership committed to safety.** August 2009. Available on http://www.jointcommission.org/assets/1/18/ SEA_43.PDF. Last accessed 6 Jan 2013.

## Chapter 23: Second Victim

- **Nurse's suicide highlights twin tragedies of medical error.** MSNBC News (June 27 2011). Available on http://www.msnbc.msn.com/id/43529641/ns/ health-health_care/t/nurses-suicide-highlights-twin-tragedies-medical-errors/. Last accessed 6 Jan 2013.
- Marx D. **Whack-a-mole: the price we pay for expecting perfection.** Plano, TX: By Your Side Studios; 2009.
- **MITSS (Medically Induced Trauma Support Services)** has a wealth of resources. Available on http://mitss.org/. Last accessed 6 Jan 2013.
- Conway J, Federico F, Stewart K, Campbell MJ. **Respectful Management of Serious Clinical Adverse Events (Second Edition).** IHI Innovation Series white paper. Cambridge, Massachusetts: Institute for Healthcare Improvement; 2011. Available on http://www.IHI.org. Last accessed 6 Jan 2013.

# Index

A. Agrawal (ed.), *Patient Safety: A Case-Based Comprehensive Guide*,
DOI 10.1007/978-1-4614-7419-7, © Springer Science+Business Media New York 2014

Printed by Printforce, the Netherlands